Integrating Welfare and Sustainability in Poultry Systems

Integrating Welfare and Sustainability in Poultry Systems

Edited by

Dana L.M. Campbell

Agriculture and Food, CSIRO (Commonwealth Scientific and Industrial Research Organisation), Armidale, NSW, Australia

Marisa A. Erasmus

Department of Animal Sciences, Purdue University, West Lafayette, IN, United States

ELSEVIER

ACADEMIC PRESS

An imprint of Elsevier

Academic Press is an imprint of Elsevier
125 London Wall, London EC2Y 5AS, United Kingdom
50 Hampshire Street, 5th Floor, Cambridge, MA 02139, United States

ISBN: 978-0-443-40594-5

For Information on all Academic Press publications
visit our website at https://www.elsevier.com/books-and-journals

Publisher: Mara Conner
Acquisitions Editor: Kelsey Connors
Editorial Project Manager: Sadhna Rajan
Production Project Manager: Adellin Bentick
Cover Designer: Raman Kumar

Printed in England by CPI Group (UK) Ltd
Croydon CR0 4YY
United Kingdom

Typeset by MPS Limited, Chennai, India

Contents

List of contributors

Carolin Adler College of Agriculture and Bioresources, University of Saskatchewan, Saskatoon, SK, Canada

Sadid Amaz Department of Poultry Science, University of Georgia, Athens, GA, United States

Jacopo Bacenetti Department of Environmental and Policy Science, Università degli Studi di Milano, Milan, Italy

Bethany Baker-Cook Institute of Agrifood Research and Technology, Animal Welfare, Monells, Spain

Darshana Bhaisare Poultry Science Division, Nagpur Veterinary Collages, MAFSU, Nagpur, Nagpur, Maharashtra, India

Dana L.M. Campbell Agriculture and Food, CSIRO (Commonwealth Scientific and Industrial Research Organisation), Armidale, NSW, Australia

Lina Elmersson Department of Applied Animal Science and Welfare, Swedish University of Agricultural Sciences (SLU), Skara, Sweden

Marisa A. Erasmus Department of Animal Sciences, Purdue University, West Lafayette, IN, United States

Giacomo Falcone Department of Agricultural Sciences, Università degli Studi Mediterranea di Reggio Calabria, Reggio Calabria, Italy

Anne C. Fanatico Sustainable Development Department, Appalachian State University, Boone, NC, United States

Gregory S. Fraley Department of Animal Sciences, Purdue University, West Lafayette, IN, United States

Sabine G. Gebhardt-Henrich Center for Proper Housing: Poultry and Rabbits (ZTHZ), Division of Animal Welfare, Vetsuisse Faculty, University of Bern, Zollikofen, Switzerland

Leonie Jacobs School of Animal Sciences, Virginia Tech, Blacksburg, VA, United States

Sai Siva Kumar Katam PHT Division, ICAR- Central Avian Research Institute, Bareilly, Uttar Pradesh, India

Lillian Lins-Taylor Auburn University, Auburn, AL, United States

Monika Madheswaran PHT Division, ICAR- Central Avian Research Institute, Bareilly, Uttar Pradesh, India

Stefano Paolo Marelli Department of Veterinary Medicine and Animal Science DIVAS, University of Milan, Lodi, Italy

Clive J.C. Phillips Curtin University Sustainability Policy (CUSP) Institute, Curtin University, Bentley, WA, Australia; Institute of Veterinary Medicine and Animal Science, Estonia University of Life Sciences, Tartu, Estonia

Geeta Pipaliya PHT Division, ICAR- Central Avian Research Institute, Bareilly, Uttar Pradesh, India

Allison Pullin Prestage Department of Poultry Science, North Carolina State University, Raleigh, NC, United States

Prafulla Regmi Department of Poultry Science, University of Georgia, Athens, GA, United States

Jaydip Rokade PHT Division, ICAR- Central Avian Research Institute, Bareilly, Uttar Pradesh, India

Attila Salamon Department of Ethology, ELTE Eötvös Loránd University, Budapest, Hungary

Jenna M. Schober Department of Animal Sciences, Purdue University, West Lafayette, IN, United States

Karen Schwean-Lardner College of Agriculture and Bioresources, University of Saskatchewan, Saskatoon, SK, Canada

Tory Shynkaruk College of Agriculture and Bioresources, University of Saskatchewan, Saskatoon, SK, Canada

Michelle Sinclair A World of Good Initiative Inc., Santa Fe, NM, United States

Inga Tiemann Precision Livestock Farming, Faculty of Agricultural Sciences and Landscape Architecture, Osnabrueck University of Applied Sciences, Osnabrueck, Germany

Stephanie Torrey Trouw Nutrition North America, Puslinch, ON, Canada

Manzoor Wani ANFT Division ICAR-Central Avian Research Institute, Bareilly, Uttar Pradesh, India

Chapter 1

Introduction: poultry welfare and sustainability in modern production

Marisa A. Erasmus[1] and Dana L.M. Campbell[2]
[1]*Department of Animal Sciences, Purdue University, West Lafayette, IN, United States,*
[2]*Agriculture and Food, CSIRO (Commonwealth Scientific and Industrial Research Organisation), Armidale, NSW, Australia*

Introduction

The global importance of poultry

Globally, poultry production represents one of the most important food systems, spanning diverse cultures and economic systems and contributing to food security. Factors that influence this contribution to food security include: (1) animal feed and the overlap between feed ingredients and human food; (2) the productive efficiency of the animals; (3) the contribution of poultry manure to crop production; (4) the contribution of poultry meat and eggs to the human diet relative to other protein sources in terms of availability and affordability; and (5) the contribution of poultry production to income (Mottet & Tempio, 2017). Poultry products are the most consumed of all animal products, with poultry meat consumption having surpassed that of pork and beef (Miller et al., 2022; Sammani et al., 2025).

The majority of poultry meat production is concentrated in the United States (U.S.) (20% of global chicken meat), China (15%), Brazil (14%), the European Union (11%), and Russia (5%) (United States Department of Agriculture, Foreign Agricultural Service, 2025). The top egg-producing countries include China (617 billion eggs), India (147 billion eggs), Indonesia (146 billion eggs), the U.S. (109 billion eggs), Brazil (65 billion eggs), and Mexico (62 billion eggs) (Ozbun, 2026). Economically, the poultry industry is a major contributor to GDP, jobs, and income in these and many other countries (Silva et al., 2026). For example, in the U.S., which is a global leader in chicken and turkey production (FAO Food and Agriculture Organization of the United Nations, 2026), poultry companies employ over 2 million people, and the poultry industry

Integrating Welfare and Sustainability in Poultry Systems. DOI: https://doi.org/10.1016/B978-0-443-40594-5.00006-7

has an economic impact of over $663 billion (U.S. Poultry and Egg Association, 2026). In Brazil, the world's largest poultry exporter, 3.6 million people are employed within the poultry industry, and poultry contributes 1.5% to the GDP (ABPA Brazilian Association of Animal Protein, 2021). In the U.K., which is not among the global leaders in poultry production, the poultry meat industry contributed £2.8 billion to the GDP in 2023 (Oxford Economics, 2025). The poultry industry is not only important for national GDP, but for smallholders as well, where poultry is a source of income and protein (Birhanu et al., 2023). In countries across sub-Saharan Africa, South Asia, and Southeast Asia, small-scale and backyard poultry keeping provide essential income for rural households, with women comprising most small-scale poultry producers (Alders & Pym, 2009; Guèye, 2000).

Animal welfare in poultry production

Various sources are projecting continued growth of and demand for poultry products globally (e.g., OECD/FAO, 2025). Furthermore, several countries have increased their poultry production and exports, notably Brazil, the U.S., the European Union, and Thailand (Miller et al., 2022). Simultaneously, citizens and consumers in various countries continue to place importance on animal welfare, as well as environmental impacts, and human health (OECD/FAO, 2025), although the demand for animal welfare (measured as willingness to pay) depends on several factors, among them people's level of income (Lagerkvist & Hess, 2011) and the level of per capita meat consumption (Parlasca et al., 2023). Citizen and consumer behavior and perceptions of animal welfare have contributed to changes affecting the housing and management of farm animals; for example, through legislation in Europe (Revision of EU animal-welfare legislation, n.d), ballot initiatives leading to state statutes in the U.S. (Farm animal housing in 2024: laws, proposals and challenges. National Agricultural Law Center, 2024) and animal welfare certification and labeling schemes in Europe and the U.S. (reviewed in Bessei, 2018). Animal welfare perceptions and concerns also influence businesses through their social license to operate (Alonso et al., 2020), whereby public approval or disapproval of an organization's actions can influence the organization's ability to continue doing business. Various studies have documented increasing concern for animal welfare in Latin America, Canada, Asia, and Australia (reviewed in Alonso et al., 2020), despite a lack of a globally consistent definition of animal welfare (Cornish et al., 2016; Alonso et al., 2020). Animal welfare is made even more complex due to the various factors influencing people's perceptions, which in turn shape how animal welfare is defined and applied in practice. Perceptions differ due to demographic factors, experience, culture and religious factors, economic status, and differences in the value that people place on certain species of animals (reviewed in De la Fuente et al., 2017).

Animal welfare and sustainability

Having a definition of animal welfare is essential for measuring and assessing animal welfare; developing animal welfare-related laws, standards and programs; enabling stakeholders to work together to address animal welfare changes and challenges; and explaining how animal welfare affects, and is affected by, human and environmental factors. Various definitions and concepts of animal welfare, as well as frameworks for assessing animal welfare, exist (see, e.g., Broom, 1986; Dawkins, 2021; Fraser & Pajor, 1997; Mellor et al., 2020; Rault et al., 2025; Reimert et al., 2023; van der Staay et al., 2025). The World Organization for Animal Health defines animal welfare as "the physical and mental state of an animal in relation to the conditions in which it lives and dies" (WOAH World Organisation for Animal Health, 2026). Increasingly, animal welfare is being recognized as critical for human health and well-being; for example, as part of the One Health (WHO World Health Organization. One Health, 2026) and One Welfare (Pinillos et al., 2016) principles, which recognize that the health and well-being, respectively, of humans, animals and the environment are inextricably linked, and that animal welfare influences human "livelihoods and economic stability," including in an agricultural context (Lymbery, 2025). Animal welfare has also been linked with the concept of sustainability (reviewed in Wawrzyniak, 2023). For example, the U.K. Animal Welfare Committee (previously the Farm Animal Welfare Committee, [FAWC]), which advises the UK Department for Environment, Food and Rural Affairs, stated that "agriculture cannot be considered sustainable if it is achieved at an unacceptable cost to animal welfare" (FAWC Farm Animal Welfare Committee, 2016).

The continual successful operation of farms that produce animal protein has become increasingly critical, as we wonder how we are going to feed an expanding human population with a finite number of resources. "Sustainability" in agriculture is salient for consumers, stakeholders, and producers alike, as society demands consideration of our changing environment. At the same time, this Green Revolution has made "sustainability" a buzzword. This brings with it a significant risk of "greenwashing" as statements are made by agricultural (and other) companies for practices being put in place to protect the environment without a genuine effort to substantiate those claims (Mendes et al., 2025). There is no denying that for poultry production to continue into the future with the proposed increased output to feed more people, change needs to occur (Kleyn & Ciacciariello, 2021). Production will need to be more sustainable, but what does this actually mean and what form will change take? Sustainability and how it applies to agriculture can often be interpreted in different ways and may take on a specific meaning depending on cultural background or individual beliefs (Jaeger et al., 2023; Lizcano-Prada et al., 2024); sustainability and animal welfare share the same challenge of being multidimensional concepts that are shaped by culture, values, and economic status.

Pillars of sustainability

The definition of sustainable agriculture, as well as the indicators used to quantify whether a farming system is sustainable or not, can vary across global regions. However, the three-pillars approach is becoming increasingly used (Bathaei & Štreimikienė, 2023), aligning with the Triple Bottom Line framework (Profit, People, Planet) coined by John Elkington in the 1990s (Elkington, 1998). The three pillars of sustainability include environmental, social, and economic, in which practices need to, for example, protect natural resources, ensure safe working environments, and still turn a profit. But when broken down into parts, and considering that those parts need to make up the whole, then sustainability becomes more complicated because meeting goals in one pillar is not necessarily going to align with meeting goals in another. If the pillars are misaligned, then the emphasis placed on one pillar versus another may depend on the personal beliefs and principles of those involved. An even larger conundrum is presented when meeting the goals of one pillar significantly reduces the ability to meet the goals of another. Thus positive in one pillar may mean negative, rather than neutral, in another.

To delve a little deeper into the three pillars, "Environmental Sustainability" is front of mind for animal production. This concerns how a particular farming system is using natural resources and what emissions or environmental destruction it may be causing. This can include the emission of greenhouse gases, degradation of the land, runoff pollutants, use of water, and loss of biodiversity through farming landscape changes. "Social Sustainability" can include food security for local communities, as well as job security alongside labor provisions such as worker health and safety or personnel availability. It also includes the social acceptance of production practices, and this is often where animal welfare is integrated prominently through social license to operate. Finally, neither environmental nor social pillars of sustainability can be independent of "Economic Sustainability" because ultimately, animal production operates as a business and if there is no profit or there is financial loss, then particular types of housing systems or management practices will not be feasibly implemented.

Animal welfare and the pillars of sustainability

There are a multitude of animal welfare factors that are interconnected with sustainability under the three pillars. Animal welfare in the poultry industry has been a high priority for many decades to ensure the birds are housed and managed to optimize welfare metrics. This is reflected in the societal expectations met through welfare assurance programs, changes in the most prevalent types of housing systems, and legislation reform that bans what are viewed as unethical practices in some countries, such as battery-cage housing and bodily mutilations. The poultry industry has adapted to the changing animal welfare landscape with much research globally to determine what types of housing systems and management practices can lead to high standards of animal welfare. However, to be secure in the future, poultry industries must also consider the changing landscapes they are faced with, as sustainability in practices becomes increasingly critical. Housing and management practices that result in higher animal welfare may not always align with goals for reduced carbon footprints, production efficiency, and reduced use of non-renewable resources. Thus it may be challenging to integrate optimal animal welfare into optimal industry sustainability. For example, slow-growing meat chickens have lower risks for several welfare issues than fast-growing birds. However, slow-growing is less efficient, and therefore not as economically or environmentally viable compared with fast-growing (Baxter et al., 2021; Leinonen, 2025). Cage-free housing systems provide more behavioral freedom for laying hens, but excessive dust can create a worker health and safety hazard (Mitchell et al., 2015). Ducks are semiaquatic animals that are motivated to bathe in water, but providing water to them in a safe way without considerable impact on water usage and environmental pollution is an ongoing challenge (Babington & Campbell, 2022). Some have argued that animal welfare should be considered an issue in its own right and should not be mixed with the concept of sustainability (Wawrzyniak, 2023).

There is a projected increasing demand for, and production of, poultry products in the coming years. This creates challenges and opportunities for advancing both poultry welfare and sustainable poultry production. Sometimes, animal welfare and sustainability objectives align and can be improved with certain interventions. However, animal welfare and some sustainability objectives often conflict, requiring trade-offs. The tensions between welfare and sustainability that exist vary significantly across global regions. For example, in Westernized regions, welfare-oriented research and practices have been occurring for decades, and there are moves toward positive affective states and a "Good Life" as we strive to give even more to the animals (Edgar et al., 2013). In other regions, the resources and infrastructure may be significantly more limited, and the focus on animal welfare is still centered around shifting away from conventional caged poultry systems or infectious disease management, where vaccine access is limited (Hartcher et al., 2023; Ouma et al., 2023).

Purpose and scope of this book

This book considers research that has guided what constitutes optimal animal welfare across a range of poultry species and how the housing and management practices may align with sustainability goals across the pillars of environmental, social, and economic sustainability. This book will address poultry welfare in the context of sustainability challenges, providing the reader with different facets of the industry that must be considered, successful sustainable housing options achieved to date, and potential research areas to address moving forward. The collection of chapters provides insights into a range of poultry species across varying types of indoor and outdoor housing systems.

Chapter 2 considers laying hen welfare alongside sustainability for birds in indoor housing systems. Commercial egg production in recent years has made a major transition away from conventional caged systems to cage-free housing such as aviaries. While these aviary systems provide birds with more behavioral opportunities and exercise to better cater to skeletal health, this exacerbates issues such as keel bone damage and flock-level abnormal and injurious behaviors such as feather pecking and smothering. There can also be higher costs associated with aviary production as well as a greater environmental footprint, driving the need for innovative, integrated solutions in these systems moving forward.

Chapter 3 focuses on broiler chicken production and how to balance welfare and sustainability for fast- and slow-growing strains. While fast-growing birds have been selected to produce one of the most feed-efficient animal products, this comes at a cost to their welfare, with skeletal, cardiovascular, and other behavioral concerns being highly prevalent. Slower-growing birds may fare better in terms of welfare, but require more time and resources to produce the equivalent protein product. Potential strategies such as lower protein diets, reducing stocking densities, or slaughtering birds at a younger age are explored as possible considerations for balancing welfare and sustainability for meat birds.

Chapter 4 moves from grower birds to their breeder parents in discussing the challenges of broiler breeder production. Current practices for maintaining stable growth and breeding incorporate feed restriction, which leads to persistent hunger and stress. Different feeding strategies, including quantitative and qualitative feed restriction, are employed, but these do not fully address chronic hunger. This chapter considers these feeding practices, as well as how more relaxed feed restriction, enrichment provision, and slower-growing birds impact both bird welfare and breeding productivity. Innovation in genetics is going to be necessary to counteract the decline in breeder bird fertility.

Chapter 5 provides an overview of duck production, including major aspects of duck housing systems that influence duck production and welfare. Duck production systems vary globally, and critical aspects of duck production include ventilation and flooring systems, which can singularly affect duck welfare and interact with other factors, such as stocking density, to further influence duck welfare. The chapter also discusses the challenges associated with providing water for ducks. Major sustainability dimensions that are discussed include greenhouse gas emissions, land use, and manure management. The chapter highlights the critical need for additional research into duck welfare and sustainability.

Chapter 6 discusses goose production across different production systems, including intensive indoor, semiintensive with outdoor access, or integrated systems such as agroforestry. These birds have water-resistant feathers, which can make them suited to a more diverse range of environments, but also present challenges in their housing requirements to meet behavioral water-related needs. This chapter considers different housing systems and their trade-offs in welfare and sustainability dimensions, as well as case study examples for challenges in waterfowl production.

Chapter 7 discusses how turkey production globally is at a smaller scale than chicken production, and this is reflected in the available welfare research for this species. Housing and management practices are often adapted from broiler chicken research, but this does not always align with species-specific needs in this larger bird that has its own behavioral and physical requirements. As global pressure increases to balance animal welfare, environmental protection, and productivity, there is a need for science-based, sustainability strategies specific to turkeys.

Chapter 8 moves away from the high-performance commercial chicken lines to consider what specialty or local breeds across the world can offer poultry production. Dual-purpose chickens and local strains are typically regionally adapted and have the potential to enable higher animal welfare status, although welfare research on these birds is currently underrepresented. When compared to the global selected hybrid lines, local breeds typically have slower growth and reduced performance, limiting their resource-economic sustainability. However, their contribution to genetic diversity, climate resilience, and niche consumer markets drives a need to include local poultry breeds in industry innovations to maintain poultry supply into the future.

Chapter 9 delves into pasture-based or free-range systems for laying hens, where the housing includes an accessible outdoor area with vegetation. These systems encompass low-intensity backyard flocks right through to large-scale, high-intensity production farms. Providing an outdoor area can facilitate key hen behaviors such as foraging and dust bathing, making pasture-based a housing system that is considered to result in higher animal welfare, from a behavioral perspective, than exclusively indoor systems. However, pasture access may lead to disease transmission among hens, including from wild birds, as well as emissions of ammonia, phosphorus, and nitrogen into the environment. Production costs can be higher in these systems, which are often passed along to the consumer in egg costs. Thus, sustainability trade-offs exist, and in a One Health approach, both positive and negative effects must be considered.

Chapter 10 also considers pasture-based systems, but focusing on the specific area of organic production where there is an emphasis on natural living, often incorporating high animal welfare standards. Organic poultry production typically aligns with consumer concerns for animal welfare, including housing and management provisions such as outdoor access, lower stocking densities, and the use of slow-growing strains. While these measures may contribute to optimized bird welfare, they can increase production costs.

The output from these systems is also typically lower than that of conventional pasture-based systems, resulting in an environmental trade-off for land use efficiency. This chapter explores key animal welfare-based features of organic production and their intersection with the economic, environmental, and social dimensions of sustainable agriculture.

Chapter 11 further builds on outdoor systems by considering the integration of silvopasture in poultry production. This concept centers around a naturalistic poultry environment that could contribute to higher bird welfare, while securing secondary use of the land, such as through timber or fruit production. This type of housing environment would bring similar welfare benefits as other pasture-based systems through outdoor access, but with additional benefits of shelter and protection from canopy cover. The system aligns with consumer demand for ethical animal production but may have high initial investment and scalability challenges, likely limiting it to niche markets. Furthermore, research on poultry welfare in this context is currently limited.

Chapter 12 presents the integration of animal welfare and sustainability in India—one of the world's largest producers of eggs and poultry meat—and the challenges faced by this region. Poultry production in this region is driven by intensive systems and rising consumer demand. Emerging initiatives, including industry partnerships and NGO-led capacity-building programmes, are supporting a gradual transition toward improved welfare practices. However, welfare research specific to the Indian context is currently limited. Strengthening regulatory frameworks, improving farmer awareness, and adopting science-based management strategies will be critical to ensuring long-term industry resilience.

Chapter 13 considers not just the birds as a resource to be preserved and capitalized on, but all natural resources that play a role in poultry production. While commercial poultry production has intensified at scale and will continue to do so to meet protein needs, it ultimately depends on natural resources. As these continue to erode, the sustainability of commercial production is at critical risk. Ecological practices such as capturing sun energy, nutrient cycling, and building biodiversity will support the health of the environment, as well as that of people and the birds we need. Social justice for humans across global regions, as well as optimized welfare for the animals, are essential for future food security. Ecology, equity, and animal welfare can build both ecological and social resiliency for a sustainable future.

Chapter 14 considers one of the most critical aspects of poultry production—feeding the birds. Chickens are typically fed maize, wheat, and soybeans, with feed often being imported. This has significant economic and environmental sustainability implications, driving a need for novel solutions. Local production of feed that is more climate-resilient and alternative feed sources could provide options for meeting chicken feed needs. Aligned with novel feed production, some plant products may enhance birds' immune responses, substituting for reliance on antibiotics. This chapter discusses the large-scale development of new feeding strategies that may enable global regions to have greater self-sufficiency, producing poultry meat and eggs from more sustainable food inputs.

Chapter 15 provides a critical review of the most recent (2021–25) literature on the application of life cycle assessment (LCA) to poultry industries, primarily chicken. LCA is a standardized framework to quantify the environmental impact of a system and is thus a key application for sustainability assessments. Poultry is highly efficient when compared to other livestock industries; however, feed production and manure management are key factors that contribute to environmental impacts. While studies applying LCA to poultry are increasing, there is a need to standardize methodologies and consider additional factors, such as animal welfare, for a more comprehensive assessment.

Chapter 16 focuses on consumer perceptions of poultry systems and how public opinions on animal welfare can drive changes in housing systems. Consumers can shape markets and agricultural practices at a global scale, highlighting the need to design systems and practices aligned with public drivers. Consumers around the world have concerns for animal welfare, preferring cage-free systems that provide more behavioral freedom to the birds. Environmental impacts of farming systems and sustainable production are viewed with high importance among young consumers representing the future generation. Cocreation of solutions that capture both animal welfare and environmental goals is critical for the economic security of poultry systems moving forward.

References

ABPA (Brazilian Association of Animal Protein). (2021). https://brazilianchicken.com.br/abpa-and-apex/abpa/?lang=en.

Alders, R. G., & Pym, R. A. E. (2009). Village poultry: Still important to millions, eight thousand years after domestication. *World's Poultry Science Journal, 65*(2), 181–190. https://doi.org/10.1017/s0043933909000117.

Alonso, M. E., González-Montaña, J. R., & Lomillos, J. M. (2020). Consumers' concerns and perceptions of farm animal welfare. *Animals, 10*(3), 385. https://doi.org/10.3390/ani10030385.

Babington, S., & Campbell, D. L. M. (2022). Water for domestic ducks: The benefits and challenges in commercial production. *Frontiers in Animal Science, 3*, 782507. https://doi.org/10.3389/fanim.2022.782507.

Bathaei, A., & Štreimikienė, D. (2023). A systematic review of agricultural sustainability indicators. *Agriculture, 13*(2), 241. https://doi.org/10.3390/agriculture13020241.

Baxter, M., Richmond, A., Lavery, U., & O'Connell, N. E. (2021). A comparison of fast growing broiler chickens with a slower-growing breed type reared on higher welfare commercial farms. *PLoS ONE, 16*(11), e0259333. https://doi.org/10.1371/journal.pone.0259333.

Bessei, W. (2018). Impact of animal welfare on worldwide poultry production. *World's Poultry Science Journal, 74*(2), 211–224. https://doi.org/10.1017/s0043933918000028.

Birhanu, M. Y., Osei-Amponsah, R., Yeboah Obese, F., & Dessie, T. (2023). Smallholder poultry production in the context of increasing global food prices: Roles in poverty reduction and food security. *Animal Frontiers, 13*(1), 17–25. https://doi.org/10.1093/af/vfac069.

Broom, D. M. (1986). Indicators of poor welfare. *British Veterinary Journal, 142*(6), 524–526. https://doi.org/10.1016/0007-1935(86)90109-0.

Cornish, A., Raubenheimer, D., & McGreevy, P. (2016). What we know about the public's level of concern for farm animal welfare in food production in developed countries. *Animals, 6*(11), 74. https://doi.org/10.3390/ani6110074.

Dawkins, M. S. (2021). *The science of animal welfare: Understanding what animals want.* Oxford Academic.

De la Fuente, M. F., Souto, A., Caselli, C., & Schiel, N. (2017). People's perception on animal welfare: Why does it matter? *Ethnobiology and Conservation, 6.* https://doi.org/10.15451/ec2017-10-6.18-1-7.

Edgar, J. L., Mullan, S. M., Pritchard, J. C., McFarlane, U. J. C., & Main, D. C. J. (2013). Towards a 'good life' for farm animals: Development of a resource tier framework to achieve positive welfare for laying hens. *Animals, 3*(3), 584–605. https://doi.org/10.3390/ani3030584.

Elkington, J. (1998). *Cannibals with forks: The triple bottom line of 21st century business.* New Society Publishers.

FAO (Food and Agriculture Organization of the United Nations). (2026). Crops and livestock products database. https://www.fao.org/faostat/en/#data/QCL.

Farm animal housing in 2024: laws, proposals and challenges. (2024). National Agricultural Law Center. https://nationalaglawcenter.org/farm-animal-housing-in-2024-laws-proposals-and-challenges/.

FAWC (Farm Animal Welfare Committee). (2016). Sustainable agriculture and animal welfare. Retrieved from https://assets.publishing.service.gov.uk/media/5a7f9365ed915d74e33f747f/Advice_about_sustainable_agriculture_and_farm_animal_welfare_-_final_2016.pdf.

Fraser, W., & Pajor, M. (1997). A scientific conception of animal welfare that reflects ethical concerns. *Animal Welfare, 6*(3), 187–205. https://doi.org/10.1017/S0962728600019795.

Guèye, E. F. (2000). The role of family poultry in poverty alleviation, food security and the promotion of gender equality in rural Africa. *Outlook on Agriculture, 29*(2), 129–136. https://doi.org/10.5367/000000000101293130.

Hartcher, K., Nuggehalli, J., Yang, Q., Catalina, M., de Luna, T., Agus, A., Ito, S., Idrus, Z., Rahayu, I. H. S., Jattuchai, J., Descovich, K., Lane, E., & Sinclair, M. (2023). Improving hen welfare on cage-free egg farms in Asia: Egg producers' perspectives. *Animal Welfare, 32*, e64. https://doi.org/10.1017/awf.2023.85.

Jaeger, S. R., Harker, F. R., & Ares, G. (2023). Consumer insights about sustainable and 'beyond organic' agriculture: A study of biodynamics in the United Kingdom, Australia, Singapore, and Germany. *Journal of Cleaner Production, 401*, 136744. https://doi.org/10.1016/j.jclepro.2023.136744.

Kleyn, F. J., & Ciacciariello, M. (2021). Future demands of the poultry industry: Will we meet our commitments sustainably in developed and developing economies? *World's Poultry Science Journal, 77*(2), 267–278. https://doi.org/10.1080/00439339.2021.1904314.

Lagerkvist, C. J., & Hess, S. (2011). A meta-analysis of consumer willingness to pay for farm animal welfare. *European Review of Agricultural Economics, 38*(1), 55–78. https://doi.org/10.1093/erae/jbq043.

Leinonen, I. (2025). Potential climate performance of modern fast- and slow-growing broiler genotypes. *Animal, 19*(5), 101499. https://doi.org/10.1016/j.animal.2025.101499.

Lizcano-Prada, J., Maestre-Matos, M., Mesias, F. J., Lami, O., Giray, H., Özçiçek Dölekoğlu, C., Bamoi, A. G. A., & Martínez-Carrasco, F. (2024). Does consumers' cultural background affect how they perceive and engage in food sustainability? A cross-cultural study. *Foods, 13*(2), 311. https://doi.org/10.3390/foods13020311.

Lymbery, P. (2025). Animal welfare as an essential element of one health. *Animal Research and One Health, 3*(4), 454–457. https://doi.org/10.1002/aro2.70021.

Mellor, D. J., Beausoleil, N. J., Littlewood, K. E., McLean, A. N., McGreevy, P. D., Jones, B., & Wilkins, C. (2020). The 2020 five domains model: Including human–animal interactions in assessments of animal welfare. *Animals, 10*(10), 1–24. https://doi.org/10.3390/ani10101870.

Mendes, J. A. J., Oliveira, A. Y., Santos, L. S., Gerolamo, M. C., & Zeidler, V. G. Z. (2025). A theoretical framework to support green agripreneurship avoiding greenwashing. *Environment, Development and Sustainability, 27*(9), 21779–21835. https://doi.org/10.1007/s10668-024-04965-z.

Miller, M., Gerval, A., Hansen, J., & Grossen, G. (2022). *Poultry expected to continue leading global meat imports as demand rises.* Amber Waves Available from: https://www.ers.usda.gov/amber-waves/2022/august/poultry-expected-to-continue-leading-global-meat-imports-as-demand-rises.

Mitchell, D., Arteaga, V., Armitage, T., Mitloehner, F., Tancredi, D., Kenyon, N., & Schenker, M. (2015). Cage versus noncage laying-hen housings: Worker respiratory health. *Journal of Agromedicine, 20*(3), 256–264. https://doi.org/10.1080/1059924X.2015.1042177.

Mottet, A., & Tempio, G. (2017). Global poultry production: Current state and future outlook and challenges. *World's Poultry Science Journal, 73*(2), 245–256. https://doi.org/10.1017/s0043933917000071.

OECD/FAO. (2025). OECD-FAO Agricultural Outlook 2025-2034. OECD Publishing. https://doi.org/10.1787/601276cd-en.

Ouma, E. A., Kankya, C., Dione, M., Kelly, T., Enahoro, D., Chiwanga, G., Abukari, Y., Msoffe, P., Kayang, B. B., & Zhou, H. (2023). Poultry health constraints in smallholder village poultry systems in Northern Ghana and Central Tanzania. *Frontiers in Veterinary Science, 10*, 1159331. https://doi.org/10.3389/fvets.2023.1159331.

Oxford Economics. (2025). The economic impact of the poultry meat industry. https://britishpoultry.org.uk/wp-content/uploads/2025/08/2025-Economic-impact-of-the-poultry-meat-industry-.pdf.

Ozbun, T. (2026). Egg production: Leading countries worldwide 2024. https://www.statista.com/statistics/263971/top-10-countries-worldwide-in-egg-production/?srsltid=AfmBOoouFbGlfdk2BnEgUhKZV62kErSXi1g1YGUhPCNKyG0Cg7SROlgU#statisticContainer.

Parlasca, M., Knößlsdorfer, I., Alemayehu, G., & Doyle, R. (2023). How and why animal welfare concerns evolve in developing countries. *Animal Frontiers, 13*(1), 26–33. https://doi.org/10.1093/af/vfac082.

Pinillos, R. G., Appleby, M. C., Manteca, X., Scott-Park, F., Smith, C., & Velarde, A. (2016). One Welfare – A platform for improving human and animal welfare. *Veterinary Record, 179*(16), 412–413. https://doi.org/10.1136/vr.i5470.

Rault, J. L., Bateson, M., Boissy, A., Forkman, B., Grinde, B., Gygax, L., Harfeld, J. L., Hintze, S., Keeling, L. J., Kostal, L., Lawrence, A. B., Mendl, M. T., Miele, M., Newberry, R. C., Sandøe, P., Špinka, M., Taylor, A. H., Webb, L. E., Whalin, L., & Jensen, M. B. (2025). A consensus on the definition of positive animal welfare. *Biology Letters, 21*(1). https://doi.org/10.1098/rsbl.2024.0382.

Reimert, I., Webb, L. E., van Marwijk, M. A., & Bolhuis, J. E. (2023). Review: Towards an integrated concept of animal welfare. *Animal, 17*, 100838. https://doi.org/10.1016/j.animal.2023.100838.

Revision of EU animal-welfare legislation. (n.d.). https://food.ec.europa.eu/animals/animal-welfare/evaluations-and-impact-assessment/revision-eu-animal-welfare-legislation_en.

Sammani, I., Yeshini, B., Siriwardhane, S., Pasindu, K., & Jayathilaka, R. (2025). Poultry consumption and sustainability: Assessing economic, trade, and climatic impacts across diverse income economies. *Sustainable Futures, 10*, 101485. https://doi.org/10.1016/j.sftr.2025.101485.

Silva, Y., Perera, N., Mendis, K., Susan, H., & Jayathilaka, R. (2026). Sustainability indicators in a globalised poultry sector: Production, consumption, trade openness, and GDP across 126 countries. *Research in Globalization, 12*, 100340. https://doi.org/10.1016/j.resglo.2026.100340.

van der Staay, F. J., Goerlich, V. C., Meijboom, F. L. B., & Arndt, S. S. (2025). Animal welfare definitions, frameworks, and assessment tools: Advancing the measurement and laying the foundation for improved animal welfare through a three-step approach. *Animal Welfare, 34*, e30. https://doi.org/10.1017/awf.2025.23.

United States Department of Agriculture. (2025). Foreign Agricultural Service, Livestock and Poultry: World Markets and Trade. https://www.fas.usda.gov/sites/default/files/2025-12/Livestock_poultry_December%209%202025.pdf.

U.S. Poultry and Egg Association 2026 https://www.uspoultry.org/economic-data/.

Wawrzyniak, D. (2023). Review: Animal husbandry and sustainable agriculture: is animal welfare (only) an issue of sustainability of agricultural production or a separate issue on its own? *Animal, 17*, 100880. https://doi.org/10.1016/j.animal.2023.100880.

WHO (World Health Organization). (2026). One Health. https://www.who.int/health-topics/one-health#tab=tab_1.

WOAH (World Organisation for Animal Health). (2026). https://www.woah.org/en/what-we-do/animal-health-and-welfare/animal-welfare/.

Chapter 2

Egg laying hens in indoor systems

Prafulla Regmi and Sadid Amaz
Department of Poultry Science, University of Georgia, Athens, GA, United States

Introduction

Housing systems and husbandry practices for laying hens have undergone continuous evolution since the emergence of the commercial egg industry in the early 20th century. Earlier periods of egg production were predominantly based on free-range and semiintensive systems that housed birds at low stocking densities ranging from 250–750 birds per hectare. Free-range systems commonly used mobile coops or slatted housing structures, whereas semi-intensive systems consisted of permanent housing with rotational access to grassed outdoor pens (Elson, 2011). During the early 1950s, rapid industry expansion and intensification led to the widespread adoption of indoor housing systems, including conventional battery cages and deep-litter floor systems. By 1980 conventional cage systems accounted for approximately 95% of global laying hen production (Elson, 2002).

The relative ease of flock management, improved biosecurity and food-safety control, and enhanced health and survivability associated with conventional cage systems facilitated large-scale industry expansion and vertical integration, contributing to the widespread availability, affordability, and global sustainability of egg production (Appleby, 2003; Holt et al., 2011; Lay et al., 2011). Despite these advantages, conventional cages have been widely criticized for ethical concerns related to the behavioral restrictions imposed on housed hens (Brambell, 1965). Conventional cages provide insufficient space per bird to allow the effective expression of evolutionarily conserved behaviors such as foraging, perching, nesting, and dustbathing. The inability to perform highly motivated behaviors can lead to frustration and compromised welfare (Appleby et al., 1992; Olsson & Keeling, 2000; Wichman & Keeling, 2009).

In addition to behavioral constraints, laying hens genetically selected for high egg production are susceptible to metabolic disorders, including

Integrating Welfare and Sustainability in Poultry Systems. DOI: https://doi.org/10.1016/B978-0-443-40594-5.00001-8

osteoporosis and fatty liver hemorrhagic syndrome, when housed in cages (Shini et al., 2019; Webster, 2004). Couch (1955) first described a condition termed "caged layer fatigue" in hens kept in battery cages, a severe form of osteoporosis characterized by structural bone loss in the vertebrae and subsequent spinal paralysis (Whitehead, 2004). Fracture prevalence of approximately 29% has been reported in end-of-lay caged hens, increasing to as high as 98% by the time carcasses reached the evisceration line (Gregory & Wilkins, 1989). As a result of these welfare concerns, conventional cages were banned in the European Union (EU) beginning in 2012 (CEC, 1999), leading to the adoption of enriched (furnished) cages and noncage systems, including single and multitier aviaries.

The transition to cage-free egg production has also accelerated across North America. As of 2025, approximately 44% of the 291 million hens in the United States are housed in cage-free systems, a proportion expected to increase further due to state-level legislative mandates (USDA-AMS. Shell Eggs: Monthly USDA Cage-Free Shell Egg Report, 2025). The transition toward more extensive indoor housing systems, such as multitier aviaries, has resulted in larger group sizes, increased space allowance, and greater environmental complexity. Cage-free systems provide resources such as feeders, drinkers, perches, scratch areas, and nest areas, distributed both vertically and horizontally according to housing manufacturer designs and to comply with regulatory guidelines. In addition, hens have access to a floor area with litter material that allows for foraging and dustbathing.

Despite these improvements, significant sustainability challenges persist in cage-free systems. Aggressive pecking and cannibalism, high prevalence of keel fractures, and greater cumulative mortality associated with opportunistic bacterial and protozoal infections remain major health and welfare concerns for laying hens (Kajlich et al., 2016; Kaukonen & Valros, 2019; Regmi, Nelson, et al., 2016). Each of these welfare issues is influenced by multiple interacting risk factors, including housing environment, genetics, nutrition, and exposure to management-related stressors such as catching, handling, thermal stress, and transportation. In addition, environmental and economic challenges, including increased production costs, higher energy and resource use, and particulate matter and ammonia emissions, also thwart the sustainability of the cage-free systems. This chapter examines social (animal welfare), environmental, and economic aspects of indoor egg production systems.

Animal welfare aspects

Skeletal and behavioral challenges that can compromise the welfare and productivity of laying hens in indoor housing systems are discussed in this section, with particular emphasis on indoor cage-free (aviary) systems, which are likely to represent the predominant housing type for commercial egg production in the near future.

Skeletal issues

Two major skeletal disorders, osteoporosis and keel bone damage, are among the most significant welfare challenges in modern egg production and are closely linked to housing design and husbandry practices for laying hens. Because hens mobilize skeletal calcium reserves, particularly medullary bone, which supplies approximately one-third of the calcium required for eggshell formation (Comar & Driggers, 1949), compromised bone quality, increased fragility, and fractures represent not only an immediate welfare concern but also a risk to flock productivity.

In the following sections, osteoporosis and keel bone damage are discussed separately with respect to their risk factors, underlying pathophysiology, and welfare implications in indoor laying hen systems.

Osteoporosis

One of the major skeletal disorders affecting laying hens housed in indoor systems, particularly conventional cages, is osteoporosis. Osteoporosis is characterized by a systemic loss of cortical and trabecular bone tissue. Because of this structural depletion, osteoporotic bones become thin, brittle, and porous, rendering them less capable of absorbing energy generated by daily mechanical loading and increasing the likelihood of fracture. Bone fractures can lead to morphological deformities of the skeleton, chronic pain, and restricted mobility in affected hens (Nasr, Murrell, et al., 2012; Nasr, Nicol, et al., 2012; Whitehead & Fleming, 2000), making osteoporosis a significant welfare concern.

Following the onset of egg laying during the early laying period, the formation of cortical and trabecular bone is markedly reduced, while the medullary cavities of long bones begin to fill with medullary bone tissue, which serves as a labile calcium reservoir for eggshell formation (Whitehead, 2004). Unlike lamellar bone, medullary bone lacks the parallel and hierarchical organization of collagen fibers, and calcium phosphate hydroxyapatite crystals are distributed randomly throughout the matrix (Dacke et al., 1993). Calcium derived from dietary sources is stored in medullary bone through osteoblastic activity, whereas during eggshell formation, medullary bone is resorbed by osteoclasts to meet the high circulating calcium demand. In addition to regulating gastrointestinal and renal mechanisms to maintain calcium homeostasis during egg production (Elaroussi et al., 1993; Tanaka et al., 1978), estrogen plays a central role by shifting osteoblastic function from lamellar bone formation toward medullary bone deposition (Whitehead, 2004). However, the initial accumulation of medullary bone within the marrow cavity may occur at the expense of cancellous bone. Moreover, osteoclast activity proceeds largely independently of estrogen concentration, allowing bone resorption to continue throughout the laying cycle. Structural bone may therefore be resorbed alongside medullary bone, particularly at exposed bone surfaces (Whitehead, 2004). Because the replacement of lost structural bone during lay is minimal, hens experience a progressive decline in skeletal strength over time. This process is further exacerbated in hens housed in conventional cages, where restricted movement promotes disuse osteoporosis (Nightingale et al., 1972). Osteoporosis in caged laying hens has been recognized since the early adoption of cage housing systems (Couch, 1955). In caged hens, progressive structural bone loss toward the end of lay can result in fracture prevalence of up to 29% (Gregory & Wilkins, 1989). The welfare consequences of osteoporosis are further intensified during preslaughter catching and handling, when weakened bones are unable to withstand externally imposed mechanical forces, leading to brittle fractures. Fracture prevalence may rise to as high as 98% during carcass processing at the slaughter plant (Gregory & Wilkins, 1989). In severe cases of osteoporosis, fracture energy can be reduced by up to 85% (Passi & Gefen, 2005), predisposing hens to spontaneous fractures. Fractures incurred at the costochondral junctions of the ribs and within the thoracic vertebrae can result in spinal paralysis (Whitehead, 2004). Moreover, injuries in these regions may compromise respiratory function and thermoregulation, thereby compounding physiological stress in affected hens (Li et al., 2024; Webster, 2004).

Unlike conventional cages, furnished cages and cage-free aviary systems allow laying hens to perform a range of locomotor and load-bearing activities, including walking, running, jumping, perching, and short bouts of flight (Campbell et al., 2016; Sandilands et al., 2009). These behaviors subject the skeleton to both low- and high-impact mechanical loading, stimulating active bone formation during the rearing (pullet) phase and attenuating bone loss during the laying (hen) phase (Regmi et al., 2015; Regmi, Smith, et al., 2016). Consequently, clinical manifestations of osteoporosis in the long bones are uncommon in laying hens housed in cage-free systems. Despite substantial evidence that load-bearing activity improves both structural and medullary bone quality in laying hens (Johnsson et al., 2023; Shipov et al., 2010), a higher proportion of hens in cage-free systems experience keel bone fractures (Wilkins et al., 2011). However, there is no empirical evidence linking keel bone fractures in cage-free hens to osteoporosis.

Keel bone damage

Keel bone deformities were reported in extensive production systems as early as the 1930s and 1940s (Buckner et al., 1949; Warren, 1937). After conventional cage systems became widespread, reports of keel bone fractures and deformities were relatively rare, and crooked keels were generally attributed to dietary deficiencies in calcium, phosphorus, and/or vitamin D. Coinciding with the shift from conventional cages to cage-free systems in several European countries in the late 1990s and early 2000s, keel bone fractures and deformities re-emerged as a major welfare issue. European studies of noncage systems report that up to 80% of laying hens may experience keel bone breakage and/or deformation, leading to severe skeletal damage and, in extreme cases, mortality (Sandilands et al., 2009; Wilkins et al., 2004, 2011). Gregory et al. (1990) conducted one of the earliest comprehensive assessments of skeletal fractures in laying hens, reporting end-of-lay fracture incidences of 25% in systems with perches and 12% in free-range systems, compared with only 5% in conventional cages; nearly 90% of fractures involved the keel and the furculum. Subsequent European studies documented keel fracture prevalences ranging from 49% to 74% across a variety of extensive housing systems (Freire et al., 2003; Nicol et al., 2006; Wilkins et al., 2004). In a comparison of 67 end-of-lay flocks in the United Kingdom, furnished cages exhibited the lowest fracture prevalence (36%), despite lower peak keel breaking strength than flocks housed in multitier systems with elevated perches (>80%; Wilkins et al., 2011). Similarly, assessments of Belgian commercial aviary systems reported a prevalence of 82.5%, with ~66% of hens exhibiting mid-keel fractures (Heerkens et al., 2016). Studies from North America indicate challenges comparable to those in Europe (Petrik et al., 2014; Regmi, Nelson, et al., 2016). A recent review across regions reported average prevalences ranging from 23.5% in conventional cages to 63% in single-tier aviaries (Rufener & Makagon, 2020).

The high incidence of keel fractures in aviaries and other extensive systems is attributed, in part, to the exposed anatomical position of the keel bone, which makes it vulnerable to traumatic collisions in complex housing environments. Approximately 21% of flights in aviary-housed hens end in failed landings (Campbell et al., 2016), which may contribute to keel fractures (Harlander-Matauschek et al., 2015; Wilkins et al., 2011), although the extent to which external trauma serves as the primary source has been debated (Thøfner et al., 2020). Accordingly, perch design and placement are key risk factors (Abrahamsson & Tauson, 1993; Sandilands et al., 2009; Wilkins et al., 2011), along with interrow spacing, flooring materials, perch access within nest boxes, and hen genetics (Heerkens et al., 2016). Housing- and husbandry-based mitigation has therefore focused on reducing collision-related trauma during movement in noncage systems. Incorporating ramps and intermediate platforms between elevated perches reduces falls by nearly 50% in multitier aviaries (Stratmann, Fröhlich, Gebhardt-Henrich, et al., 2015). Correspondingly, keel fractures assessed by palpation decline during early and mid-lay (Heerkens et al., 2016; Stratmann, Fröhlich, Gebhardt-Henrich, et al., 2015), although these benefits were not maintained at end-of-lay when fractures were assessed postmortem (66 weeks; Stratmann, Fröhlich, Gebhardt-Henrich, et al., 2015). Covering metal perches with polyurethane also shows moderate success in reducing collision-associated fractures (Stratmann, Fröhlich, Harlander-Matauschek, et al., 2015). Even so, these interventions typically lower overall fracture prevalence by only 20%–40% under controlled conditions, and approximately 40% of hens in well-managed cage-free systems continue to suffer keel fractures (Heerkens et al., 2016; Stratmann, Fröhlich, Gebhardt-Henrich, et al., 2015).

In parallel with housing and management interventions, genetic and genomic approaches have been explored to reduce keel fracture incidence. Substantial genetic variation in bone quality and fracture susceptibility has been reported (Bishop et al., 2000; Clark et al., 2008; Fleming et al., 2004; Käppeli et al., 2011; Vits et al., 2005). Although keel bone damage occurs across most contemporary commercial lines, prevalence and severity differ among genotypes. Modern brown strains, including Hy-Line Brown and Hy-Line Silver Brown, exhibit more severe keel deformities than Barred Rock hens (Regmi, Nelson, et al., 2016), and fracture prevalence is higher in ISA Brown than in Lohmann Brown Classic hens (50.6% vs 36.4%; Heerkens et al., 2016). These differences, however, are not consistently observed when data are pooled across housing systems (Wilkins et al., 2011). Quantitative genetic analyzes show that genetics explains over 40% of the variation in keel bone phenotypes, with heritability estimates for palpation scores ranging from 0.15 to 0.30 depending on the genetic line (Andersson et al., 2017; Candelotto et al., 2017). Importantly, keel status appears largely independent of shell strength, egg weight, and body weight, suggesting selection could improve keel integrity without major detriment to these traits. However, an inverse genetic relationship has been reported between keel bone integrity and early egg number (genetic correlations from −0.24 to −0.54), and in some lines, later onset of lay is genetically associated with improved bone quality and reduced keel damage (Andersson et al., 2017).

Age and production stage also influence keel fracture outcomes. Fracture prevalence is minimal at the end of the pullet phase, irrespective of rearing condition, but increases rapidly during early lay and typically plateaus after ~50 weeks of age (Casey-Trott et al., 2017; Petrik et al., 2014). The keel bone differs from long bones of the appendicular skeleton in that its ossification proceeds more slowly; 7%–18% of the keel remains unossified at the end of the pullet phase (Casey-Trott et al., 2017; Mutch et al., 2016). This developmental pattern may help explain the timing of fracture occurrence, particularly during early lay. Consistent with this, an ex vivo collision study found no age-related differences in fracture risk when similar impact energies were applied to birds aged 31, 45, and 65 weeks, suggesting functional keel maturity is achieved by approximately 30 weeks of age (Toscano et al., 2013).

Finally, individual behavioral strategies within aviaries contribute to heterogeneity in fracture risk and welfare outcomes. Variation in vertical space use and resource occupancy reflects how hens cope with physical (e.g., height transitions and perch geometry) and social challenges, with fracture risk varying by tier preference and production stage (Casey-Trott et al., 2017; Harlander-Matauschek et al., 2015). Hens that sustain severe keel injuries preferentially occupy upper tiers and elevated resources, whereas birds remaining predominantly in lower tiers tend to show poorer plumage condition—an indication of alternative welfare trade-offs linked to space use (Ali et al., 2020; Rufener et al., 2019; Yin et al., 2024).

In summary, keel bone fractures and deviations in laying hens have a multifactorial etiology involving housing design and use, collision risk, keel bone developmental characteristics, age and production stage, and genetics. Therefore genetic selection combined with favorable housing and nutritional interventions holds strong potential to reduce keel damage.

Behavioral issues

Indoor laying hen housing systems, particularly large cage-free and aviary environments, can create social and environmental conditions that can lead to behavioral issues within a flock. While hens form stable dominance hierarchies in small groups (Siegel, 1976), large commercial groups limit individual recognition, resulting in unstable social organization and increased competition for key resources such as feeders, perches, and nest boxes (Ali et al., 2020; D'Eath & Keeling, 2003; Grethen et al., 2023; Lorentz et al., 2022). This social instability and resource competition can promote maladaptive behaviors, including injurious feather pecking, toe pecking, piling, and smothering, as well as crowding and floor-egg laying (mislaid eggs). Increased social contact in these systems may further facilitate the spread of damaging behaviors throughout the flock, particularly under suboptimal rearing, housing, or management conditions. Common behavioral issues, as well as common causes of mortality in indoor housing systems for laying hens, are discussed below.

Feather pecking

Behavioral abnormalities, including feather pecking and cannibalism, are observed in both domestic and wild avian species and remain persistent challenges in commercial poultry production. Within the laying hen industry, feather pecking represents a major welfare concern due to its association with feather damage, skin injuries, and, in severe cases, mortality resulting from cannibalism (Duncan & Hughes, 1972). Feather pecking is defined as the pecking, pulling, and sometimes consumption of feathers from conspecifics (Blokhuis & Arkes, 1984). In addition to its welfare implications, this behavior leads to substantial economic losses through increased mortality, impaired production performance, and elevated management costs.

Pecking behaviors in poultry are commonly categorized into five types: aggressive pecking, which involves forceful pecking without necessarily causing injury or feather loss; gentle feather pecking, characterized by pecking at feather tips and generally considered a normal social behavior; severe feather pecking, which consists of repetitive forceful pecking and feather removal that often results in skin damage or bleeding; denuded area tissue pecking; and vent pecking (Savory, 1995). While gentle feather pecking is typically viewed as a benign social interaction, severe feather pecking can escalate into tissue pecking, resulting in feather loss and exposure of the skin. These injuries may subsequently trigger cannibalistic pecking, leading to open wounds, hemorrhage, and, in some cases death. Vent pecking represents a particularly damaging form of cannibalism, in which birds peck at the vent region and may extract internal tissues. Cannibalistic pecking can occur even in well-feathered birds and is sometimes observed near the onset of lay (Newberry, 2004).

Feather pecking is widely regarded as a multifactorial behavioral issue influenced by hormonal regulation, environmental conditions, nutritional factors, and genetic predisposition (Rodenburg et al., 2013). Severe feather pecking has frequently been linked to frustration arising from thwarted behavioral needs. Bird genotypes selected for high feather pecking behavior have been shown to exhibit more scratching behavior than low feather pecking genotypes, suggesting increased motivation for foraging or dust-bathing activities that may become redirected toward conspecifics in suboptimal environments (Rodenburg et al., 2005). During the rearing phase, high stocking density and poor litter quality have been identified as major risk factors for the development of severe feather pecking. During lay, additional risk factors include limited access to range areas, inadequate or poor-quality litter, insufficient environmental enrichment, and plumage damage acquired during rearing (Schwarzer et al., 2022). To reduce the physical consequences of injurious pecking, beak trimming is commonly implemented in commercial systems, often at one day of age, despite increasing concerns regarding its welfare implications.

The prevalence and severity of feather pecking vary across housing systems, including free-range, indoor cage-free, and conventional cage systems (Rehman et al., 2018). Although feather pecking occurs in all systems, it remains a significant welfare issue in cage-free environments. The rearing environment exerts long-lasting effects on pecking behavior in both pullets and adult laying hens (Gilani et al., 2013). Surveys of commercial housing systems have reported feather pecking prevalence ranging from 40% to 80% in cage-free flocks (Blokhuis et al., 2007). Feather and skin damage caused by feather pecking are strongly correlated with mortality due to cannibalism, with correlation coefficients reported between 0.60 and 0.80. Birds that die as a result of cannibalistic pecking typically exhibit significantly poorer plumage condition than birds that die from other causes (Kjaer & Sørensen, 2002). Marked differences among genotypes have been observed in plumage damage, skin injury, and pecking-related mortality, whereas some environmental variables, such as dietary methionine and cysteine levels, light intensity during rearing, and age at first access to outdoor range, have shown limited or inconsistent effects on pecking behavior (Kjaer & Sørensen, 2002).

In free-range systems, cannibalism prevalence has been reported to reach 8.8% at 61 weeks of age, while severe feather pecking affected approximately

23.8% of birds (Coton et al., 2019). Additional risk factors influencing severe feather pecking include lighting programs, genotype, farm location, and house age, as well as the extent of outdoor range use in indoor–outdoor systems. A survey of Canadian laying hen flocks reported a feather damage prevalence of approximately 26% in noncage systems, indicating that feather pecking remains problematic under alternative housing conditions (Decina et al., 2019). The authors suggested that litter provision and management practices, including manure removal and its subsequent effects on air quality, may play important roles in influencing feather pecking prevalence.

Nutrition is another important factor affecting feather pecking behavior in cage-free laying hens. Feeding barley silage has been shown to improve plumage condition by reducing aggressive and feather pecking behaviors, without negatively affecting egg production or egg quality (Johannson et al., 2016). Other nutritional factors associated with injurious pecking include feed restriction, low-fiber and high-energy diets, and deficiencies in essential amino acids and minerals (Van Krimpen et al., 2005). Dietary strategies that promote prolonged feeding and foraging activity are therefore considered important components of pecking prevention programs.

Feather pecking behavior has frequently been explained using the redirected foraging motivation hypothesis, which proposes that the behavior arises when foraging and exploratory motivations cannot be adequately expressed and are redirected toward conspecific feathers (Aerni et al., 2000; Blokhuis, 1986). Consequently, a wide range of environmental enrichment strategies aimed at stimulating foraging and exploration, such as provision of litter, roughage, and manipulable substrates, have been implemented to reduce feather pecking behavior to some extent (Dixon et al., 2010; van Staaveren et al., 2021). However, environmental enrichment is generally only one aspect of broader management interventions recommended for effectively reducing feather pecking and damage (van Staaveren et al., 2021).

Recent mechanistic research has expanded this behavioral framework by identifying roles for stress physiology, inflammation, and bidirectional gut–brain interactions in the development of feather pecking. Birds experimentally challenged with *Clostridium perfringens* showed reduced activity, including decreased feeding and gentle feather pecking, alongside increases in sitting and sleeping behavior (Calefi et al., 2016). Outbreaks of severe feather pecking have also been reported following *Escherichia coli* infections and during cases of chronic enteritis (Brunberg et al., 2016). Evidence for microbial involvement has been further supported by microbiota transfer studies comparing high feather pecking and low feather pecking genetic lines, in which high feather pecking birds showed greater abundance of Clostridiales, whereas low feather pecking birds exhibited higher levels of Lactobacillales (van der Eijk et al., 2020). Following microbiota transfer, behavioral differences were observed, including increased vocalization and reduced fear responses in birds receiving high feather pecking microbiota.

More recently, supplementation with *Lactobacillus rhamnosus* has been shown to reduce stress-induced feather pecking behavior in laying hens, accompanied by increased numbers of regulatory T cells, suggesting an immunomodulatory mechanism influencing behavior (Mindus et al., 2021). Collectively, these findings indicate that indoor cage-free systems, when improperly managed, may expose hens to social, environmental, and pathogenic stressors that elevate the risk of injurious feather pecking. Effective mitigation thus requires an integrated approach combining appropriate housing design, environmental enrichment, nutritional management, genetic selection, and health control strategies.

Toe pecking

Toe pecking is classified as a form of cannibalistic allo-pecking behavior in poultry and represents an important welfare concern. It is generally considered to be initiated by the presence of feather injury or blood on the feet, which attracts pecking from conspecifics. Toe-pecking victims exhibited a short latency to step down from perches, suggesting that conspecific hens targeting exposed toes at higher elevations may initiate or maintain toe-pecking behavior (Krause et al., 2011). Once initiated, toe pecking often becomes self-perpetuating with toe injuries drawing further attention from flock mates, intensifying pecking behavior (Leonard et al., 1995), and affected birds may even peck at their own injured toes (Hughes & Whitehead, 2007). In addition to pecking, toe and foot injuries may also result from contact with barn equipment, which can act as an initial trigger for subsequent pecking behavior (Rørvang et al., 2019).

Toe injuries are associated with reduced mobility and impaired walking ability, which pose particular challenges in noncage housing systems. In these systems, hens must navigate complex three-dimensional environments to access essential resources such as feed, water, nest boxes, and perches. Reduced mobility can therefore exacerbate welfare issues by limiting access to resources and increasing vulnerability to further pecking. Severe toe pecking injuries often develop rapidly and may be accompanied by hemorrhage and tissue loss, substantially compromising animal welfare (Buitenhuis et al., 2003). In addition to physical damage, toe pecking has been linked to altered stress physiology and the expression of fearful behavior in affected birds (Krause et al., 2011). Toe pecking has been identified as one of the most critical welfare issues in high-mortality genetic lines, although it has been reported as relatively infrequent in small furnished cage systems in Denmark (Rørvang et al., 2019).

Although toe pecking occurs less frequently than feather pecking or cloacal cannibalism (Leonard et al., 1995), it can still result in substantial mortality under certain conditions. Toe pecking has been documented in Swiss cage-free laying hen flocks, where it contributed to significant losses (Gebhardt-Henrich et al., 2023). In cage-free systems, toe pecking has been observed primarily during periods of heightened stress, such as the onset of lay following relocation to a new barn between 20 and 30 weeks of age, as well as during peak egg production between 26 and 50 weeks of age (Gebhardt-Henrich et al., 2023).

Associations between toe pecking, productivity, and physical condition have also been reported. In cage-free housing systems, plumage condition has been shown to correlate positively with egg production proxies, whereas toe injuries exhibit a negative correlation, suggesting that toe pecking may adversely affect both welfare and production outcomes (Toscano et al., 2025). Similar to feather pecking, toe pecking appears to be multifactorial and shares many of the same risk factors. Reduced light exposure has been suggested as a potential mitigation strategy (Coton et al., 2019), and the behavior has been reported more frequently in white hybrid strains (Gebhardt-Henrich et al., 2023). Light flicker may further contribute to toe pecking by inducing stress, as birds possess highly developed visual systems that are sensitive to subtle changes in light quality and intensity (Lisney et al., 2011).

Physiological and behavioral differences between toe-pecking victims and unaffected birds provide additional evidence of the stress associated with this behavior. Birds subjected to toe pecking have exhibited enlarged adrenal glands and a reduced latency to exit an elevated platform in step-down tests, indicating heightened stress responsiveness. The increased relative adrenal gland weight observed in toe-pecked birds is an indicator of chronic stress (Carsia & Weber, 2000).

Effective prevention and control in indoor systems depend on an integrated management approach, including housing and equipment design that minimizes toe injuries, appropriate lighting regimes to reduce visual stress, and broader strategies aimed at mitigating injurious pecking and stress-related behaviors (EFSA AHAW Panel (EFSA Panel on Animal Health and Animal Welfare), et al., 2023; Rodenburg et al., 2013).

Piling and smothering

Piling and smothering are anomalous behaviors observed in commercial laying hens. Piling occurs when hens congregate in tightly packed clusters, whereas smothering refers to fatalities that result from piling events. Although smothering is always preceded by piling, not all piling incidents result in smothering (Campbell et al., 2016). Commercial cage-free laying hens are particularly susceptible to smothering due to high bird densities within confined indoor spaces. While piling and smothering can occur across all housing systems, smothering in commercial cage-free flocks represents a significant welfare concern and a source of economic loss due to increased mortality (Bright & Johnson, 2011).

Piling and smothering behaviors can be broadly categorized into three types: fear-induced piling (hysteria or panic), nest-box piling, and recurrent (or creeping) piling (Bright & Johnson, 2011). Fear-induced piling and smothering are typically isolated events triggered by sudden disturbances and are often associated with high mortality, sometimes exceeding 20 birds. Nest-box piling and smothering occur when a large number of hens occupy the same nest box simultaneously. These events may result in mortality ranging from a few birds (1–2) to large numbers (>20) and are readily identifiable due to their specific location. The third type, recurrent or creeping piling and smothering, is characterized by slowly developing, repetitive piling behavior in apparently undisturbed flocks that can occur throughout the production period. This form of smothering generally results in lower mortality rates (Bright & Johnson, 2011; Giersberg et al., 2019; Hansen, 1976; Richards et al., 2012). Smothering events (excluding nest-box smothering) typically occur at unpredictable times and locations within the housing system (Richards et al., 2012). In laying hens, piling behavior was frequently observed when experimental cues to induce piling were not presented and appeared to be more prevalent around midday, particularly when hens were engaged in diverse activities in pen corners (Winter et al., 2022). Gray et al. (2020) proposed four hypotheses to explain piling behavior across different temporal scales: (1) hens approaching or avoiding an attractant or repellent, (2) social influences within the flock, (3) the effects of early life experiences, and (4) piling as a maladaptive form of collective behavior.

Although piling and smothering can occur across different housing systems, their welfare and economic impacts are greater in cage-free flocks due to higher cumulative mortality associated with larger flock sizes. Several environmental and management-related factors have been identified as potential contributors to piling in cage-free laying hens, including physical space, environmental enrichment, ambient temperature, lighting, human–animal interactions, and the overall farm management system (Mazocco et al., 2024). Cage-free hens often engage in synchronized behaviors such as dustbathing, which can result in localized overcrowding within litter areas. The open litter space may therefore promote piling during peak lay, mid lay, and end of lay stages of production (Campbell et al., 2016). Insufficient environmental stimulation may further contribute to piling behavior, as hens may adopt a socially oriented positioning strategy in the absence of competitive or exploratory cues (Gebhardt-Henrich & Stratmann, 2016). Daily temperature fluctuations have also been identified as a likely contributing factor to piling and the associated risk of suffocation (Bright & Johnson, 2011).

Effective husbandry management is generally more challenging in cage-free systems than in conventional cages. Stockpersons must devote additional time to managing multiple variables, including guiding hens to underutilized nest boxes—particularly during the early laying period—collecting floor eggs, and actively dispersing birds when abnormal behaviors such as piling are observed (Gray et al., 2020). Inadequate feeding management, including insufficient feeding frequency during lay and limited feed availability throughout the day, can increase competition among birds, elevating stress levels and potentially triggering piling behavior (Waldroup & Hellwig, 2000).

Commercial layer flocks (n = 12) in the United Kingdom were observed to pile more than four times per day on average, with each event lasting approximately 44 minutes (Armstrong et al., 2023). In the same study, peak piling activity typically occured during early afternoon and was associated with reduced egg production on the following day; specifically, an average of four piling events per day corresponded to a reduction of 7.35 eggs per 1000 birds (Armstrong et al., 2023). Considerable variation in piling duration has been documented among flocks. Notably, sister flocks originating from the same hatch and reared together exhibited similar piling distributions, as did flocks housed within the same shed or managed by the same personnel (Armstrong et al., 2023).

In another study, 48 piling events were recorded, with a maximum density of 187.93 birds/m^2 and up to 1204 birds involved in a single pile. Piling occurred repeatedly at the same location on 33 of 34 observation days, providing early evidence of consistent, site-specific piling behavior (Herbert et al., 2021). This consistency may be related to factors such as the water-to-feed ratio or the temperature conditions recorded on those days. In a cage-free Swiss layer flock, piling events were most often preceded by solitary hen movements (77.9%) or nonhysterical group movements (7.6%). The greatest accumulation of birds in piles was observed in both white and brown flocks more than 5 to 10 hours after lights-on. Compared with 30 weeks of age, flocks at 20 weeks exhibited fewer piling events, but involved more birds per event and longer piling durations (Winter et al., 2021). These patterns may reflect response facilitation, individual stimulus responses, or antipredator behavior. Additionally, piling behavior may be associated with diurnal activities such as dustbathing and floor egg laying in the litter area. These findings suggest that various factors, such as genetics, early life experiences, and husbandry practices may play important roles in shaping piling behavior.

In indoor cage-free environments, hens express a wide range of natural behaviors, provided that stocking density, flooring materials, system maintenance, and resource availability are appropriate. Adequate provision of enclosed nest boxes and sufficient perch space is essential to support foraging, ground scratching, and dustbathing behaviors. Hen movement data have shown promise for predicting piling events, as well as broader welfare and health outcomes, particularly when movement patterns are consistent over time (Sibanda et al., 2020). To optimize aviary utilization and reduce the risk of competition and suffocation, it is critical to analyze hen movement patterns and space use within the shed. For flocks that predominantly occupy upper tiers, installing ramps may encourage use of lower tiers and outdoor ranges. Additionally, placing supplementary resources—such as drinker lines or nest boxes—on higher feeder tiers may promote a more uniform spatial distribution of hens and help mitigate piling and smothering events.

Common causes of mortality

In commercial laying hen operations mortality is routinely monitored on a daily, weekly, or cumulative basis. Veterinarians often perform necropsy on mortality cases and use the flock history to identify underlying causes of mortality. Identifying mortality patterns at an early stage allows for timely corrective actions, improved disease control, and the recognition of underlying management or environmental shortcomings. In indoor cage-free systems, cumulative mortality is comparatively higher than in conventional cages and ranges from 5% to 12% and increases with age, often reaching 9% or more by 72 weeks (Schuck-Paim et al., 2021; Weeks et al., 2016). In the United States, laying hen flocks are frequently depopulated when cumulative mortality exceeds levels recommended in breeder management guidelines, highlighting the substantial economic consequences of elevated mortality during the production cycle (U.S. Department of Agriculture, Animal and Plant Health Inspection Service. Layers 2013, Part I: Reference of health and management practices on table-egg farms in the United States, 2013, 2014).

Mortality in indoor cage-free systems arises from a combination of infectious and noninfectious causes that often interact. Major infectious contributors include enteric diseases such as coccidiosis and secondary necrotic enteritis, *Escherichia coli* infections leading to septicemia, salpingitis, and egg peritonitis, and bacterial diseases such as fowl cholera, spotty liver disease, and emerging *Gallibacterium anatis* infections (Fulton, 2017; Gretarsson et al., 2023). Viral infections, particularly infectious bronchitis, contribute to mortality largely through reproductive tract damage and increased susceptibility to secondary bacterial infections (Fossum et al., 2009). Reproductive disorders, especially salpingitis, are among the most common pathological findings associated with death in laying hens (Fulton, 2017; Gretarsson et al., 2023).

Noninfectious causes of mortality are equally important in cage-free systems. Mortality associated with behavioral problems such as piling and smothering, injurious pecking, and cannibalism was discussed above. Metabolic disorders, including fatty liver hemorrhagic syndrome, account for a substantial proportion of deaths in older hens, whereas hypocalcemia can cause sudden death in actively laying birds with minimal gross lesions (Fulton, 2017; Gretarsson et al., 2023). Chronic under-conditioning and emaciation reflect prolonged health, nutritional, or access-to-resource issues. Overall, mortality in cage-free laying hens is multifactorial, and effective reduction depends on routine monitoring, consistent necropsy, and integrated management strategies addressing health, nutrition, behavior, and housing conditions throughout the production cycle.

Environmental and economic aspects

Increased global production and consumption trends have established eggs as the most affordable and environmentally efficient animal protein source (FAO,

IFAD, UNICEF, WFP and WHO, 2021). However, balancing environmental impacts with social and economic considerations remains central to achieving sustainable egg production.

Economic aspects

The transition from conventional cages to alternative housing systems has affected the profitability of egg production across Europe and the United States. Increased production costs in cage-free systems are primarily associated with feed, labor, and greater space allowances for hens (Dekker et al., 2011; Horne & Bondt, 2022; Matthews & Sumner, 2015). In smaller flocks, the transition from cages to cage-free housing can increase production costs by as much as 19% (Kritsa et al., 2024).

Conventional cages generally have the lowest variable costs, whereas single-tier and multitier aviary systems incur higher variable costs. In a controlled US study comparing conventional cages, enriched colony cages, and multitier aviaries within the same location, aviary systems had higher feed costs per dozen eggs than caged systems. This difference was largely attributed to higher cumulative mortality in aviaries, which resulted in lower hen-housed egg production rates (Matthews & Sumner, 2015).

Labor costs were also substantially higher in aviary systems, ranging from 6 to 8 cents per dozen eggs, compared with 2 to 5 cents per dozen in caged systems, and these costs increased over the production cycle. Labor associated with egg collection in aviaries accounted for up to 50% of total labor costs (Matthews & Sumner, 2015). Pullet costs further contributed to higher production expenses in aviaries. The cost of pullets per dozen eggs was compounded by higher cumulative mortality, with aviaries experiencing 13.3% mortality compared with 5% in caged systems. Although aviary systems may require more energy to maintain thermal comfort during colder months, energy use costs did not differ significantly among housing types (Matthews & Sumner, 2015).

Despite higher production costs, the premium paid for eggs produced in aviary systems can offset these expenses. This price premium has resulted in higher gross margins per laying hen per production cycle, with increases of up to 13% reported in some cases, particularly within European production systems (Dekker et al., 2011; Kritsa et al., 2024).

Environmental aspects

Within indoor egg production systems, factors such as air quality, greenhouse gas (GHG) emissions, and land and energy use are important determinants of overall environmental impact. As the industry transitions from caged to cage-free systems, stakeholders have experienced trade-offs between environmental performance and animal welfare outcomes (de Olde et al., 2020). These trade-offs are often driven by differences in housing design, egg production per hen, feed formulation, and management practices. Public perceptions of sustainability outcomes across egg production systems have further renewed interest in quantifying their environmental impacts. For example, surveys of consumers and the general public indicate that cage-free aviary systems are perceived as more environmentally sustainable than conventional cages, despite published literature suggesting otherwise (Ochs et al., 2018). This perception gap underscores the importance of aligning environmental sustainability goals with the animal welfare benefits associated with cage-free systems.

Across housing types, the environmental footprint per unit of egg produced is consistently lower in conventional cage systems than in single or multitiered aviaries. Life cycle assessments conducted on caged and indoor cage-free systems in the Netherlands reported approximately 20% greater global warming potential per kilogram of eggs produced in cage-free systems compared with conventional cages (Dekker et al., 2011). Emissions of GHGs such as carbon dioxide and nitrous oxide were lower in conventional cages due to improved feed conversion ratios, reduced fossil energy use during production and transport, and lower nitrous oxide emissions associated with manure management (Dekker et al., 2011).

Poultry production also contributes substantially to ammonia emissions, with implications for worker and bird health and a role in acid rain formation (Xin et al., 2011). Nearly 20% of total ammonia emissions in the United States are attributed to poultry production (Battye et al., 1994). Aviary hen houses have been reported to emit approximately 36% more ammonia per hen per day than conventional cage systems (Shepherd et al., 2015). Elevated ammonia concentrations can exacerbate particulate matter generation, resulting in poor indoor air quality. Farm-level emission rates of particulate matter, including PM_{10} and $PM_{2.5}$, have been reported to be 6 to 10-fold higher in aviary systems compared with conventional cages (Shepherd et al., 2015; Zhao et al., 2015). The presence of floor litter, increased hen activity, and reduced ventilation efficiency during colder months are primary contributors to air quality challenges in aviary housing (Zhao et al., 2015).

Energy use follows a similar pattern to gaseous and particulate emissions, with conventional cages requiring less energy per kilogram of eggs produced than single or multitiered aviary systems. Feed concentrate, litter production and transport account for approximately one-third to one-half of total energy use across systems (Dekker et al., 2011). The lower carbon footprint of conventional cages is further reflected in their reduced land use requirements relative to alternative indoor systems (Dekker et al., 2011). Additionally, lower cumulative mortality rates contribute to the environmental sustainability of conventional cages. In commercial egg production, mortality losses are not replaced, leading to increased energy and resource inputs per animal unit in systems with higher cumulative mortality. An increase in cumulative mortality from 0% to 10% within a single production cycle can raise energy demand and

GHG emissions by approximately 15% (Weeks et al., 2016). Conversely, reducing cumulative mortality to 5% can lower GHG emissions by as much as 25% (Weeks et al., 2016).

Despite these differences, the housing system alone does not solely determine environmental outcomes. A comprehensive life cycle assessment of the US egg supply chain demonstrated that differences in feed formulation, particularly the inclusion of animal-derived ingredients, can outweigh differences in resource-use efficiency between cage and cage-free systems (Turner et al., 2025). Consequently, dietary interventions aimed at reducing crude protein levels with supplementation of synthetic amino acids have been adopted to mitigate environmental impacts. A reduction in dietary crude protein by 2 percentage points has been shown to decrease global warming potential by 5% and energy use by 2% through reduced nitrogen excretion (Setoguchi et al., 2025). Additional mitigation strategies include improved heat exchangers and air filtration systems, increased litter removal frequency, and manure treatment using adsorbents, acidifiers, and urease inhibitors (de Olde et al., 2020; Xin et al., 2011). Moreover, reducing litter surface area has been shown to decrease litter use in cage-free systems without adversely affecting hen comfort or maintenance behaviors (Gonzalez-Mora et al., 2021).

Stakeholders have consistently emphasized the need for integrated solutions to address sustainability challenges in egg production systems. Desired approaches include innovation in emission reduction technologies, outcome-oriented regulatory frameworks, and the adoption of real-time emission monitoring systems (de Olde et al., 2020).

AI disclosure

During the preparation of this work, the author(s) used Microsoft CoPilot in order to address grammatical and syntactic errors. After using this tool/service, the author(s) reviewed and edited the content as needed and take(s) full responsibility for the content of the publication.

References

Abrahamsson, P., & Tauson, R. (1993). Effect of perches at different positions in conventional cages for laying hens of two different strains. *Acta Agriculturae Scandinavica A: Animal Sciences, 43*(4), 228–235. https://doi.org/10.1080/09064709309410171.

Aerni, V., El-Lethey, H., & Wechsler, B. (2000). Effect of foraging material and food form on feather pecking in laying hens. *British Poultry Science, 41*(1), 16–21. https://doi.org/10.1080/00071660086349.

Ali, A. B. A., Campbell, D. L. M., & Siegford, J. M. (2020). A risk assessment of health, production, and resource occupancy for 4 laying hen strains across the lay cycle in a commercial-style aviary system. *Poultry Science, 99*(10), 4672–4684. https://doi.org/10.1016/j.psj.2020.05.057.

Andersson, B., Icken, W., Kaufmann, F., & Schmutz, M. (2017). Genetic aspects of keel bone deformities and fractures determined by palpation in laying hens. *Lohmann Information, 51*(2), 33–40. https://lohmann-breeders.com/media/2020/08/Andersson.pdf.

Appleby, M. C. (2003). The European Union ban on conventional cages for laying hens: History and prospects. *Journal of Applied Animal Welfare Science, 6*(2), 103–121. https://doi.org/10.1207/S15327604JAWS0602_03.

Appleby, M. C., Hughes, B. O., & Elson, H. A. (1992). Poultry production systems: Behaviour, management and welfare. CAB International, Wallingford, UK.

Armstrong, D., Asher, L., Rayner, A., Ngidda, H., Sharma, B., & Gray, H. (2023). The effect of piling behavior on the production and mortality of free-range laying hens. *Poultry Science, 102*(10), 102989. https://doi.org/10.1016/j.psj.2023.102989.

Battye, R., Battye, W., Overcash, C., & Fudge, S. (1994). Development and selection of ammonia emission factors. EPA Contract No. 68-D3-0034, Work Assignment 0-3. Office of Research and Development, U.S. Environmental Protection Agency, Washington, DC.

Bishop, S. C., Fleming, R. H., McCormack, H. A., Flock, D. K., & Whitehead, C. C. (2000). Inheritance of bone characteristics affecting osteoporosis in laying hens. *British Poultry Science, 41*(1), 33–40. https://doi.org/10.1080/00071660086376.

Blokhuis, H. J. (1986). Feather-pecking in poultry: Its relation with ground-pecking. *Applied Animal Behaviour Science, 16*(1), 63–67. https://doi.org/10.1016/0168-1591(86)90040-7.

Blokhuis, H. J., & Arkes, J. G. (1984). Some observations on the development of feather-pecking in poultry. *Applied Animal Behaviour Science, 12*(1-2), 145–157. https://doi.org/10.1016/0168-1591(84)90104-7.

Blokhuis, H. J., Van Niekerk, T. F., Bessei, W., Elson, A., Guémené, D., Kjaer, J. B., Maria Levrino, G. A., Nicol, C. J., Tauson, R., Weeks, C. A., & Van De Weerd, H. A. (2007). The LayWel project: Welfare implications of changes in production systems for laying hens. *World's Poultry Science Journal, 63*(1), 101–114. https://doi.org/10.1017/s0043933907001328.

Brambell, F. W. (1965). Report of the technical committee to enquire into the welfare of animals kept under intensive livestock husbandry systems. Cmnd., 2836. Her Majesty's Stationery Office.

Bright, & Johnson (2011). Short communications: Smothering in commercial free-range laying hens: A preliminary investigation. *Veterinary Record, 168*(19), 512–512 https://doi.org/10.1136/vr.c7462.

Brunberg, E. I., Bas Rodenburg, T., Rydhmer, L., Kjaer, J. B., Jensen, P., & Keeling, L. J. (2016). Omnivores going astray: A review and new synthesis of abnormal behavior in pigs and laying hens. *Frontiers in Veterinary Science, 3*, 57. https://doi.org/10.3389/fvets.2016.00057.

Buckner, G. D., Insko, W. M., Henry, A. H., & Wachs, E. F. (1949). Rate of growth and calcification of the sternum of male and female new hampshire chickens having crooked keels. *Poultry Science, 28*(2), 289–292. https://doi.org/10.3382/ps.0280289.

Buitenhuis, A. J., Rodenburg, T. B., Siwek, M., Cornelissen, S. J. B., Nieuwland, M. G. B., Crooijmans, R. P. M. A., Groenen, M. A. M., Koene, P., Bovenhuis, H., & Van Der Poel, J. J. (2003). Identification of quantitative trait loci for receiving pecks in young and adult laying hens. *Poultry Science, 82*(11), 1661–1667. https://doi.org/10.1093/ps/82.11.1661.

Calefi, A. S., Da Silva Fonseca, J. G., Cohn, D. W. H., Honda, B. T. B., Costola-De-Souza, C., Tsugiyama, L. E., Quinteiro-Filho, W. M., Piantino Ferreira, A. J., & Palermo-Neto, J. (2016). The gut-brain axis interactions during heat stress and avian necrotic enteritis. *Poultry Science, 95*(5), 1005–1014. https://doi.org/10.3382/ps/pew021.

Campbell, D. L. M., Goodwin, S. L., Makagon, M. M., Swanson, J. C., & Siegford, J. M. (2016). Failed landings after laying hen flight in a commercial aviary over two flock cycles. *Poultry Science, 95*(1), 188–197. https://doi.org/10.3382/ps/pev270.

Candelotto, L., Stratmann, A., Gebhardt-Henrich, S. G., Rufener, C., Van De Braak, T., & Toscano, M. J. (2017). Susceptibility to keel bone fractures in laying hens and the role of genetic variation. *Poultry Science, 96*(10), 3517–3528. https://doi.org/10.3382/ps/pex146.

Carsia, R. V., & Weber, H. (2000). Dietary protein restriction stress in the domestic chicken (*Gallus gallus* domesticus) induces remodeling of adrenal steroidogenic tissue that supports hyperfunction. *General and Comparative Endocrinology, 120*(1), 99–107. https://doi.org/10.1006/gcen.2000.7534.

Casey-Trott, T. M., Korver, D. R., Guerin, M. T., Sandilands, V., Torrey, S., & Widowski, T. M. (2017). Opportunities for exercise during pullet rearing, Part I: Effect on the musculoskeletal characteristics of pullets. *Poultry Science, 96*(8), 2509–2517. https://doi.org/10.3382/ps/pex059.

Clark, W. D., Cox, W. R., & Silversides, F. G. (2008). Bone fracture incidence in end-of-lay high-producing, noncommercial laying hens identified using radiographs. *Poultry Science, 87*(10), 1964–1970. https://doi.org/10.3382/ps.2008-00115.

Comar, C. L., & Driggers, J. C. (1949). Secretion of radioactive calcium in the hen's egg. *109*(2829), 282. https://doi.org/10.1126/science.109.2829.282.

Commission of the European Communities (CEC). (1999). Council Directive 1999/74/EC laying down minimum standards for the protection of laying hens. Official Journal of European Communities. L203. 53–57.

Coton, J., Guinebretière, M., Guesdon, V., Chiron, G., Mindus, C., Laravoire, A., Pauthier, G., Balaine, L., Descamps, M., Bignon, L., Huneau-Salaün, A., & Michel, V. (2019). Feather pecking in laying hens housed in free-range or furnished-cage systems on French farms. *British Poultry Science, 60*(6), 617–627. https://doi.org/10.1080/00071668.2019.1639137.

Couch, J. R. (1955). Cage layer fatigue. *Feed Age, 5*, 55–57.

Dacke, C. G., Arkle, S., Cook, D. J., Wormstone, I. M., Jones, S., Zaidi, M., & Bascal, Z. A. (1993). Medullary bone and avian calcium regulation. *Journal of Experimental Biology, 184*(1), 63–88. https://doi.org/10.1242/jeb.184.1.63.

de Olde, E. M., van der Linden, A., olde Bolhaar, L. D., & de Boer, I. J. M. (2020). Sustainability challenges and innovations in the Dutch egg sector. *Journal of Cleaner Production, 258*, 120974. https://doi.org/10.1016/j.jclepro.2020.120974.

D'Eath, R. B., & Keeling, L. J. (2003). Social discrimination and aggression by laying hens in large groups: From peck orders to social tolerance. *Applied Animal Behaviour Science, 84*(3), 197–212. https://doi.org/10.1016/j.applanim.2003.08.010.

Decina, C., Berke, O., Van Staaveren, N., Baes, C. F., Widowski, T. M., & Harlander-Matauschek, A. (2019). A cross-sectional study on feather cover damage in Canadian laying hens in non-cage housing systems. *BMC Veterinary Research, 15*(1), 435. https://doi.org/10.1186/s12917-019-2168-2.

Dekker, S. E. M., de Boer, I. J. M., Vermeij, I., Aarnink, A. J. A., & Koerkamp, P. W. G. G. (2011). Ecological and economic evaluation of Dutch egg production systems. *Livestock Science, 139*(1-2), 109–121. https://doi.org/10.1016/j.livsci.2011.03.011.

Dixon, L. M., Duncan, I. J. H., & Mason, G. J. (2010). The effects of four types of enrichment on feather-pecking behaviour in laying hens housed in barren environments. *Animal Welfare, 19*(4), 429–435. https://doi.org/10.1017/s0962728600001913.

Duncan, I. J. H., & Hughes, B. O. (1972). The influence of strain and environmental factors upon feather pecking and cannibalism in fowls. *British Poultry Science, 13*(6), 525–547. https://doi.org/10.1080/00071667208415981.

Elaroussi, M. A., Forte, L. R., Eber, S. L., & Biellier, H. V. (1993). Adaptation of the kidney during reproduction: Role of estrogen in the regulation of responsiveness to parathyroid hormone. *Poultry Science, 72*(8), 1548–1556. https://doi.org/10.3382/ps.0721548.

Elson, H. A. (2002). Half a century of egg production – The industry's development from 1962 to 2012. *Poultry International, 41*(5), 8–16.

Elson, H. A. (2011). Housing and husbandry of laying hens: Past, present, and future. *Lohmann Information, 46*, 16–24.

FAO, IFAD, UNICEF, WFP and WHO. 2021. The State of Food Security and Nutrition in the World 2021. Transforming food systems for food security, improved nutrition and affordable healthy diets for all. Rome, FAO.https://doi.org/10.4060/cb4474en.

Fleming, R. H., McCormack, H. A., McTeir, L., & Whitehead, C. C. (2004). Incidence, pathology and prevention of keel bone deformities in the laying hen. *British Poultry Science, 45*(3), 320–330. https://doi.org/10.1080/00071660410001730815.

Fossum, O., Jansson, D. S., Etterlin, P. E., & Vgsholm, I. (2009). Causes of mortality in laying hens in different housing systems in 2001 to 2004. *Acta Veterinaria Scandinavica, 51*(1). https://doi.org/10.1186/1751-0147-51-3.

Freire, R., Wilkins, L. J., Short, F., & Nicol, C. J. (2003). Behaviour and welfare of individual laying hens in a non-cage system. *British Poultry Science, 44*(1), 22–29. https://doi.org/10.1080/0007166031000085391.

Fulton, R. M. (2017). Causes of normal mortality in commercial egg-laying chickens. *Avian Diseases, 61*(3), 289–295. https://doi.org/10.1637/11556-120816-RegR.

Gebhardt-Henrich, S. G., Mueller, S., Zanini, L., & Toscano, M. J. (2023). A survey about the welfare problem "toe pecking" in Swiss laying hens. *Applied Animal Behaviour Science, 259*, 105854. https://doi.org/10.1016/j.applanim.2023.105854.

Gebhardt-Henrich, S. G., & Stratmann, A. (2016). What is causing smothering in laying hens? *Veterinary Record, 179*(10), 250–251. https://doi.org/10.1136/vr.i4618.

Giersberg, M. F., Kemper, N., & Spindler, B. (2019). Pecking and piling: The behaviour of conventional layer hybrids and dual-purpose hens in the nest. *Applied Animal Behaviour Science, 214*, 50–56. https://doi.org/10.1016/j.applanim.2019.02.016.

Gilani, A. M., Knowles, T. G., & Nicol, C. J. (2013). The effect of rearing environment on feather pecking in young and adult laying hens. *Applied Animal Behaviour Science, 148*(1-2), 54–63. https://doi.org/10.1016/j.applanim.2013.07.014.

Gonzalez-Mora, A. F., Larios, A. D., Rousseau, A. N., Godbout, S., Morin, C., Palacios, J. H., Grenier, M., & Fournel, S. (2021). Assessing environmental control strategies in cage-free egg production systems: Effect on spatial occupancy and natural behaviors. *Animals, 11*(1), 1–13. https://doi.org/10.3390/ani11010017.

Gray, H., Davies, R., Bright, A., Rayner, A., & Asher, L. (2020). Why do hens pile? Hypothesizing the causes and consequences. *Frontiers in Veterinary Science, 7*, 616836. https://doi.org/10.3389/fvets.2020.616836.

Gregory, N. G., & Wilkins, L. J. (1989). Broken bones in domestic fowl: Handling and processing damage in end-of-lay battery hens. *British Poultry Science, 30*(3), 555–562. https://doi.org/10.1080/00071668908417179.

Gregory, N. G., Wilkins, L. J., Eleperuma, S. D., Ballantyne, A. J., & Overfield, N. D. (1990). Broken bones in domestic fowls: Effect of husbandry system and stunning method in end-of-lay hens. *British Poultry Science, 31*(1), 59–69. https://doi.org/10.1080/00071669008417231.

Gretarsson, P., Kittelsen, K., Moe, R. O., Vasdal, G., & Toftaker, I. (2023). End of lay postmortem findings in aviary housed laying hens. *Poultry Science, 102*(2), 102332. https://doi.org/10.1016/j.psj.2022.102332.

Grethen, K. J., Gómez, Y., & Toscano, M. J. (2023). Coup in the coop: Rank changes in chicken dominance hierarchies over maturation. *Behavioural Processes, 210*, 104904. https://doi.org/10.1016/j.beproc.2023.104904.

Hansen, R. S. (1976). Nervousness and hysteria of mature female chickens. *Poultry Science, 55*(2), 531–543. https://doi.org/10.3382/ps.0550531.

Harlander-Matauschek, A., Rodenburg, T. B., Sandilands, V., Tobalske, B. W., & Toscano, M. J. (2015). Causes of keel bone damage and their solutions in laying hens. *World's Poultry Science Journal, 71*(3), 461–472. https://doi.org/10.1017/s0043933915002135.

Heerkens, J. L. T., Delezie, E., Rodenburg, T. B., Kempen, I., Zoons, J., Ampe, B., & Tuyttens, F. A. M. (2016). Risk factors associated with keel bone and foot pad disorders in laying hens housed in aviary systems. *Poultry Science, 95*(3), 482–488. https://doi.org/10.3382/ps/pev339.

Herbert, G. T., Redfearn, W. D., Brass, E., Dalton, H. A., Gill, R., Brass, D., Smith, C., Rayner, A. C., & Asher, L. (2021). Extreme crowding in laying hens during a recurrent smothering outbreak. *Veterinary Record, 188*(12). https://doi.org/10.1002/vetr.245.

Holt, P. S., Davies, R. H., Dewulf, J., Gast, R. K., Huwe, J. K., Jones, D. R., Waltman, D., & Willian, K. R. (2011). The impact of different housing systems on egg safety and quality. *Poultry Science, 90*(1), 251–262. https://doi.org/10.3382/ps.2010-00794.

Horne, P. L. M., & Bondt, N. (2022). *Competitiveness of the EU egg sector, base year 2021: International comparison of production costs of eggs and egg products* (Report No. 52). Wageningen Economic Research, Wageningen, the Netherlands. https://doi.org/10.18174/583668.

Hughes, B. O., & Whitehead, C. C. (2007). Sodium deprivation, feather pecking and activity in laying hens. *British Poultry Science, 15*(5), 435–439. https://doi.org/10.1080/00071667408416131.

Johannson, S. G., Raginski, C., Schwean-Lardner, K., Classen, H. L., & Plaizier, J. (2016). Providing laying hens in group-housed enriched cages with access to barley silage reduces aggressive and feather-pecking behaviour. *Canadian Journal of Animal Science, 96*(2), 161–171. https://doi.org/10.1139/cjas-2015-0133.

Johnsson, M., Wall, H., Lopes Pinto, F. A., Fleming, R. H., McCormack, H. A., Benavides-Reyes, C., Dominguez-Gasca, N., Sanchez-Rodriguez, E., Dunn, I. C., Rodriguez-Navarro, A. B., Kindmark, A., & de Koning, D. J. (2023). Genetics of tibia bone properties of crossbred commercial laying hens in different housing systems. *G3: Genes, Genomes, Genetics, 13*(2). https://doi.org/10.1093/g3journal/jkac302.

Kajlich, A. S., Shivaprasad, H. L., Trampel, D. W., Hill, A. E., Parsons, R. L., Millman, S. T., & Mench, J. A. (2016). Incidence, severity, and welfare implications of lesions observed postmortem in laying hens from commercial noncage farms in California and Iowa. *Avian Diseases, 60*(1), 8–15. https://doi.org/10.1637/11247-080415-Reg.1.

Käppeli, S., Gebhardt-Henrich, S. G., Fröhlich, E., Pfulg, A., Schäublin, H., & Stoffel, M. H. (2011). Effects of housing, perches, genetics, and 25-hydroxycholecalciferol on keel bone deformities in laying hens. *Poultry Science, 90*(8), 1637–1644. https://doi.org/10.3382/ps.2011-01379.

Kaukonen, E., & Valros, A. (2019). Feather pecking and cannibalism in non-beak-trimmed laying hen flocks—Farmers' perspectives. *Animals, 9*(2), 43. https://doi.org/10.3390/ani9020043.

Kjaer, J. B., & Sørensen, P. (2002). Feather pecking and cannibalism in free-range laying hens as affected by genotype, dietary level of methionine + cystine, light intensity during rearing and age at first access to the range area. *Applied Animal Behaviour Science, 76*(1), 21–39. https://doi.org/10.1016/S0168-1591(01)00209-X.

Krause, E. T., Petow, S., & Kjaer, J. B. (2011). A note on the physiological and behavioural consequences of cannibalistic toe pecking in laying hens (*Gallus gallus* domesticus). *European Poultry Science, 75*(2), 140–143. https://doi.org/10.1016/s0003-9098(25)00768-4.

Kritsa, M. Z., Tsiboukas, K., Sossidou, E. N., Simitzis, P. E., & Goliomytis, M. (2024). Partial budget analysis of laying hens' transition from cages to production systems of improved welfare: A case study in Greece. *British Poultry Science, 65*(5), 503–512. https://doi.org/10.1080/00071668.2024.2360631.

Lay, D. C., Fulton, R. M., Hester, P. Y., Karcher, D. M., Kjaer, J. B., Mench, J. A., Mullens, B. A., Newberry, R. C., Nicol, C. J., O'Sullivan, N. P., & Porter, R. E. (2011). Hen welfare in different housing systems. *Poultry Science, 90*(1), 278–294. https://doi.org/10.3382/ps.2010-00962.

Leonard, M. L., Horn, A. G., & Fairfull, R. W. (1995). Correlates and consequences of allopecking in white leghorn chickens. *Applied Animal Behaviour Science, 43*(1), 17–26. https://doi.org/10.1016/0168-1591(94)00551-O.

Li, X., Cai, X., Wang, X., Zhu, L., Yan, H., Yao, J., & Yang, C. (2024). Understanding the causes of keel bone damage and its effects on the welfare of laying hens. *Animals, 14*(24), 3655. https://doi.org/10.3390/ani14243655.

Lisney, T. J., Rubene, D., Rózsa, J., Løvlie, H., Håstad, O., & Ödeen, A. (2011). Behavioural assessment of flicker fusion frequency in chicken *Gallus gallus* domesticus. *Vision Research, 51*(12), 1324–1332. https://doi.org/10.1016/j.visres.2011.04.009.

Lorentz, B., Neupane, N., Robison, C., & Regmi, P. (2022). Leveraging RFID tracking to monitor hen movement in cage-free aviaries. In Proceedings of the Poultry Science Association Annual Meeting (No. 101, E-supplement). San Antonio, TX.

Matthews, W. A., & Sumner, D. A. (2015). Effects of housing system on the costs of commercial egg production. *Poultry Science, 94*(3), 552–557. https://doi.org/10.3382/ps/peu011.

Mazocco, C. C., de Castro Júnior, S. L., Silveira, R. M. F., Poletto, R., & da Silva, I. J. O. (2024). Laying hens: Why smothering and not surviving?—A literature review. *Animals, 14*(11), 1518. https://doi.org/10.3390/ani14111518.

Mindus, C., van Staaveren, N., Bharwani, A., Fuchs, D., Gostner, J. M., Kjaer, J. B., Kunze, W., Mian, M. F., Shoveller, A. K., Forsythe, P., & Harlander-Matauschek, A. (2021). Ingestion of *Lactobacillus rhamnosus* modulates chronic stress-induced feather pecking in chickens. *Scientific Reports, 11*(1), 17119. https://doi.org/10.1038/s41598-021-96615-x.

Mutch, V. Q., Regmi, P., Robison, C. I., & Karcher, D. M. (2016). Influence of different housing systems on laying hen keel bone. In Proceedings of the Poultry Science Association Annual Meeting (No. 95, E-supplement). New Orleans, Louisiana.

Nasr, M. A. F., Murrell, J., Wilkins, L. J., & Nicol, C. J. (2012). The effect of keel fractures on egg-production parameters, mobility and behaviour in individual laying hens. *Animal Welfare, 21*(1), 127–135. https://doi.org/10.7120/096272812799129376.

Nasr, M. A. F., Nicol, C. J., & Murrell, J. C. (2012). Do laying hens with keel bone fractures experience pain? *PLoS One, 7*(8), e42420. https://doi.org/10.1371/journal.pone.0042420.

Newberry, R. C. (2004). Welfare of the laying hen. In G. C. Perry (Ed.). Cannibalism. Wallingford, UK: CABI.

Nicol, C. J., Brown, S. N., Glen, E., Pope, S. J., Short, F. J., Warriss, P. D., Zimmerman, P. H., & Wilkins, L. J. (2006). Effects of stocking density, flock size and management on the welfare of laying hens in single-tier aviaries. *British Poultry Science, 47*(2), 135–146. https://doi.org/10.1080/00071660600610609.

EFSA AHAW Panel (EFSA Panel on Animal Health and Animal Welfare), Nielsen, S. S., Alvarez, J., Bicout, D. J., Calistri, P., Canali, E., Drewe, J. A., Garin-Bastuji, B., Rojas, J. L. G., Schmidt, C. G., Herskin, M., Chueca, M.Á. M., Padalino, B., Pasquali, P., Roberts, H. C., Spoolder, H., Stahl, K., Velarde, A., Viltrop, A., Winckler, C., ... Michel, V. (2023). Welfare of laying hens on farm. EFSA Journal, 21(2), e07789. https://doi.org/10.2903/j.efsa.2023.7789

Nightingale, T. E., Littlefield, L. H., & Merkley, L. W. (1972). Osteoporosis induced by unilateral wing immobilization. *Poultry Science, 51*, 1844–1845.

Ochs, D. S., Wolf, C. A., Widmar, N. J. O., & Bir, C. (2018). Consumer perceptions of egg-laying hen housing systems. *Poultry Science, 97*(10), 3390–3396. https://doi.org/10.3382/ps/pey205.

Olsson, I. A. S., & Keeling, L. J. (2000). Night-time roosting in laying hens and the effect of thwarting access to perches. *Applied Animal Behaviour Science, 68*(3), 243–256. https://doi.org/10.1016/S0168-1591(00)00097-6.

Passi, N., & Gefen, A. (2005). Trabecular bone contributes to strength of the proximal femur under mediolateral impact in the avian. *Journal of Biomechanical Engineering, 127*(1), 198–203. https://doi.org/10.1115/1.1835366.

Petrik, M. T., Guerin, M. T., & Widowski, T. M. (2014). On-farm comparison of keel fracture prevalence and other welfare indicators in conventional cage and floor-housed laying hens in Ontario, Canada. *Poultry Science, 94*(4), 579–585. https://doi.org/10.3382/ps/pev039.

Regmi, P., Deland, T. S., Steibel, J. P., Robison, C. I., Haut, R. C., Orth, M. W., & Karcher, D. M. (2015). Effect of rearing environment on bone growth of pullets. *Poultry Science, 94*(3), 502–511. https://doi.org/10.3382/ps/peu041.

Regmi, P., Nelson, N., Steibel, J. P., Anderson, K. E., & Karcher, D. M. (2016). Comparisons of bone properties and keel deformities between strains and housing systems in end-of-lay hens. *Poultry Science, 95*(10), 2225–2234. https://doi.org/10.3382/ps/pew199.

Regmi, P., Smith, N., Nelson, N., Haut, R. C., Orth, M. W., & Karcher, D. M. (2016). Housing conditions alter properties of the tibia and humerus during the laying phase in Lohmann white Leghorn hens. *Poultry Science, 95*(1), 198–206. https://doi.org/10.3382/ps/pev209.

Rehman, M. S., Mahmud, A., Mehmood, S., Pasha, T. N., Khan, M. T., & Hussain, J. (2018). Assessing behavior in Aseel pullets under free-range, part-time free-range, and cage system during growing phase. *Poultry Science, 97*(3), 725–732. https://doi.org/10.3382/ps/pex355.

Richards, G. J., Brown, S. N., Booth, F., Toscano, M. J., & Wilkins, L. J. (2012). Panic in free-range laying hens. *Veterinary Record, 170*(20), 519–519. https://doi.org/10.1136/vr.100685.

Rodenburg, T. B., Koene, P., Bokkers, E. A. M., Bos, M. E. H., Uitdehaag, K. A., & Spruijt, B. M. (2005). Can short-term frustration facilitate feather pecking in laying hens? *Applied Animal Behaviour Science, 91*(1-2), 85–101. https://doi.org/10.1016/j.applanim.2004.08.023.

Rodenburg, T. B., Van Krimpen, M. M., De Jong, I. C., De Haas, E. N., Kops, M. S., Riedstra, B. J., Nordquist, R. E., Wagenaar, J. P., Bestman, M., & Nicol, C. J. (2013). The prevention and control of feather pecking in laying hens: Identifying the underlying principles. *World's Poultry Science Journal, 69*(2), 361–374. https://doi.org/10.1017/s0043933913000354.

Rørvang, M. V., Hinrichsen, L. K., & Riber, A. B. (2019). Welfare of layers housed in small furnished cages on Danish commercial farms: The condition of keel bone, feet, plumage and skin. *British Poultry Science, 60*(1), 1–7. https://doi.org/10.1080/00071668.2018.1533632.

Rufener, C., Baur, S., Stratmann, A., & Toscano, M. J. (2019). Keel bone fractures affect egg laying performance but not egg quality in laying hens housed in a commercial aviary system. *Poultry Science, 98*(4), 1589–1600. https://doi.org/10.3382/ps/pey544.

Rufener, C., & Makagon, M. M. (2020). Keel bone fractures in laying hens: A systematic review of prevalence across age, housing systems, and strains. *Journal of Animal Science, 98*, S36–S51. https://doi.org/10.1093/JAS/SKAA145.

Sandilands, V., Moinard, C., & Sparks, N. H. C. (2009). Providing laying hens with perches: Fulfilling behavioural needs but causing injury? *British Poultry Science, 50*(4), 395–406. https://doi.org/10.1080/00071660903110844.

Savory, C. J. (1995). Feather pecking and cannibalism. *World's Poultry Science Journal, 51*(2), 215–219. https://doi.org/10.1079/WPS19950016.

Schuck-Paim, C., Negro-Calduch, E., & Alonso, W. J. (2021). Laying hen mortality in different indoor housing systems: A meta-analysis of data from commercial farms in 16 countries. *Scientific Reports, 11*(1). https://doi.org/10.1038/s41598-021-81868-3.

Schwarzer, A., Rauch, E., Bergmann, S., Kirchner, A., Lenz, A., Hammes, A., Erhard, M., Reese, S., & Louton, H. (2022). Risk factors for the occurrence of feather pecking in non-beak-trimmed pullets and laying hens on commercial farms. *Applied Sciences, 12*(19), 9699. https://doi.org/10.3390/app12199699.

Setoguchi, A., Oishi, K., Ogino, A., & Hirooka, H. (2025). Life cycle assessment of egg production: Effects of breed and feeding a low-protein diet with amino acids during laying period. *Animal Science Journal, 96*(1), e70129. https://doi.org/10.1111/asj.70129.

Shepherd, T. A., Zhao, Y., Li, H., Stinn, J. P., Hayes, M. D., & Xin, H. (2015). Environmental assessment of three egg production systems- Part II. Ammonia, greenhouse gas, and particulate matter emissions. *Poultry Science, 94*(3), 534–543. https://doi.org/10.3382/ps/peu075.

Shini, A., Shini, S., & Bryden, W. L. (2019). Fatty liver haemorrhagic syndrome occurrence in laying hens: Impact of production system. *Avian Pathology, 48*(1), 25–34. https://doi.org/10.1080/03079457.2018.1538550.

Shipov, A., Sharir, A., Zelzer, E., Milgram, J., Monsonego-Ornan, E., & Shahar, R. (2010). The influence of severe prolonged exercise restriction on the mechanical and structural properties of bone in an avian model. *Veterinary Journal, 183*(2), 153–160. https://doi.org/10.1016/j.tvjl.2008.11.015.

Sibanda, T. Z., Welch, M., Schneider, D., Kolakshyapati, M., & Ruhnke, I. (2020). Characterising free-range layer flocks using unsupervised cluster analysis. *Animals, 10*(5), 855. https://doi.org/10.3390/ani10050855.

Siegel, P. B. (1976). Social behavior of the fowl. *Poultry Science, 55*(1), 5–13. https://doi.org/10.3382/ps.0550005.

Stratmann, A., Fröhlich, E. K. F., Gebhardt-Henrich, S. G., Harlander-Matauschek, A., Würbel, H., & Toscano, M. J. (2015). Modification of aviary design reduces incidence of falls, collisions and keel bone damage in laying hens. *Applied Animal Behaviour Science, 165*, 112–123. https://doi.org/10.1016/j.applanim.2015.01.012.

Stratmann, A., Fröhlich, E. K. F., Harlander-Matauschek, A., Schrader, L., Toscano, M. J., Würbel, H., & Gebhardt-Henrich, S. G. (2015). Soft perches in an aviary system reduce incidence of keel bone damage in laying hens. *PLoS One, 10*(3), e0122568. https://doi.org/10.1371/journal.pone.0122568.

Tanaka, Y., Castillo, L., Wineland, M. J., & Deluca, H. F. (1978). Synergistic effect of progesterone, testosterone, and estradiol in the stimulation of chick renal 25-hydroxyvitamin D3-lα-hydroxylase. *Endocrinology, 103*(6), 2035–2039. https://doi.org/10.1210/endo-103-6-2035.

Thøfner, I., Hougen, H. P., Villa, C., Lynnerup, N., & Christensen, J. P. (2020). Pathological characterization of keel bone fractures in laying hens does not support external trauma as the underlying cause. *PLoS One, 15*(3), 0229735. https://doi.org/10.1371/journal.pone.0229735.

Toscano, M. J., Neuditschko, M., & Gebhardt-Henrich, S. G. (2025). Use of location tracking to link egg production with toe pecking injuries, feather quality, body mass in laying hens. *Journal of Applied Poultry Research, 34*(4), 100603. https://doi.org/10.1016/j.japr.2025.100603.

Toscano, M. J., Wilkins, L. J., Millburn, G., Thorpe, K., & Tarlton, J. F. (2013). Development of an ex vivo protocol to model bone fracture in laying hens resulting from collisions. *PLoS One, 8*(6), e66215. https://doi.org/10.1371/journal.pone.0066215.

Turner, I., Arulnathan, V., Ibarburu, M., Gates, R., & Pelletier, N. (2025). Differences in feed formulation are more determining of life cycle environmental impacts in U.S. egg production systems than are resource use efficiency and housing system choice. *Journal of Cleaner Production, 533*, 147032. https://doi.org/10.1016/j.jclepro.2025.147032.

U.S. Department of Agriculture, Animal and Plant Health Inspection Service. (2014). Layers 2013, Part I: Reference of health and management practices on table-egg farms in the United States, 2013. Report number 685.0614. USDA–APHIS–VS–CEAH–NAHMS, For Collins, CO, USA. https://www.aphis.usda.gov/sites/default/files/layers2013-dr-parti.pdf.

USDA-AMS. (2025). Shell eggs: Monthly USDA cage-free shell egg report. https://www.ams.usda.gov/mnreports/pymcagefree.pdf.

van der Eijk, J. A. J., Rodenburg, T. B., de Vries, H., Kjaer, J. B., Smidt, H., Naguib, M., Kemp, B., & Lammers, A. (2020). Early-life microbiota transplantation affects behavioural responses, serotonin and immune characteristics in chicken lines divergently selected on feather pecking. *Scientific Reports, 10*(1), 2750. https://doi.org/10.1038/s41598-020-59125-w.

Van Krimpen, M. M., Kwakkel, R. P., Reuvekamp, B. F. J., Van Der Peet-Schwering, C. M. C., Den Hartog, L. A., & Verstegen, M. W. A. (2005). Impact of feeding management on feather pecking in laying hens. *World's Poultry Science Journal, 61*(4), 663–686. https://doi.org/10.1079/wps200478.

van Staaveren, N., Ellis, J., Baes, C. F., & Harlander-Matauschek, A. (2021). A meta-analysis on the effect of environmental enrichment on feather pecking and feather damage in laying hens. *Poultry Science, 100*(2), 397–411. https://doi.org/10.1016/j.psj.2020.11.006.

Vits, A., Weitzenbürger, D., Hamann, H., & Distl, O. (2005). Production, egg quality, bone strength, claw length, and keel bone deformities of laying hens housed in furnished cages with different group sizes. *Poultry Science, 84*(10), 1511–1519. https://doi.org/10.1093/ps/84.10.1511.

Waldroup, P. W., & Hellwig, H. M. (2000). The potential value of morning and afternoon feeds for laying hens. *Journal of Applied Poultry Research. 9*(1), 98–110. https://doi.org/10.1093/japr/9.1.98.

Warren, D. E. (1937). Physiologic and genetic studies of crooked keels in chickens. *Kansas Agricultural Experimental Station Technical Bulletin, 44*. https://extension.k-state.edu/historicpublications/pubs/STB044.PDF.

Webster, A. B. (2004). Welfare implications of avian osteoporosis. *Poultry Science, 83*(2), 184–192. https://doi.org/10.1093/ps/83.2.184.

Weeks, C. A., Lambton, S. L., & Williams, A. G. (2016). Implications for welfare, productivity and sustainability of the variation in reported levels of mortality for laying hen flocks kept in different housing systems: A meta-analysis of ten studies. *PLoS One, 11*(1), e0146394. https://doi.org/10.1371/journal.pone.0146394.

Whitehead, C. C. (2004). Overview of bone biology in the egg-laying hen. *Poultry Science, 83*(2), 193–199. https://doi.org/10.1093/ps/83.2.193.

Whitehead, C. C., & Fleming, R. H. (2000). Osteoporosis in cage layers. *Poultry Science, 79*(7), 1033–1041. https://doi.org/10.1093/ps/79.7.1033.

Wichman, A., & Keeling, L. J. (2009). The influence of losing or gaining access to peat on the dustbathing behaviour of laying hens. *Animal Welfare, 18*(2), 149–157. https://doi.org/10.1017/s0962728600000294.

Wilkins, L. J., Brown, S. N., Zimmerman, P. H., Leeb, C., & Nicol, C. J. (2004). Investigation of palpation as a method for determining the prevalence of keel and furculum damage in laying hens. *Veterinary Record, 155*(18), 547–549. https://doi.org/10.1136/vr.155.18.547.

Wilkins, L. J., McKinstry, J. L., Avery, N. C., Knowles, T. G., Brown, S. N., Tarlton, J., & Nicol, C. J. (2011). Influence of housing system and design on bone strength and keel bone fractures in laying hens. *Veterinary Record, 169*(16), 414. https://doi.org/10.1136/vr.d4831.

Winter, J., Toscano, M. J., & Stratmann, A. (2021). Piling behaviour in Swiss layer flocks: Description and related factors. *Applied Animal Behaviour Science, 236*. https://doi.org/10.1016/j.applanim.2021.105272.

Winter, J., Toscano, M. J., & Stratmann, A. (2022). The potential of a light spot, heat area, and novel object to attract laying hens and induce piling behaviour. *Animal, 16*(8), 100567. https://doi.org/10.1016/j.animal.2022.100567.

Xin, H., Gates, R. S., Green, A. R., Mitloehner, F. M., Moore, P. A., & Wathes, C. M. (2011). Environmental impacts and sustainability of egg production systems. *Poultry Science, 90*(1), 263–277. https://doi.org/10.3382/ps.2010-00877.

Yin, P., Tong, Q., Li, B. M., Zheng, W. C., Wang, Y., Peng, H. Q., Xue, X. L., & Wei, S. Q. (2024). Spatial distribution, movement, body damage, and feather condition of laying hens in a multi-tier system. *Poultry Science, 103*(1), 103202. https://doi.org/10.1016/j.psj.2023.103202.

Zhao, Y., Shepherd, T. A., Li, H., & Xin, H. (2015). Environmental assessment of three egg production systems–Part I: Monitoring system and indoor air quality. *Poultry Science, 94*(3), 518–533. https://doi.org/10.3382/ps/peu076.

Chapter 3

Broiler chickens: slow and fast growing

Bethany Baker-Cook[1] and Lillian Lins-Taylor[2]
[1]*Institute of Agrifood Research and Technology, Animal Welfare, Monells, Spain,*
[2]*Auburn University, Auburn, AL, United States*

Meat chickens, also known as broiler chickens, are an important source of animal protein globally. Throughout the past few decades, the demand for chicken meat has steadily increased, and a 50% increase in broiler production by 2050 has been forecast to meet this increasing demand (FAO, 2018). As broiler production continues to increase, it is faced with the challenge of meeting the global demand for an affordable protein source whilst ensuring the welfare of the chickens and sustainable production practices as environmental resources become scarcer.

Welfare of broilers

Throughout the years, genetic selection for feed efficiency, rapid growth, and body conformation, combined with advances in nutrition and management practices, has resulted in a bird that is highly efficient at converting feed into muscle and reaches slaughter weight within record time. However, these advances in nutrition, management, and growth rate have led to various welfare issues within the modern conventional broiler. For example, the rapid growth rate of conventional broilers places a high demand on the cardiovascular system to deliver oxygen to the rapidly growing tissue; this high demand can cause cardiovascular strain and has led to an increased prevalence of metabolic disorders such as pulmonary hypertension and sudden death syndrome (Bessei, 2006). Similarly, the rapid weight gain has also led to skeletal problems and locomotor disorders (both infectious and noninfectious) (Riber & Wurtz, 2024). Included disorders are tibial dyschondroplasia, femoral head necrosis, and osteomyelitis (Kwon et al., 2024). These can cause locomotor issues and lameness, which often causes pain and discomfort to the bird. Furthermore, lameness reduces the birds' ability to reach feed and water and prevents the birds from performing other important and desired behaviors, and can lead to death (Weeks et al., 2000), as do the leg health

Integrating Welfare and Sustainability in Poultry Systems. DOI: https://doi.org/10.1016/B978-0-443-40594-5.00005-5

issues due to rapid growth, body conformation, and particularly large breast size (Dixon, 2020; Rayner et al., 2020). Rapid weight gain has also been linked to poor immune function, leading birds to be more susceptible to infections and any suffering that may be associated with this (Forseth et al., 2023). Heat stress is another concern with modern conventional broilers, due to the high metabolic rates associated with fast growth. Birds produce a lot of metabolic heat and, especially in warm and humid environments, can struggle to dissipate this heat, often leading to heat stress and, in extreme situations, mortality. As climate variability increases, broilers are also experiencing more frequent and prolonged periods of high environmental temperatures, which further heightens the risk of heat-related welfare issues. This is something that is likely to become a larger issue as temperatures continue to rise due to climate change.

Management issues such as high stocking density and poor litter quality can also lead to welfare issues. One such issue is contact dermatitis, often seen in the form of foot pad dermatitis, hock burn, or breast blister. These are inflamed, necrotic lesions that form on the skin in areas that have had prolonged exposure to litter with high levels of ammonia and moisture (Bessei, 2006; EFSA, 2023). These conditions are painful and can prevent birds from moving and performing desired behaviors (Weber Wyneken et al., 2015). High stocking densities and the resulting overcrowding also create competition for space and resources (Shynkaruk et al., 2023). When space is limited, it prevents birds from being able to move around comfortably and hinders the performance of comfort, exploratory, and foraging behaviors (EFSA, 2023). Particularly as these aforementioned behaviors have social aspects to them, birds often perform them synchronously. Overcrowding can also prevent birds from resting and sleeping comfortably, as the close presence of active and resting birds often leads to disturbance. Rest disturbance can negatively impact bird health and affective states (Malleau et al., 2007). The limited space also means birds walk over one another or pile, and this can lead to scratches, wounds, and other injuries, which are painful and may get infected and cause further suffering (EFSA, 2023), in addition to carcass downgrading at processing. Understimulation is a significant issue for broilers raised indoors in conventional settings, regardless of genotype. The barns are often barren, containing only litter, feed, and waterlines for interaction (EFSA, 2023). This lack of stimulation can negatively impact the mental and physical welfare of the birds (Anderson et al., 2021; Campbell et al., 2023; EFSA, 2023).

Slow-growing broiler chickens

Concerns about bird welfare, particularly health issues linked to rapid growth and management practices, have led to increased pressure from the media, NGOs, and consumers for higher-welfare production systems. This movement has led to a market for birds with slower growth rates (Forseth et al., 2023).

Slower-growing broilers grow at a slower rate than conventional broilers, which have growth rates of around 60 g/day (Nicol et al., 2024). However, the definition of a slow-growing broiler varies, as growth rates within this group range widely from less than 25 g/day to just below 50 g/day (Nicol et al., 2024). This variation in slow-growing strains complicates drawing clear research conclusions. Nevertheless, multiple studies across experimental and commercial conditions report improved welfare aspects for slow-growing broilers compared to conventional broilers.

For example, slow-growing birds have been found to have improved leg and foot health, with studies in both experimental and commercial settings finding that slow growers are more mobile and have better gait scores than conventional or faster growing broilers (Dixon, 2020; Rayner et al., 2020; Abeyesinghe et al., 2021; Baxter et al., 2021; Averós et al., 2022; Nicol et al., 2024). Similarly, many studies found that slower-growing broilers had less foot pad dermatitis (Dixon, 2020; Rayner et al., 2020; Abeyesinghe et al., 2021; Nicol et al., 2024). Hock burn was found to be lower in slow-growing broilers under experimental conditions (Dixon, 2020; Weimer et al., 2020), but this was not found in a study conducted on commercial farms (Averós et al., 2022). In addition, no difference was found between fast and slow-growing broilers in the latency to lie test (Baxter et al., 2021). Slow-growing broilers have been found to have fewer heart defects, tendon degeneration, rotated tibias, and scoliosis (Bokkers & Koene, 2003), as well as improved feather cover and cleanliness (Dixon, 2020; Averós et al., 2022). Slow-growing broilers have been found to have better immunity (Snyder et al., 2022) and are less susceptible to infection and disease (Fanatico et al., 2008; Rayner et al., 2020). Several studies have found mortality rates to be lower for slow-growing broilers than conventional broilers (Fanatico et al., 2008; Dixon, 2020; Rayner et al., 2020; Abeyesinghe et al., 2021; Averós et al., 2022). There are also fewer condemnations of slow-growing birds compared to conventional broilers at slaughter (Rayner et al., 2020; Forseth et al., 2023). There are some welfare outcomes that have been found to be worse in slow-growing broilers. For example, higher incidence of keel bone deviations (Bokkers & Koene, 2003), breast blisters (Allain et al., 2009), scratches and skin lesions (De Jong et al., 2022), as well as higher incidences of toe pecking (Weimer et al., 2020) have been reported in studies comparing slower and faster growing broilers. Furthermore, agonistic interactions have been found to occur more with slow-growing broilers (Bokkers & Koene, 2003), as well as more reports of incidences of feather pecking and cannibalism.

There are also differences in behavior between slow-growing and conventional broilers. Slow growers have been observed performing more active, comfort, and exploratory behaviors, which are indicative of better welfare, whilst conventional broilers spent more time inactive (Fanatico et al., 2008; Dixon, 2020; Abeyesinghe et al., 2021; Dawson et al., 2021; De Jong et al., 2021; Van der Eijk et al., 2023). No consistent and clear difference has been found between fast and slow-growing birds in terms of the performance of play behaviors (Baxter et al., 2021; Averós et al., 2022; Van der Eijk et al., 2023). Various studies have assessed fear levels, using a variety of different tests with differing results (Lindholm et al., 2016; Rayner et al., 2020; Baxter et al., 2021), with some finding no difference in fear level, some finding an increased fear level in slow growers (Lindholm et al., 2016; Baxter et al., 2021), and others finding a decrease in fear level in slow growers (Lindholm et al., 2016; Van der Eijk et al., 2023). It is important to note that genetic selection on both conventional and slow-growing strains is continuous. Some recent selections for welfare traits are improving leg health in conventional strains (Neeteson et al., 2023), whilst some strains of slow-growing birds have started to have higher daily weight gain (Nicol et al., 2024).

Sustainability and broiler production

The advances in feed conversion, nutrition, and management have resulted in chicken being a relatively efficient and environmentally sustainable form of animal protein in comparison with other forms of animal protein (although other forms of nonanimal protein are more sustainable) (Vaarst et al., 2015). However, there are still several serious issues regarding sustainability within broiler production. For further discussion on some of the environmental impacts, see Chapter 13, Chapter 14, and Chapter 15.

Carbon footprint or greenhouse gas (GHG) emissions are one of the main topics featuring in discussions about the environmental sustainability of poultry production. The main GHGs associated with poultry production are methane, nitrous oxide, and carbon dioxide (CO_2), which arise from multiple stages of broiler production. Among these, feed production is the primary contributor to GHG emissions, with as much as 70%–80% of GHG emissions from broiler production being attributed to feed production (Beal et al., 2023; Attia et al., 2024). Growing, harvesting, transporting, and milling feed crops requires significant energy, largely from fossil fuels (the process of which releases CO_2 into the atmosphere). A major component of broiler feed is soy meal, often imported from regions such as South America. Soy cultivation often leads to changes in land use and deforestation, which not only emits a lot of CO_2 but also reduces the land's carbon sequestration capacity, therefore amplifying the climatic impact (Leinonen & Kyriazakis, 2016). Beyond GHG emissions, the feed crop cultivation also contributes to eutrophication and acidification, as the oversupply of nutrients to crop fields enters the water system via leaching, runoff, and atmospheric deposition (Leinonen & Kyriazakis, 2016; Attia et al., 2024). Furthermore, broiler diets often overfeed protein to meet the birds' amino acid requirements. Not all the nitrogen in the diet is retained within the body; the excess nitrogen gets excreted into the environment, further intensifying these impacts (Vaarst et al., 2015; Leinonen & Kyriazakis, 2016). Additionally, direct farm energy use is another source of emissions, largely related to heating and ventilation; however, this is relatively small compared to the emissions from feed production (Pelletier, 2008; Leinonen & Kyriazakis, 2016). Often overlooked, the production of chicks that will be reared on farms for meat is another large contributor to GHG (Beal et al., 2023). This is often excluded from the analysis of the environmental impact of broiler production.

Another way broiler production impacts the environment is through manure and its management. The manure is a source of GHG, such as CO_2, methane (a very potent GHG), and nitrous oxide (Leinonen & Kyriazakis, 2016; Mostert et al., 2022). Additionally, manure is high in nitrogen and phosphorus, and when poorly managed, these nutrients can leach into water bodies and cause eutrophication (Leinonen & Kyriazakis, 2016). Eutrophication is the process in which excess nutrients in the water lead to the occurrence of algal blooms, which cause oxygen depletion in water bodies and lead to the death of aquatic life. Alongside nutrient leaching, pathogens and heavy metals in the manure can also leach into the groundwater, causing environmental and human safety issues. Manure is also responsible for acidification (Leinonen & Kyriazakis, 2016). The microbial decomposition of the nitrogen in manure leads to the production of ammonia, which volatilizes into the atmosphere. Once airborne, ammonia and its derivatives contribute to the formation of acidifying compounds. These acidifying compounds then return to Earth in either the form of dust or acid rain and lower the pH of the soil and water bodies. This reduces nutrient availability for plants and harms aquatic life. Moreover, it can also have an additive effect with eutrophication, as the nitrogen deposition can act as a fertilizer.

Transportation and global trade also make significant contributions to the environmental impact of poultry production. The transportation of feed and chicks to the farm, and chickens to the slaughterhouse, plays a contributing role in this. Notably, a larger impact is the transportation of poultry feed ingredients and poultry meat across the globe. Feed ingredients, such as corn and soy, are transported great distances from the country of production, often in South America, to other countries around the globe to feed chickens (Mostert et al., 2022). Similarly, poultry meat is also transported around the globe, with over 15 million metric tonnes of poultry meat being traded globally, particularly between Asia, North and South America, and the European Union (Vaarst et al., 2015). The use of fossil fuels in this transportation generates GHG emissions and increases acidification.

Poultry production also uses a lot of water, the direct consumption by the birds themselves, in the production of feed, in cooling and cleaning the housing, and in processing. One study calculated that 4300 L of water is needed to produce every kilogram of chicken (Attia et al., 2024). Most of this water usage is attributed to the production of concentrated poultry feed (Vaarst et al., 2015). As water is vital to birds, it is important that water management is optimized, especially as water scarcity is a global issue, and will continue to increase as the earth's temperature continues to rise (Attia et al., 2024).

Biodiversity loss is another aspect of sustainability, and this is also greatly impacted by poultry production. As with GHG emissions, the production of ingredients for broiler feed is one of the main drivers of biodiversity loss. A lot of biodiversity loss is associated with the production of monocrops, such as soy for animal feed production, as deforestation and conversion of natural habitats to cropland cause habitat loss and disrupt ecosystems, leading to loss of diversity in both flora and fauna. This includes the loss of pollinators, which are an important component of food production (Vaarst et al., 2015). As previously discussed, the high nitrogen and phosphorus content in poultry manure runoff can lead to eutrophication, causing loss of aquatic species and dead zones within the ecosystem. Similarly, acidification alters the soil chemistry, impacting both the flora and microbial diversity.

Another aspect of sustainability is social sustainability, or the human impact of poultry production, particularly on the quality of life of humans. There are several quality-of-life concerns for humans involved in poultry production, such as health risks from dust and ammonia exposure, high injury risk, and high risk of infection by antibiotic-resistant bacteria leading to hospitalizations (Vaarst et al., 2015), and zoonotic diseases. Furthermore, depending on the location and production system, another issue is the exploitation of contract farmers (Vaarst et al., 2015), and the fact that the large-scale global trade of poultry products is undermining the livelihood of local farmers in places such as Asia and Africa (Queenan et al., 2021).

Welfare and sustainability in broiler chickens

Slow-growing broilers gain weight more slowly and therefore take longer to reach their final slaughter weight (this length differs greatly with strain). This means that to produce the same amount of meat, more birds are required (estimated to be 19%–47% depending on country and strain), and more land is needed to raise the birds (estimated to be 20%–60% depending on country and bird strain) (Chan et al., 2022; Khire & Ryba, 2025). Slow-growing broilers are also less feed efficient, that is, they cannot use energy within the diet for growth as efficiently (Dixon, 2020; Weimer et al., 2020; Abeyesinghe et al., 2021; Singh et al., 2021) as conventional broilers. This means that they require more feed to reach the same slaughter weight. As more birds need to be raised, housed, and fed, there will be an increase in GHG emissions and energy usage (Leinonen & Kyriazakis, 2016; Karlsson et al., 2025). Additionally, more water would also be required due to a higher number of birds being raised and more feed being produced. However, some of the aforementioned environmental impacts could be offset by slow-growing birds having a lower protein requirement (Kyriazakis et al., 2025) and being able to use alternative protein sources more effectively (Leinonen & Kyriazakis, 2016). As feed makes up a major component of the environmental impact of broiler production, less protein in the diets and more local alternative protein sources could have a big impact. Lower-protein diets would reduce the need for monocrop soy production and the global transport of feed ingredients, which would reduce GHG emissions and biodiversity loss. The reduction in protein intake can reduce nitrogen excretion, thereby reducing acidification and eutrophication, potentially reducing biodiversity loss. However, further research is needed to understand this.

Raising more birds also means more manure is produced, and unless properly treated, this will also negatively impact eutrophication and acidification. There is little information on whether the nitrogen emissions differ between fast- and slow-growing strains. One possible solution to this is utilizing manure produced for other things, such as fertilizer, biogas, or as an energy source. Particularly, using litter, that is, manure and bedding material, as fuel to generate energy (Williams et al., 2016) can reduce the need for fossil fuels as an energy source and reduce emissions, eutrophication, and acidification (Leinonen & Kyriazakis, 2016). Another possible option is the use of litter amendments or treatments, which are typically added to the litter as a powder or spray and reduce ammonia emissions from the manure, thereby aiding in reducing eutrophication and acidification. A reduction in the amount of ammonia in the litter would also be beneficial for bird welfare, as it would likely reduce the incidence of contact dermatitis, such as foot pad dermatitis, hock burn, breast blisters, and reduce the ammonia burn to the eyes and nasal passageways.

Meat production also differs between slow-growing and conventional broilers. Some studies reported that conventional birds had higher carcass weights and breast weights, whereas slow-growing birds had higher thigh, drumstick, and wing weights (Singh et al., 2021; Karlsson et al., 2025). Different bird strains have different body conformations (Weimer et al., 2020) and, due to genetic selection, are better at depositing muscle in different regions. This may mean that to meet the demand for certain cuts of meat, more slow-growing broilers are needed compared to conventional broilers. However, conventional broilers have been found to have higher incidences of muscle myopathies, such as white striping, woody breast, and spaghetti meat (Barbut et al., 2024; Kyriazakis et al., 2025), a higher number of condemnations at the slaughterhouse (Rayner et al., 2020; Forseth et al., 2023), and higher mortality rates (Fanatico et al., 2008; Dixon, 2020; Rayner et al., 2020; Abeyesinghe et al., 2021; Averós et al., 2022). The condemnation of the carcass, postmortem meat loss, and mortality means a loss of resources, such that there is an environmental and economic impact without any product. The higher resilience of slow-growing birds and fewer condemnations may mean that the reduction in loss of resources may offset some of the environmental impact. However, condemnations and mortality are left out of most sustainability analyses, requiring further research to quantify the impacts thereof. Additionally, slow growers are less susceptible to heat stress due to a lower metabolism than faster-growing birds (De Jong et al., 2012; EFSA, 2023). This means that less cooling is required during the hot seasons, which could have the benefit of requiring less on-farm energy usage. Furthermore, as the climate becomes more extreme, heat stress is becoming more common, and broilers' ability to tolerate heat better may become more important.

Interestingly, the emissions from slow-growing broiler breeders are lower than those of fast-growing broiler breeders and partially offset the higher emissions of the slow-growing broilers (Karlsson et al., 2025). This is because fewer breeders are needed, as slow-growing breeders have higher fertility and longer laying periods, meaning more broiler chicks per breeder (Arrazola et al., 2022; Mostert et al., 2022). Furthermore, slow-growing breeders are also under less extreme feed restriction, which is a significant improvement in breeder welfare.

It is also worth mentioning that although slow-growing broilers suffer from less extreme welfare issues, that is, leg and cardiovascular health, the longer production time means that if these birds are raised under the same environmental and management conditions, they will still be exposed to other, less extreme welfare issues, that is, barren environments, boredom, lack of stimulation, and be exposed to them for a longer period of time (Khire & Ryba, 2025). Furthermore, as slow-growing broilers become more common in production, we will likely start seeing other welfare issues specific to slower-growing birds. One example that is already being noticed is the presence of feather pecking or cannibalism in slow-growing flocks. These problems are not usually an issue in fast-growing broilers but are starting to be reported in slow-growing broilers (B. B-C. Pers. Comm.).

It is also important to remember that poultry are sentient beings, and part of ensuring sustainable production is ensuring not only that their mental and physical needs are fulfilled but also that they get to experience positive welfare. Although reducing the growth rate of broiler chickens seems to reduce some of the welfare issues within broiler production, there are still many aspects of welfare that are not being met simply with a movement towards slow-growing broilers. Therefore we need to find other solutions to help improve the welfare of broilers. One possibility could be reducing stocking densities, as a reduction in stocking density has been shown to be beneficial for the welfare of the birds (regardless of the strain used) (Van der Eijk et al., 2022; Van der Eijk et al., 2023), and the environmental impact has been found to only increase marginally (Karlsson et al., 2025) or be improved (Mostert et al., 2022). Another alternative is to slaughter the birds at a younger age; this would improve their welfare, as many of the welfare issues worsen as the birds get older (Nicol et al., 2024). This may also be beneficial for the environment, as GHG emissions are higher with heavier slaughter weights (Kyriazakis et al., 2025). There are still many aspects of welfare and how we can improve it for broilers kept in conventional housing that need further examination.

Overall, with land and water resources becoming scarcer, the ultimate solution for both the environment and animal welfare may be to increase our consumption and production of crops and reduce our reliance on animal protein and focus on producing less but higher-welfare animal protein (i.e., eat less but better meat).

References

Abeyesinghe, S. M., Chancellor, N. M., Hernandez Moore, D., Chang, Y.-M., Pearce, J., Demmers, T., & Nicol, C. J. (2021). Associations between behaviour and health outcomes in conventional and slow-growing breeds of broiler chicken. *Animal, 15*(7), 100261. https://doi.org/10.1016/j.animal.2021.100261.

Allain, V., Mirabito, L., Arnould, C., Colas, M., Le Bouquin, S., Lupo, C., & Michel, V. (2009). Skin lesions in broiler chickens measured at the slaughterhouse: Relationships between lesions and between their prevalence and rearing factors. *British Poultry Science, 50*(4), 407–417. https://doi.org/10.1080/00071660903110901.

Anderson, M. G., Campbell, A. M., Crump, A., Arnott, G., & Jacobs, L. (2021). Environmental complexity positively impacts affective states of broiler chickens. *Scientific Reports, 11*(1), 16966. https://doi.org/10.1038/s41598-021-95280-4.

Arrazola, A., Widowski, T. M., & Torrey, S. (2022). In pursuit of a better broiler: Welfare and productivity of slower-growing broiler breeders during lay. *Poultry Science, 101*(8), 101917. https://doi.org/10.1016/j.psj.2022.101917.

Attia, Y. A., Aldhalmi, A. K., Youssef, I. M., Bovera, F., Tufarelli, V., El-Hack, M. E. A., El-Kholy, K. H., & Shukry, M. (2024). Climate change and its effects on poultry industry and sustainability. *Discover Sustainability, 5*(1), 397. https://doi.org/10.1007/s43621-024-00627-2.

Averós, X., Nazar, F. N., & Estevez, I. (2022). Animal welfare assessment: Quantifying differences among commercial medium and fast growth broiler flocks. *Frontiers in Animal Science, 3*, 868851. https://doi.org/10.3389/fanim.2022.868851.

Barbut, S., Mitchell, R., Hall, P., Bacon, C., Bailey, R., Owens, C. M., & Petracci, M. (2024). Review: Myopathies in broilers: Supply chain approach to provide solutions to challenges related to raising fast growing birds. *Poultry Science, 103*(8), 103801. https://doi.org/10.1016/j.psj.2024.103801.

Baxter, M., Richmond, A., Lavery, U., & O'Connell, N. E. (2021). A comparison of fast growing broiler chickens with a slower-growing breed type reared on Higher Welfare commercial farms. *PLoS One, 16*(11), e0259333. https://doi.org/10.1371/journal.pone.0259333.

Beal, C. M., Robinson, D. M., Smith, J., Gerber Van Doren, L., Tabler, G. T., Rochell, S. J., Kidd, M. T., Bottje, W. G., & Lei, X. (2023). Economic and environmental assessment of U.S. broiler production: Opportunities to improve sustainability. *Poultry Science, 102*(10), 102887. https://doi.org/10.1016/j.psj.2023.102887.

Bessei, W. (2006). Welfare of broilers: A review. *World's Poultry Science Journal, 62*(03), 455–466.

Bokkers, E. A. M., & Koene, P. (2003). Behaviour of fast- and slow growing broilers to 12 weeks of age and the physical consequences. *Applied Animal Behaviour Science, 81*(1), 59–72. https://doi.org/10.1016/S0168-1591(02)00251-4.

Campbell, A. M., Anderson, M. G., & Jacobs, L. (2023). Measuring chronic stress in broiler chickens: Effects of environmental complexity and stocking density on immunoglobulin-A levels. *Animals, 13*(13), 2058. https://doi.org/10.3390/ani13132058.

Chan, I., Franks, B., & Hayek, M. N. (2022). The 'sustainability gap' of US broiler chicken production: Trade-offs between welfare, land use and consumption. *Royal Society Open Science, 9*(6), 210478. https://doi.org/10.1098/rsos.210478.

Dawson, L. C., Widowski, T. M., Liu, Z., Edwards, A. M., & Torrey, S. (2021). In pursuit of a better broiler: A comparison of the inactivity, behavior, and enrichment use of fast- and slower growing broiler chickens. *Poultry Science, 100*(12), 101451. https://doi.org/10.1016/j.psj.2021.101451.

De Jong, I. C., Blaauw, X. E., Van der Eijk, J. A. J., Souza da Silva, C., Van Krimpen, M. M., Molenaar, R., & Van den Brand, H. (2021). Providing environmental enrichments affects activity and performance, but not leg health in fast- and slower-growing broiler chickens. *Applied Animal Behaviour Science, 241*, 105375. https://doi.org/10.1016/j.applanim.2021.105375.

De Jong, I. C., Bos, B., Van Harn, J., Mostert, P., & Beest, D. T. (2022). Differences and variation in welfare performance of broiler flocks in three production systems. *Poultry Science, 101*(7), 101933. https://doi.org/10.1016/j.psj.2022.101933.

De Jong, I., Berg, C., Butterworth, A., & Estevéz, I. (2012). Scientific report updating the EFSA opinions on the welfare of broilers and broiler breeders. *Supporting Publications 2012*. **EN-295**. https://doi.org/10.2903/sp.efsa.2012.EN-295.

Dixon, L. M. (2020). Slow and steady wins the race: The behaviour and welfare of commercial faster growing broiler breeds compared to a commercial slower growing breed. *PLoS One, 15*(4), e0231006. https://doi.org/10.1371/journal.pone.0231006.

EFSA (2023). EFSA scientific opinion: Welfare of broilers on farm. *EFSA Journal, 21*. https://doi.org/10.2903/j.efsa.2023.7788.

Fanatico, A. C., Pillai, P. B., Hester, P. Y., Falcone, C., Mench, J. A., Owens, C. M., & Emmert, J. L. (2008). Performance, livability, and carcass yield of slow- and fast-growing chicken genotypes fed low-nutrient or standard diets and raised indoors or with outdoor access. *Poultry Science, 87*(6), 1012–1021. https://doi.org/10.3382/ps.2006-00424.

FAO. (2018). The future of food and agriculture - Alternative pathways to 2050. Summary Version. Rome. https://www.fao.org/3/CA1553EN/ca1553en.pdf.

Forseth, M., Moe, R. O., Kittelsen, K., Skjerve, E., & Toftaker, I. (2023). Comparison of carcass condemnation causes in two broiler hybrids differing in growth rates. *Scientific Reports, 13*(1), 4195. https://doi.org/10.1038/s41598-023-31422-0.

Karlsson, L., Keeling, L., & Röös, E. (2025). What is a better chicken? Exploring trade-offs between animal welfare and greenhouse gas emissions in higher-welfare broiler systems. *Sustainable Production and Consumption, 55*, 203–216. https://doi.org/10.1016/j.spc.2025.02.015.

Khire, I., & Ryba, R. (2025). Are slow-growing broiler chickens actually better for animal welfare? Shining light on a poultry welfare concern using a farm-scale economic model. *British Poultry Science, 66*(3), 290–298. https://doi.org/10.1080/00071668.2024.2432926.

Kwon, B. Y., Park, J., Kim, D. H., & Lee, K. W. (2024). Assessment of welfare problems in broilers: Focus on musculoskeletal problems associated with their rapid growth. *Animals, 14*(7), 1116. https://doi.org/10.3390/ani14071116.

Kyriazakis, I., Dokou, S., Taylor, J., Giannenas, I., & Murphy, E. (2025). A meta-analysis of the sources of variation in the environmental impacts of different broiler production systems. *British Poultry Science, 66*(2), 193–205. https://doi.org/10.1080/00071668.2024.2409192.

Leinonen, I., & Kyriazakis, I. (2016). How can we improve the environmental sustainability of poultry production? *Proceedings of the Nutrition Society, 75*(3), 265–273. https://doi.org/10.1017/S0029665116000094.

Lindholm, C., Karlsson, L., Johansson, A., & Altimiras, J. (2016). Higher fear of predators does not decrease outdoor range use in free-range Rowan Ranger broiler chickens. *Acta Agriculturae Scandinavica A: Animal Sciences, 66*(4), 231–238. https://doi.org/10.1080/09064702.2017.1337214.

Malleau, A. E., Duncan, I. J. H., Widowski, T. M., & Atkinson, J. L. (2007). The importance of rest in young domestic fowl. *Applied Animal Behaviour Science, 106*(1-3), 52–69. https://doi.org/10.1016/j.applanim.2006.06.017.

Mostert, P. F., Bos, A. P., Van Harn, J., & De Jong, I. C. (2022). The impact of changing toward higher welfare broiler production systems on greenhouse gas emissions: A Dutch case study using life cycle assessment. *Poultry Science, 101*(12), 102151. https://doi.org/10.1016/j.psj.2022.102151.

Neeteson, A. M., Avendaño, S., Koerhuis, A., Duggan, B., Souza, E., Mason, J., Ralph, J., Rohlf, P., Burnside, T., Kranis, A., & Bailey, R. (2023). Evolutions in commercial meat poultry breeding. *Animals, 13*(19), 3150. https://doi.org/10.3390/ani13193150.

Nicol, C. J., Abeyesinghe, S. M., & Chang, Y. M. (2024). An analysis of the welfare of fast-growing and slower-growing strains of broiler chicken. *Frontiers in Animal Science, 5*, 1374609. https://doi.org/10.3389/fanim.2024.1374609.

Pelletier (2008). Environmental performance in the US broiler poultry sector: Life cycle energy use and greenhouse gas, ozone depleting, acidifying and eutrophying emissions. *Agricultural Systems, 98*(2), 67–73. https://doi.org/10.1016/j.agsy.2008.03.007.

Queenan, K., Cuevas, S., Mabhaudhi, T., Chimonyo, M., Slotow, R., & Häsler, B. (2021). A qualitative analysis of the commercial broiler system, and the links to consumers' nutrition and health, and to environmental sustainability: A South African case study. *Frontiers in Sustainable Food Systems, 5*, 650469. https://doi.org/10.3389/fsufs.2021.650469.

Rayner, A. C., Newberry, R. C., Vas, J., & Mullan, S. (2020). Slow-growing broilers are healthier and express more behavioural indicators of positive welfare. *Scientific Reports, 10*(1), 15151. https://doi.org/10.1038/s41598-020-72198-x.

Riber, A. B., & Wurtz, K. E. (2024). Impact of growth rate on the welfare of broilers. *Animals, 14*(22), 3330. https://doi.org/10.3390/ani14223330.

Shynkaruk, T., Long, K., LeBlanc, C., & Schwean-Lardner, K. (2023). Impact of stocking density on the welfare and productivity of broiler chickens reared to 34 d of age. *Journal of Applied Poultry Research, 32*(2), 100344. https://doi.org/10.1016/j.japr.2023.100344.

Singh, M., Lim, A. J., Muir, W. I., & Groves, P. J. (2021). Comparison of performance and carcass composition of a novel slow-growing crossbred broiler with fast-growing broiler for chicken meat in Australia. *Poultry Science, 100*(3), 100966. https://doi.org/10.1016/j.psj.2020.12.063.

Snyder, A. M., Riley, S. P., Robison, C. I., Karcher, D. M., Wickware, C. L., Johnson, T. A., & Weimer, S. L. (2022). Behavior and immune response of conventional and slow-growing broilers to *Salmonella Typhimurium. Frontiers in Physiology, 13*, 890848. https://doi.org/10.3389/fphys.2022.890848.

Vaarst, M., Steenfeldt, S., & Horsted, K. (2015). Sustainable development perspectives of poultry production. *World's Poultry Science Journal, 71*(4), 609–620. https://doi.org/10.1017/s0043933915002433.

Van der Eijk, J. A. J., Gunnink, H., Melis, S., Van Riel, J. W., & De Jong, I. C. (2022). Reducing stocking density benefits behaviour of fast- and slower-growing broilers. *Applied Animal Behaviour Science, 257*, 105754. https://doi.org/10.1016/j.applanim.2022.105754.

Van der Eijk, J. A. J., Van Harn, J., Gunnink, H., Melis, S., Van Riel, J. W., & De Jong, I. C. (2023). Fast- and slower-growing broilers respond similarly to a reduction in stocking density with regard to gait, hock burn, skin lesions, cleanliness, and performance. *Poultry Science, 102*(5), 102603. https://doi.org/10.1016/j.psj.2023.102603.

Weber Wyneken, C., Sinclair, A., Veldkamp, T., Vinco, L. J., & Hocking, P. M. (2015). Footpad dermatitis and pain assessment in turkey poults using analgesia and objective gait analysis. *British Poultry Science, 56*(5), 522–530. https://doi.org/10.1080/00071668.2015.1077203.

Weeks, C. A., Danbury, T. D., Davies, H. C., Hunt, P., & Kestin, S. C. (2000). The behaviour of broiler chickens and its modification by lameness. *Applied Animal Behaviour Science, 67*(1-2), 111–125. https://doi.org/10.1016/s0168-1591(99)00102-1.

Weimer, S. L., Mauromoustakos, A., Karcher, D. M., & Erasmus, M. A. (2020). Differences in performance, body conformation, and welfare of conventional and slow-growing broiler chickens raised at 2 stocking densities. *Poultry Science, 99*(9), 4398–4407. https://doi.org/10.1016/j.psj.2020.06.009.

Williams, A. G., Leinonen, I., & Kyriazakis, I. (2016). Environmental benefits of using Turkey litter as a fuel instead of a fertiliser. *Journal of Cleaner Production, 113*, 167–175. https://doi.org/10.1016/j.jclepro.2015.11.044.

Chapter 4

Broiler breeders

Stephanie Torrey
Trouw Nutrition North America, Puslinch, ON, Canada

Broiler breeder production can be described as "more art than science." Success depends on many variables—management, flock, nutritional program, breed, season, and housing—and small changes can affect behavior, welfare, mortality, efficiency, fertility, hatchability, and productivity. In this chapter, literature is reviewed that highlights the challenges of feeding modern broiler breeders, especially concerning feed restriction and fertility. The focus will be primarily on female broiler breeders, given that they comprise 90% of the population, and there are few studies on male broiler breeders. The chapter concludes with a discussion on potential solutions to these challenges and their relation to sustainability in the poultry industry.[1]

Broiler breeder production

Over the past 50–60 years, broiler breeder growth trajectories have changed dramatically (Renema et al., 2007). This change is primarily due to selective breeding for greater efficiency and faster growth rates in broiler chickens (Richards et al., 2010; Tallentire et al., 2016). Consequently, feed restriction has become necessary to prevent both mortality (Katanbaf et al., 1989a; Fig. 4.1) and obesity-related infertility (Buzała et al., 2015; Siegel & Dunnington, 1985). The degree of feed restriction required has increased over time.

In 1979, the target body weight of 6-week-old breeder cockerels was about 50% of that of male broilers at the same age. By 2005, this target dropped; breeder cockerels were about a third of male broiler body weights at the same age (Renema et al., 2007; Fig. 4.2). More recently, modern breeder body weight targets are around 22% of same-age broilers (Carney et al., 2022). Most modern breeds eat *ad libitum* for the first 2 to 4 weeks after which both pullets and cockerels are feed restricted for life. This chronic restriction causes poor welfare, as indicated by both behavioral and physiological signs of chronic

1. The views in this chapter do not necessarily represent those of my employer.

Integrating Welfare and Sustainability in Poultry Systems. DOI: https://doi.org/10.1016/B978-0-443-40594-5.00003-1

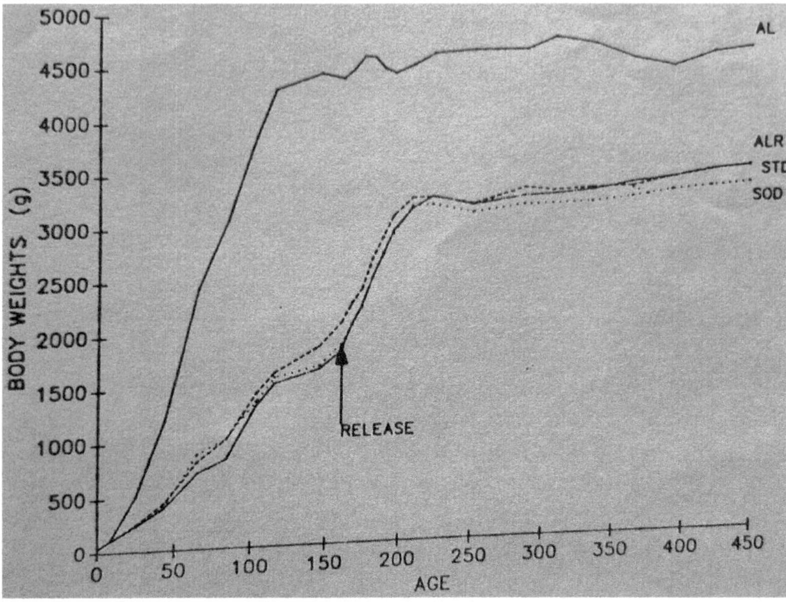

FIGURE 4.1 **Cumulative mortality of breeders fed with different feeding strategies.** AL = *ad libitum*; ALR = restricted. SOD = skip-one-day; STD = skip-two-days. Release denotes when SOD and STD regimens were changed to ALR regimen (Katanbaf et al., 1989a).
Reproduced from Katanbaf et al., 1989a. Restricted feeding in early and late-feathering chickens. 1. Growth and physiological responses. Poultry Science, 68(3), 344–351. https://doi.org/10.3382/ps.0680344.

hunger. Feed-restricted broiler breeders display high feeding motivation even right after feeding, high levels of feed-seeking behavior, redirected oral behavior, and feather pecking (de Jong et al., 2003; D'Eath et al., 2009; Mens & de Jong, van Riel, et al., 2022; Morrissey et al., 2014a; Zukiwsky et al., 2021). Chronic hunger has also been associated with altering metabolic rates (as indicated by glucose: nonesterified fatty acids (NEFA) levels; de Jong et al., 2003) and causing fault bars in feathers (Arrazola & Torrey, 2019; Wurtz et al., 2025). Still, not restricting feed would also harm welfare as full-fed breeders rapidly become obese and develop all the physiological issues associated with obesity (e.g., lameness, cardiovascular health problems, sudden death). Thus some say this is a paradox with no clear solution (Decuypere et al., 2006; Renema & Robinson, 2004; Richards et al., 2010).

While some degree of feed restriction is necessary for modern, commercial broiler breeders to reduce mortality and infertility, breeder fertility has been decreasing despite this practice. As a result of successful selection for lean growth in broiler chickens to meet consumer demands, modern breeders have significantly less abdominal fat pad weight than those in previous generations, which contributes to lower fertility (Carney et al., 2022), an issue that may be

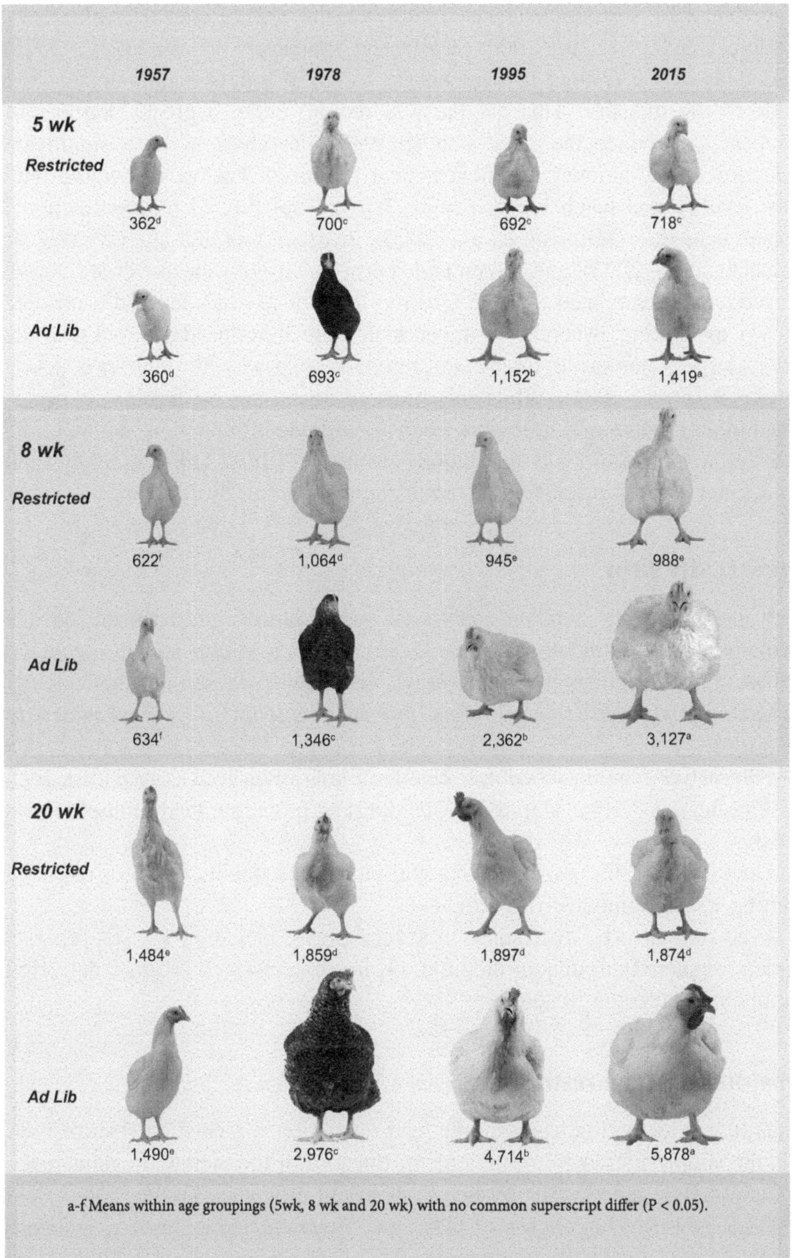

FIGURE 4.2 Age and feeding changes in bird size. Changes in size for two meat bird strains under *ad libitum* or restricted feeding across age (Carney et al., 2022).
Reproduced from Carney, V. L., Anthony, N. B., Robinson, F. E., Reimer, B. L., Korver, D. R., Zuidhof, M. J., & Afrouziyeh, M. (2022). Evolution of maternal feed restriction practices over 60 years of selection for broiler productivity. Poultry Science, 101(10), 101957. https://doi.org/10.1016/j.psj.2022.101957.

exacerbated by restriction levels (Fattori et al., 1991; Renema et al., 1999; Romero-Sanchez et al., 2008). Cash and colleagues (Cash et al., 2025) predicted a 60% decline in hatchability by 2050 unless major changes are made to the industry. This decline will require either a greater number of breeders to produce the same number of broiler chickens or a significant reduction in the amount of chicken meat produced. Factors associated with poor fertility and hatchability include, but are not limited to, dysfunctional sexual behavior (Millman et al., 2000), ovarian morphology (Weaver & Ramachandran, 2023), and semen and sperm quality (Ayeneshet et al., 2024). To reduce issues related to male fertility, the industry has adopted a practice known as spiking, whereby younger males are introduced to flocks, while older, larger, nonfertile males are removed mid-way through production (Wolanski et al., 2004). While this practice does improve fertility, it brings about other issues surrounding biosecurity (with the introduction of new birds) and welfare (male-to-male aggression) (Sabah & Yilmaz Dikmen, 2023), and it does not address problems related to female ovarian dysfunction.

Feed restriction

Feed restriction is a common practice in broiler breeder management, but the methods used can vary significantly. Each approach affects growth, behavior, welfare, and productivity in different ways. The main feed restriction methods are:

Quantitative restriction: limiting the daily amount of feed provided to each bird.

Alternative feeding schedules: changing how often feed is provided, such as skipping days (e.g., "skip-a-day" feeding) or providing feed multiple times per day.

Qualitative restriction: adjusting the nutritional quality of feed, often by diluting it with nonnutritive ingredients.

Feed may also be distributed in several ways, including via sex-separate feeders, scattered throughout the litter, or, in some research settings, delivered via precision feeding systems.

Quantitative feed restriction

Quantitative restriction has been the feed restriction approach most studied, as it is the simplest strategy to implement. Because of the incredible gains made in quantitative genetics for broiler growth and efficiency (Havenstein et al., 2003; Siegel & Dunnington, 1985), feed restriction has been a standard practice since the 1970s (Carney et al., 2022). Broiler breeder pullets and cockerels are routinely feed restricted throughout their life, starting from 2 to 4 weeks of age. This restriction can be severe, with breeder guidelines recommending pullets receive 25%–33% of the amount that *ad libitum*-fed birds eat at a similar age (De Jong & Guémené, 2011). Primary breeding companies publish recommended feed allotments and growth curves to maximize egg production. These feed allotments typically require some level of daily feed restriction starting within the first few weeks after hatch.

Over the years, the necessary level of restriction has increased, with current modern breeds now receiving less than 30% of what they would choose to eat at the same age (Carney et al., 2022; De Jong & Guémené, 2011). This restriction is most severe during rearing (de Jong et al., 2003; Hocking et al., 1993), as moderated growth from 7 to 15 weeks of age is critical for maximizing reproductive outcomes (Bruggeman et al., 1999). However, indicators of poor welfare generally relate to the degree of restriction (de Jong et al., 2003; Savory & Lariviere, 2000). Increasing levels of restriction have resulted in higher physiological and behavioral signs of chronic hunger and stress (Bowling et al., 2018; de Jong et al., 2003) and altered the growth patterns and immune responses in broiler offspring (Bowling et al., 2018).

Alternative feeding schedules

In commercial production, chains are used to distribute feed throughout long, narrow feeders. While newer systems can distribute feed quite quickly (e.g., Big Dutchman at 120' (36.5 m) per minute Chain Feeding | Champion® Chain Feeding System | Big Dutchman; Fig. 4.3), older systems may take up to 5 minutes to distribute the feed to the entire house (van Emous, 2022). This delayed distribution can lead to intense competition at the feeders, particularly when birds are receiving a small, daily allotment of feed. This competition at

FIGURE 4.3 **Breeder management: the world's first chain feeding system with the entire chain loop in just one trough—Big Dutchman.** Image of broiler breeder chain feeding system © *Photo Courtesy Big Dutchman.*

the feeders reduces body weight uniformity, which then negatively affects productivity. Because of the competition at the feeders and resulting lack of body-weight uniformity, nondaily feeding schedules have been used during rearing since at least the 1970s (Voitle et al., 1970). However, in some countries, nondaily feeding schedules are not permitted as a regular practice (e.g., Sweden; Lindholm et al. (2018)).

Early adopted nondaily feeding schedules included skip-a-day (SKIP) and skip-two-days (STD), where birds received two or more times their daily feed allotment every other or every third day. On farm, producers may also consider a 4:3 schedule (4 days with feed, three nonconsecutive days without feed per week) as SKIP, since it enables them to set the days of the week with and without feed. Research has shown that compared to SKIP, STD was less efficient (Bartov et al., 1988) and did not appear to confer any advantages over SKIP (Katanbaf et al., 1989a, 1989b).

Numerous researchers have compared SKIP to daily feeding schedules and quantified production and welfare. Compared to daily feeding schedules, SKIP or 4:3 was consistently less efficient (Arrazola et al., 2019b; Carneiro et al., 2019; de Beer & Coon, 2007; Zuidhof et al., 2015), and inconsistently improved body weight uniformity (improved: Arrazola & Widowski, Guerin, et al., 2019b; Richards et al., 2010) no difference: (Avila et al., 2023; de Beer & Coon, 2007; Morrissey et al., 2014b; Zuidhof et al., 2015). In contrast, the impact of SKIP or 4:3 on behavior and welfare is unclear. Some improvements in welfare indicators with 4:3 compared to daily feeding have been found (feather condition: Morrissey et al., 2014a); feeding motivation, fault bars: (Arrazola, Mosco, et al., 2019a) but other studies have found daily feeding to yield better outcomes (physiological indicators of stress: (de Beer et al., 2008); feeding motivation: Arrazola et al., 2019b).

In practice, nondaily feeding schedules may include more feeding days per week than SKIP. Producers utilize 6:1 (6 days with feed, one day without feed), 5:2 (5 days with feed, two nonconsecutive days without feed), and graduated (changing the number of days without feed depending on the pullets' age) feeding schedules. However, the few studies to compare performance and welfare among different nondaily feeding schedules found no strong evidence in favor of one schedule over another (Arrazola et al., 2019a; Arrazola & Widowski, Guerin, et al., 2019b; Carneiro et al., 2019; Taylor et al., 2025).

More recently, researchers have studied whether feeding broiler breeders more than once a day could maintain production and improve welfare. While impractical on farm, twice or thrice a day feeding during rearing (van Emous et al., 2021), or after photostimulation (Gholami-Soltanmoradi et al., 2024; Moradi et al., 2013; Spradley et al., 2008; Taherkhani et al., 2010), improved egg production compared to daily feeding. Fewer studies have assessed the welfare implications of more frequent feeding. Spradley and colleagues (Spradley et al., 2008) found that twice-a-day feeding during production increased mortality compared to once daily feeding. Compared to SKIP, multiple feedings per day during rearing (via a precision feeding station) reduced some indicators of feeding frustration but did not eliminate them (Girard et al., 2017), indicating that the breeders remained hungry. Similarly, others have reported some transient positive effects of more frequent feeding (Mens & van Emous, 2022; van Emous et al., 2021). Overall, providing a limited quantity of feed once or twice a day or on different schedules cannot supersede the fact that quantitative feed restriction reduces welfare.

Qualitative feed restriction

When fed a standard ration daily, broiler breeders consume their feed allotment within minutes of feed distribution. They may or may not experience any satiety (Richards et al., 2010), although de Jong et al. (2003) theorized that their lack of satiety may not mean they are continuously hungry. To increase time spent eating and reduce signs of chronic hunger, researchers have assessed whether reducing the quality of the feed to increase the quantity of feed was effective.

One strategy that has been tested in various iterations is the addition of insoluble fiber or other nonnutritive, bulking agents to the diet. When insoluble fibre is provided to pullets during rearing, the results are mixed. While numerous researchers reported decreases in feeding motivation (Arrazola et al., 2019a; Mens et al., 2022; Savory & Lariviere, 2000; Tahamtani et al., 2020; Tahamtani & Riber, 2020), others indicated that some aspects of welfare were worse with fibre than without (litter quality: Riber & Tahamtani, 2020); foot pad lesion scores: (Tahamtani et al., 2020). In addition, fibre added at a level to reduce feeding motivation (30%–70%) was not effective at controlling growth rates when used without restricting quantity (Savory & Lariviere, 2000), and even up to 400% increased volume only slightly increased the time to consume all feed (Zuidhof et al., 2015). There is also the practical concern with storing large quantities of nonnutritive feedstuffs on farm, from the logistics of feed bin storage capacity to sourcing or growing large quantities of fibre.

Since fibre on its own is not an effective way to control growth rates while improving breeder welfare, researchers have explored the efficacy of a combination of fibre and calcium propionate, an anorectic agent sometimes added to feedstuffs as a mold inhibitor. Pinchasov and colleagues (Pinchasov et al., 1993) described the appetite-suppressing capabilities of calcium propionate when added to breeder diets. However, without some level of quantitative or qualitative feed restriction, calcium propionate cannot suppress feed intake enough to limit growth (Pinchasov et al., 1993; Savory & Lariviere, 2000).

Later studies explored the combination of fibre (oat hulls or soybean hulls) and calcium propionate fed *ad libitum* on breeder welfare. Sandilands and colleagues (Sandilands et al., 2005; Sandilands et al., 2006) added calcium propionate and oat hulls to breeder diets and reported reductions in stereotypic object pecking in rearing and lower feeding motivation. While this combination of added fibre and calcium propionate conferred some minor benefits (Morrissey et al., 2014a; Sandilands et al., 2006), there was no strong evidence that the combined use of these with restricted diets measurably improved welfare (Buckley et al., 2012). There were also real concerns that the level of acid needed to suppress appetite was aversive (Tolkamp et al., 2005). Indeed, breeders learned to associate calcium propionate with negative affective states (Arrazola & Torrey, 2019).

Other researchers have investigated altering protein and energy levels to provide breeders with a larger quantity of feed without influencing nutrient specifications. During early rearing, lower density diets prolonged the time spent feeding and induced more restful behavior but did not change physiological indicators of hunger (de Jong et al., 2005). Van Krimpen and De Jong (2014) reviewed nutritional and management aspects of feeding broiler breeders and concluded that, while some minor improvements in welfare can be realized, results are often transient and no study since then has found otherwise.

Solutions to welfare issues

While adjusting feeding schedules or adding fiber to the diet can offer some welfare improvements, these changes do not fully resolve the broiler breeder paradox. Three strategies show more promise: lessening the degree of restriction, using alternative genetics, and providing enrichment.

Twenty years ago, Bruggeman et al. (2005) experimented with providing breeders with 55% of the allotment consumed by those fed *ad libitum*, but only until 15 weeks; thereafter, they were fed *ad libitum*. The moderate feed allotment resulted in poorer laying production than standard feed restriction but improved laying persistency more than *ad libitum* feeding. More recently, van Emous et al. (2015) increased the feed allocation to achieve a 9% heavier body weight at 20 weeks. With the increased allotment, hens had higher fertility, lower early embryonic mortality, and tended to lay more male broilers, which is advantageous to the broiler industry since males are more feed efficient. The heavier pullets spent more time sitting and eating than pullets fed to the standard (Van Emous et al., 2014). However, there was no effect of the extra allotment on stereotypic object and feather pecking (Van Emous et al., 2014).

During lay, Bowling et al. (2018) provided hens with up to an extra 20 g of feed per day. Those provided with the highest allotment had better immune parameters and showed fewer behavioral signs of chronic hunger. Like the finding by van Emous et al. (2015), the heavy hens also laid a greater proportion of male broilers. In a series of studies, a team from the University of Alberta used a precision feeding system to incrementally increase feed allotment, aiming for up to 22.5% extra body weight gain (Afrouziyeh et al., 2021; Afrouziyeh et al., 2023; Zukiwsky et al., 2021). Even at the highest incremental increase, reproductive outcomes were not negatively affected, and the relaxed restriction program yielded higher returns on investment than the standard program (Afrouziyeh et al., 2023). However, breeders still showed signs of feed-seeking behavior at the lowest level of restriction (Zukiwsky et al., 2021).

Other researchers have taken a different strategy to improving breeder welfare—slower-growing genetics. As part of the Broiler Breeder Paradox European Project (see: Decuypere et al., 2010), a series of studies were conducted comparing conventional to slower-growing breeds (dwarf, label rouge or experimental). Slower-growing breeds can maintain a good laying rate even when fed *ad libitum* (Decuypere et al., 2010; Heck et al., 2004). However, when feed was restricted, dwarf breeders displayed similar physiological signs of chronic hunger as did conventional breeders (de Jong et al., 2003). More recently, the welfare and productivity of three strains of breeders fed to their breed specifications were compared (Arrazola & Torrey, 2021). While some level of restriction was still required, the slower strains exhibited few signs of frustration and hunger and, in lay, a significant proportion of pens had feed remaining hours after feeding (Arrazola et al., 2022). The slower strains also had high levels of productivity and used less water than conventional breeds.

While providing enrichment to farm animals in commercial environments is not a new concept (Newberry, 1995), it has largely been overlooked in broiler breeder production (Riber et al., 2017). To improve an animal's welfare, an enrichment should prevent or reduce stereotypic or abnormal behavior, increase species-specific behavior, and/or increase the animal's ability to cope with stressors (Leone & Estévez, 2008). Newberry (1995) provided cover panels to breeders in production and found that, compared to flocks without them, flocks with the panels had better egg production due to better fertility and hatchability. The authors hypothesized that the hens experienced lower levels of stress due to their ability to better control their environment (and their interactions with males), although behavioral observations were not performed. Other research studies have tested perches and pecking objects for their influence on breeder welfare. When they are available, broiler breeders use perches and raised platforms (Baxter and O'Connell, 2025; Gebhardt-Henrich et al., 2020; Mens & van Emous, 2022; Vasdal et al., 2022a, 2022b), which increases species-specific behavior and animals' control over their environment. Gebhardt-Henrich and colleagues (Gebhardt-Henrich et al., 2020) found that perches reduced forced matings without negatively influencing reproductive performance. Other pecking objects, including bales of nonnutritive substrates (Hocking & Jones, 2006) and hanging balls (Baxter and O'Connell, 2025), elicit pecking behavior, although Hocking & Jones (2006) did not find them to be effective at eliminating behavioral signs of chronic hunger.

For all three strategies mentioned above, much more research is needed to better understand if and to what extent they can improve welfare within commercial production systems. In addition, these strategies need to be considered in conjunction with sustainability, as will be discussed below.

Sustainability of the broiler breeder industry

The above-mentioned "solutions" do not eliminate the paradox related to modern broiler breeder production. Lessening feed restriction for conventional breeders is possible without negatively affecting productivity, but it will have a limited effect on welfare. Using alternative, slow-growing strains reduces the necessity of feed restriction and would improve the welfare of breeders and broilers, but would require an acceptance of slower-growing strains of broiler chickens, which can be at odds with environmental sustainability (Karlsson et al., 2025). Yet, we cannot not consider animal welfare, and trade-offs are not necessarily inevitable (Boyle & Stevenson, 2025; Keeling, 2025).

While the traditional definition of sustainability focused primarily on environmental sustainability and the reduction of greenhouse gas (GHG) emissions, it has expanded in recent years to include social and economic sustainability, with animal welfare becoming part of the solution (Buller et al., 2018). In 2015, the United Nations put forth 17 Sustainable Development Goals, with the understanding that advancement toward one goal must account for the effects on other goals (UN, Sustainable Development Goals | United Nations Development Programme). This creates another paradox for the broiler breeder industry, as recommendations that promote higher welfare (meeting social sustainability goals) may result in higher costs (missing economic sustainability goals) and a greater environmental footprint (missing environmental sustainability goals) (Dawkins & Layton, 2012). Indeed, growing only slower-growing broilers (which would enable lower levels of feed restriction for their parent stock) would use significantly more land (Chan et al., 2022). However, other variables need to be considered, as explained below.

Because of the rapid rate of decline in breeder fertility anticipated through 2050, continuing with the status quo will have a large environmental impact as significantly more breeder hens will be required to produce the same number of broiler chickens (Cash et al., 2025). Primary breeding companies already put significant resources into understanding and improving breeder fertility in concert with welfare and production but competing selection goals and the multiyear lag between great-grandparent pedigree selections and breeder production mean that changes are incremental (Arango & Fulton, 2026) (Burnside & Neeteson, 2025), although advances in genotyping, phenotyping, computing and artificial intelligence will surely increase the rate of progress (Neeteson et al., 2023).

While some recent comparisons of broiler breeder production systems consider differing mortality rates (Cash et al., 2025; Mostert et al., 2022), not all do (Chan et al., 2022). With climate change increasing the risk of heat stress, conventional breeders will be at higher risk than slower-growing breeders for morbidity and mortality (Brugaletta et al., 2021), due to disproportionate resource allocation (Rothschild, 2019). Due to their efficiency and reproductive capabilities, breeders have limited physiological resources remaining to allocate to their resilience (Huber, 2018) after accounting for growth, reproduction, and maintenance.

Mostert and colleagues (Mostert et al., 2022) estimated the GHG emissions from three broiler systems (differing in growth rates), considering the breeder flocks, hatchery, feed, transportation, and energy, among other factors. They found that using an intermediate broiler breed resulted in lower GHG emissions than a conventional breed (with a broiler production period 11 days shorter), in part due to the efficiencies realized in the breeder rearing, laying, and egg hatching systems. Although all systems could be improved by shifting the source of soybean products to countries with low land use change (Mostert et al., 2022). In their case study report, Dunkley et al. (2015) reported that breeder houses have higher emissions than either broiler or layer houses, with the difference due to nonmechanical sources of emissions, such as manure management. Improvements to these systems would undoubtedly improve the sustainability of even slower- growing broiler breeders.

In her most recent publication, Keeling (2025) outlined the potential synergies between sustainable development goals and general farm animal welfare. However, for broiler breeders, a multipronged approach, incorporating growth rates, feed production, housing systems, and economics, would be required, and conflicts would almost certainly remain (Karlsson et al., 2025). Because of its obvious influence on broiler production, changes to broiler breeder production cannot be considered in a vacuum. The use of intermediate, slower-growing strains can reduce some of the conflict between sustainability and welfare for breeders (Karlsson et al., 2025) but would increase the trade-offs for broilers. Regenerative agricultural practices and a circular economy approach could both be used to mitigate GHG losses realized from using slower-growing strains (Boyle & Stevenson, 2025; Karlsson et al., 2025), while using specialty feed ingredients, including crystalline amino acids and phytase, can significantly improve environmental sustainability without negatively impacting production (Kebreab et al., 2016). Consumers have indicated that they place value in products or practices that prioritize animal welfare, even over environmental sustainability (Ammann et al., 2024). Coupling the two in marketing could improve both over time (Ammann et al., 2025), but only if the consumer is willing to pay.

Since the late 1940s,when the American grocery store chain A&P and the United States Department of Agriculture cocreated a competition for the "Chicken of Tomorrow," a meat-type chicken that could compete with beef and pork on Americans' dinner plates, the global poultry industry has undergone remarkable changes, most notably the creation of and divergence between the meat and egg industries (Shrader, 1952). As is evident in Fig. 4.2, advances in poultry science, primarily due to genetics but also nutrition, husbandry, and veterinary science, among others, have changed the very definition of a chicken. Even more advances in poultry science will be needed to strike a balance between broiler breeder production, welfare, and environmental and economic sustainability.

References

Afrouziyeh, Carney, & Zuidhof (2023). Is relaxing broiler breeder growth restriction cost-effective? *Poultry Science, 102*(11), 103082. https://doi.org/10.1016/j.psj.2023.103082.

Afrouziyeh, M., Zukiwsky, N. M., & Zuidhof, M. J. (2021). Timing of growth affected broiler breeder feeding motivation and reproductive traits. *Poultry Science, 100*(9), 101375. https://doi.org/10.1016/j.psj.2021.101375.

Ammann, J., Mack, G., El Benni, N., & Saleh, R. (2025). Consumers would rather buy a product with a levy for enhancing animal welfare than for environmental sustainability. *Sustainable Production and Consumption, 53*, 99–108. https://doi.org/10.1016/j.spc.2024.11.024.

Ammann, J., Mack, G., El Benni, N., Jin, S., Newell-Price, P., Tindale, S., Hunter, E., Vicario-Modroño, V., Gallardo-Cobos, R., Sánchez-Zamora, P., Miškolci, S., & Frewer, L. J. (2024). Consumers across five European countries prioritise animal welfare above environmental sustainability when buying meat and dairy products. *Food Quality and Preference, 117*, 105179. https://doi.org/10.1016/j.foodqual.2024.105179.

Arango, J., & Fulton, J. E. (2026). Commercial breeding for disease resistance in poultry: The breeder's perspective. *Poultry Science, 105*(1), 106188. https://doi.org/10.1016/j.psj.2025.106188.

Arrazola, & Torrey (2019). The development of fault bars in domestic chickens *(Gallus gallus domesticus*) increases with acute stressors and individual propensity: Implications for animal welfare. *Animal Welfare, 28*(3), 279–286. https://doi.org/10.7120/109627286.28.3.279.

Arrazola, A., Mosco, E., Widowski, T. M., Guerin, M. T., Kiarie, E. G., & Torrey, S. (2019a). The effect of alternative feeding strategies for broiler breeder pullets: 1. Welfare and performance during rearing. *Poultry Science, 98*(9), 3377–3390. https://doi.org/10.3382/ps/pez170.

Arrazola, A., Widowski, T. M., Guerin, M. T., Kiarie, E. G., & Torrey, S. (2019b). The effect of alternative feeding strategies for broiler breeder pullets: 2. Welfare and performance during lay. *Poultry Science, 98*(12), 6205–6216. https://doi.org/10.3382/ps/pez447.

Arrazola, A., & Torrey, S. (2021). Welfare and performance of slower growing broiler breeders during rearing. *Poultry Science, 100*(11). https://doi.org/10.1016/j.psj.2021.101434.

Arrazola, A., & Torrey, S. (2019). Conditioned place avoidance using encapsulated calcium propionate as an appetite suppressant for broiler breeders. *PLoS ONE, 14*(7), e0206271. https://doi.org/10.1371/journal.pone.0206271.

Arrazola, A., Widowski, T. M., & Torrey, S. (2022). In pursuit of a better broiler: Welfare and productivity of slower-growing broiler breeders during lay. *Poultry Science, 101*(8), 101917. https://doi.org/10.1016/j.psj.2022.101917.

Avila, L. P., Sweeney, K. M., Evans, C. R., White, D. L., Kim, W. K., Regmi, P., Williams, S. M., Nicholds, J., & Wilson, J. L. (2023). Body composition, gastrointestinal, and reproductive differences between broiler breeders fed using everyday or skip-a-day rearing programs. *Poultry Science, 102*(9), 102853. https://doi.org/10.1016/j.psj.2023.102853.

Ayeneshet, B., Taye, M., Esatu, W., & Tsefa, A. (2024). Comparative analysis of semen quality and fertility in diverse rooster breeds: A systematic review. *World's Poultry Science Journal,* (3), 947–975. https://doi.org/10.1080/00439339.2024.2338347.

Bartov, I., Bornstein, S., Lev, Y., Pines, M., & Rosenberg, J. (1988). Feed restriction in broiler breeder pullets: skip-a-day versus skip-two-days. *Poultry Science, 67*(5), 809–813. https://doi.org/10.3382/ps.0670809.

Baxter, M., & O'Connell, N. E. (2025). Developing appropriate environmental enrichment strategies for broiler breeders. *Scientific Reports, 15*(1), 9738. https://doi.org/10.1038/s41598-025-89691-w.

Bowling, M., Forder, R., Hughes, R. J., Weaver, S., & Hynd, P. I. (2018). Effect of restricted feed intake in broiler breeder hens on their stress levels and the growth and immunology of their offspring. *Translational Animal Science, 2*(3), 263–271. https://doi.org/10.1093/tas/txy064, https://academic.oup.com/tas.

Boyle, L., & Stevenson, P. (2025). Sustainability of livestock production: Raising the profile of animal welfare and the risks with "techno-fixes. *Animal Frontiers, 15*(5), 47–57. https://doi.org/10.1093/af/vfaf034.

Brugaletta, G., Greene, E., Tabler, T., Orlowski, S., Sirri, F., & Dridi, S. (2021). Effect of cyclic heat stress on feeding-related hypothalamic neuropeptides of three broiler populations and their ancestor jungle fowl. *Frontiers in Physiology, 12*, 809341. https://doi.org/10.3389/fphys.2021.809341.

Bruggeman, V., Onagbesan, O., D'Hondt, E., Buys, N., Safi, M., Vanmontfort, D., Berghman, L., Vandesande, F., & Decuypere, E. (1999). Effects of timing and duration of feed restriction during rearing on reproductive characteristics in broiler breeder females. *Poultry Science, 78*(10), 1424–1434. https://doi.org/10.1093/ps/78.10.1424.

Bruggeman, V., Onagbesan, O., Ragot, O., Metayer, S., Cassy, S., Favreau, F., Jego, Y., Trevidy, J. J., Tona, K., Williams, J., Decuypere, E., & Picard, M. (2005). Feed allowance-genotype interactions in broiler breeder hens. *Poultry Science, 84*(2), 298–306. https://doi.org/10.1093/ps/84.2.298.

Buckley, L. A., Sandilands, V., Hocking, P. M., Tolkamp, B. J., & D'eath, R. B. (2012). The use of conditioned place preference to determine broiler preferences for quantitative or qualitative dietary restriction. *British Poultry Science, 53*(3), 291–306. https://doi.org/10.1080/00071668.2012.698727.

Buller, H., Blokhuis, H., Jensen, P., & Keeling, L. (2018). Towards farm animal welfare and sustainability. *Animals, 8*(6), 81. https://doi.org/10.3390/ani8060081.

Burnside, T., & Neeteson, A.-M. (2025). Developments in welfare of parent stock and commercial broilers. *Poultry Science, 104*(2), 104732. https://doi.org/10.1016/j.psj.2024.104732.

Buzała, M., Janicki, B., & Czarnecki, R. (2015). Consequences of different growth rates in broiler breeder and layer hens on embryogenesis, metabolism and metabolic rate: A review. *Poultry Science, 94*(4), 728–733. https://doi.org/10.3382/ps/pev015.

Carneiro, P. R. O., Lunedo, R., Fernandez-Alarcon, M. F., Baldissera, G., Freitas, G. G., & Macari, M. (2019). Effect of different feed restriction programs on the performance and reproductive traits of broiler breeders. *Poultry Science, 98*(10), 4705–4715. https://doi.org/10.3382/ps/pez181.

Carney, V. L., Anthony, N. B., Robinson, F. E., Reimer, B. L., Korver, D. R., Zuidhof, M. J., & Afrouziyeh, M. (2022). Evolution of maternal feed restriction practices over 60 years of selection for broiler productivity. *Poultry Science, 101*(10), 101957. https://doi.org/10.1016/j.psj.2022.101957.

Cash, C., Witherspoon, K., & Athrey, G. (2025). How concerned should we be about broiler breeder fertility declines? *Poultry Science, 104*(5), 104992. https://doi.org/10.1016/j.psj.2025.104992.

Chan, I., Franks, B., & Hayek, M. N. (2022). The 'sustainability gap' of US broiler chicken production: Trade-offs between welfare, land use and consumption. *Royal Society Open Science, 9*(6), 210478. https://doi.org/10.1098/rsos.210478.

D'Eath, R. B., Tolkamp, B. J., Kyriazakis, I., & Lawrence, A. B. (2009). Freedom from hunger' and preventing obesity: The animal welfare implications of reducing food quantity or quality. *Animal Behaviour, 77*(2), 275–288. https://doi.org/10.1016/j.anbehav.2008.10.028.

Dawkins, M. S., & Layton, R. (2012). Breeding for better welfare: Genetic goals for broiler chickens and their parents. *Animal Welfare, 21*(2), 147–155. https://doi.org/10.7120/09627286.21.2.147.

de Beer, M., & Coon, C. N. (2007). The effect of different feed restriction programs on reproductive performance, efficiency, frame size, and uniformity in broiler breeder hens. *Poultry Science, 86*(9), 1927–1939. https://doi.org/10.1093/ps/86.9.1927.

de Beer, M., McMurtry, J. P., Brocht, D. M., & Coon, C. N. (2008). An examination of the role of feeding regimens in regulating metabolism during the broiler breeder grower period. 2. Plasma hormones and metabolites. *Poultry Science, 87*(2), 264–275. https://doi.org/10.3382/ps.2007-00196.

de Jong, I. C., Enting, H., van Voorst, A., & Blokhuis, H. J. (2005). Do low-density diets improve broiler breeder welfare during rearing and laying? *Poultry Science, 84*(2), 194–203. https://doi.org/10.1093/ps/84.2.194.

de Jong, I. C., van Voorst, A. S., & Blokhuis, H. J. (2003). Parameters for quantification of hunger in broiler breeders. *Physiology & Behavior, 78*(4-5), 773–783. https://doi.org/10.1016/s0031-9384(03)00058-1.

De Jong, I. C., & Guémené, D. (2011). Major welfare issues in broiler breeders. *World's Poultry Science Journal, 67*(1), 73–82. https://doi.org/10.1017/s0043933911000067.

Decuypere, E., Hocking, P. M., Tona, K., Onagbesan, O., Bruggeman, V., Jones, E. K. M., Cassy, S., Rideau, N., Metayer, S., Jego, Y., Putterflam, J., Tesseraud, S., Collin, A., Duclos, M., Trevidy, J. J., & Williams, J. (2006). Broiler breeder paradox: A project report. *World's Poultry Science Journal, 62*(03), 443–453. https://doi.org/10.1017/S0043933906001073.

Decuypere, Bruggeman, Everaert, Li, Y., Boonen, de Tavernier, Janssens, & Buys (2010). The broiler breeder paradox: Ethical, genetic and physiological perspectives, and suggestions for solutionsy. *British Poultry Science, 51*(5), 569–579. https://doi.org/10.1080/00071668.2010.519121.

Dunkley, C. S., Fairchild, B. D., Ritz, C. W., Kiepper, B. H., & Lacy, M. P. (2015). Carbon footprint of poultry production farms in South Georgia: A case study. *Journal of Applied Poultry Research, 24*(1), 73–79. https://doi.org/10.3382/japr/pfu005.

Fattori, T. R., Wilson, H. R., Harms, R. H., & Miles, R. D. (1991). Response of broiler breeder females to feed restriction below recommended levels. 1. *Growth and reproductive performance. Poultry Science, 70*(1), 26–36. https://doi.org/10.3382/ps.0700026.

Gebhardt-Henrich, S. G., Jordan, A., Toscano, M. J., & Würbel, H. (2020). The effect of perches and aviary tiers on the mating behaviour of two hybrids of broiler breeders. *Applied Animal Behaviour Science, 233*, 105145. https://doi.org/10.1016/j.applanim.2020.105145.

Gholami-Soltanmoradi, M., Solka, M., Marchewka, J., Seidavi, A., Cooper, R. G., Dadashbeiki, M., Zigo, F., Sztandarski, P., Jaszczyk, A., & Horbańczuk, J. O. (2024). Feeding frequency can affect the morphology of reproductive tract in broiler breeder hens. *Poultry Science, 103*(12), 104307. https://doi.org/10.1016/j.psj.2024.104307.

Girard, M. T. E., Zuidhof, M. J., & Bench, C. J. (2017). Feeding, foraging, and feather pecking behaviours in precision-fed and skip-a-day-fed broiler breeder pullets. *Applied Animal Behaviour Science, 188*, 42–49. https://doi.org/10.1016/j.applanim.2016.12.011.

Havenstein, G. B., Ferket, P. R., & Qureshi, M. A. (2003). Growth, livability, and feed conversion of 1957 versus 2001 broilers when fed representative 1957 and 2001 broiler diets. *Poultry Science, 82*(10), 1500–1508. https://doi.org/10.1093/ps/82.10.1500.

Heck, A., Onagbesan, O., Tona, K., Metayer, S., Putterflam, J., Jego, Y., Trevidy, J. J., Decuypere, E., Williams, J., Picard, M., & Bruggeman, V. (2004). Effects of ad libitum feeding on performance of different strains of broiler breeders. *British Poultry Science, 45*(5), 695–703. https://doi.org/10.1080/00071660400006537.

Hocking, P. M., & Jones, E. K. M. (2006). On-farm assessment of environmental enrichment for broiler breeders. *British Poultry Science, 47*(4), 418–425. https://doi.org/10.1080/00071660600825074.

Hocking, P. M., Maxwell, M. H., & Mitchell, M. A. (1993). Welfare assessment of broiler breeder and layer females subjected to food restriction and limited access to water during rearing. *British Poultry Science, 34*(3), 443–458. https://doi.org/10.1080/00071669308417600.

Huber, K. (2018). Invited review: Resource allocation mismatch as pathway to disproportionate growth in farm animals – prerequisite for a disturbed health. *Animal, 12*(3), 528–536. https://doi.org/10.1017/s1751731117002051.

Karlsson, L., Keeling, L., & Röös, E. (2025). What is a better chicken? Exploring trade-offs between animal welfare and greenhouse gas emissions in higher-welfare broiler systems. *Sustainable Production and Consumption, 55*, 203–216. https://doi.org/10.1016/j.spc.2025.02.015.

Katanbaf, M. N., Dunnington, E. A., & Siegel, P. B. (1989a). Restricted feeding in early and late-feathering chickens. *1. Growth and physiological responses. Poultry science, 68*(3), 344–351. https://doi.org/10.3382/ps.0680344.

Katanbaf, M. N., Dunnington, E. A., & Siegel, P. B. (1989b). Restricted feeding in early and late-feathering chickens. *2. Reproductive responses. Poultry science, 68*(3), 352–358. https://doi.org/10.3382/ps.0680352.

Kebreab, E., Liedke, A., Caro, D., Deimling, S., Binder, M., & Finkbeiner, M. (2016). Environmental impact of using specialty feed ingredients in swine and poultry production: A life cycle assessment. *Journal of Animal Science, 94*(6), 2664–2681. https://doi.org/10.2527/jas.2015-9036.

Keeling, L. J. (2025). Animal welfare: Part of the solution, not part of the problem in the move toward achieving sustainable development in animal agriculture. *Animal Frontiers, 15*(2), 8–11. Available from https://doi.org/10.1093/af/vfaf009.

Leone, E. H., & Estévez, I. (2008). Economic and welfare benefits of environmental enrichment for broiler breeders. *Poultry Science, 87*(1), 14–21. https://doi.org/10.3382/ps.2007-00154.

Lindholm, Johansson, Middelkoop, Lees, Yngwe, Berndtson, Cooper, & Altimiras (2018). The quest for welfare-friendly feeding of broiler breeders: Effects of daily vs. 5:2 feed restriction schedules. *Poultry Science, 97*(2), 368–377. https://doi.org/10.3382/ps/pex326.

Mens, A. J. W., de Jong, I. C., van Riel, J. W., Gunnink, H., van Hattum, T., & van Emous, R. A. (2022). Diet dilution and feeding frequency have only minor effects on the behaviour of broiler breeder pullets. *Applied Animal Behaviour Science, 253*, 105678. https://doi.org/10.1016/j.applanim.2022.105678.

Mens, A. J. W., & van Emous, R. A. (2022). Broiler breeders roosted more on slats than on perches during the laying period. *Applied Animal Behaviour Science, 246*, 105531. https://doi.org/10.1016/j.applanim.2021.105531.

Millman, S. T., Duncan, I. J., & Widowski, T. M. (2000). Male broiler breeder fowl display high levels of aggression toward females. *Poultry Science, 79*(9), 1233–1241. https://doi.org/10.1093/ps/79.9.1233.

Moradi, S., Zaghari, M., Shivazad, M., Osfoori, R., & Mardi, M. (2013). Response of female broiler breeders to qualitative feed restriction with inclusion of soluble and insoluble fiber sources. *Journal of Applied Poultry Research, 22*(3), 370–381. https://doi.org/10.3382/japr.2012-00504.

Morrissey, K. L. H., Widowski, T., Leeson, S., Sandilands, V., Arnone, A., & Torrey, S. (2014a). The effect of dietary alterations during rearing on feather condition in broiler breeder females. *Poultry Science, 93*(7), 1636–1643. https://doi.org/10.3382/ps.2013-03822.

Morrissey, K. L. H., Widowski, T., Leeson, S., Sandilands, V., Arnone, A., & Torrey, S. (2014b). The effect of dietary alterations during rearing on growth, productivity, and behavior in broiler breeder females. *Poultry Science, 93*(2), 285–295. https://doi.org/10.3382/ps.2013-03265.

Mostert, P. F., Bos, A. P., van Harn, J., & de Jong, I. C. (2022). The impact of changing toward higher welfare broiler production systems on greenhouse gas emissions: A Dutch case study using life cycle assessment. *Poultry Science, 101*(12), 102151. https://doi.org/10.1016/j.psj.2022.102151.

Neeteson, A.-M., Avendaño, S., Koerhuis, A., Duggan, B., Souza, E., Mason, J., Ralph, J., Rohlf, P., Burnside, T., Kranis, A., & Bailey, R. (2023). Evolutions in commercial meat poultry breeding. *Animals, 13*(19), 3150. https://doi.org/10.3390/ani13193150.

Newberry, R. C. (1995). Environmental enrichment: Increasing the biological relevance of captive environments. *Applied Animal Behaviour Science, 44*(2-4), 229–243. https://doi.org/10.1016/0168-1591(95)00616-Z.

Pinchasov, Y., Galili, D., Yonash, N., & Klandorf, H. (1993). Effect of feed restriction using self-restricting diets on subsequent performance of broiler breeder females. *Poultry Science, 72*(4), 613–619. https://doi.org/10.3382/ps.0720613.

Renema, R. A., & Robinson, F. E. (2004). Defining normal: Comparison of feed restriction and full feeding of female broiler breeders. *World's Poultry Science Journal, 60*(4), 508–522. https://doi.org/10.1079/wps200434.

Renema, R. A., Robinson, F. E., Newcombe, M., & McKay, R. I. (1999). Effects of body weight and feed allocation during sexual maturation in broiler breeder hens. 1. Growth and carcass characteristics. *Poultry Science, 78*(5), 619–628. https://doi.org/10.1093/ps/78.5.619.

Renema, R. A., Rustad, M. E., & Robinson, F. E. (2007). Implications of changes to commercial broiler and broiler breeder body weight targets over the past 30 years. *World's Poultry Science Journal, 63*(3), 457–472. https://doi.org/10.1017/s0043933907001572.

Riber, A. B., & Tahamtani, F. M. (2020). Motivation for feeding in broiler breeder pullets fed different types of restricted high-fibre diets. *Applied Animal Behaviour Science, 230*, 105678. https://doi.org/10.1016/j.applanim.2020.105048.

Riber, A. B., de Jong, I. C., van de Weerd, H. A., & Steenfeldt, S. (2017). Environmental enrichment for broiler breeders: An undeveloped field. *Frontiers in Veterinary Science, 4*, 86. https://doi.org/10.3389/fvets.2017.00086.

Richards, M. P., Rosebrough, R. W., Coon, C. N., & McMurtry, J. P. (2010). Feed intake regulation for the female broiler breeder: In theory and in practice. *Journal of Applied Poultry Research, 19*(2), 182–193. https://doi.org/10.3382/japr.2010-00167.

Romero-Sanchez, H., Plumstead, P. W., Leksrisompong, N., Brannan, K. E., & Brake, J. (2008). Feeding broiler breeder males. 4. Deficient feed allocation reduces fertility and broiler progeny body weight. *Poultry Science, 87*(4), 805–811. https://doi.org/10.3382/ps.2007-00285.

Rothschild, D. (2019). *A comparison of immune response and organ weights in broiler chickens (MSc Thesis)*. Guelph: University of Guelph. Available from http://hdl.handle.net/10214/17512.

Sabah, S., & Yilmaz Dikmen, B. (2023). Spiking applications in broiler breeders and its effect on reproduction performance. *World's Poultry Science Journal, 79*(3), 581–591. https://doi.org/10.1080/00439339.2023.2208386.

Sandilands, V., Tolkamp, B., & Kyriazakis, I. (2005). Behaviour of food restricted broilers during rearing and lay—Effects of an alternative feeding method. *Physiology & Behavior, 85*(2), 115–123. https://doi.org/10.1016/j.physbeh.2005.03.001.

Sandilands, V., Tolkamp, B. J., Savory, C. J., & Kyriazakis, I. (2006). Behaviour and welfare of broiler breeders fed qualitatively restricted diets during rearing: Are there viable alternatives to quantitative restriction? *Applied Animal Behaviour Science, 96*(1-2), 53–67. https://doi.org/10.1016/j.applanim.2005.04.017.

Savory, C. J., & Lariviere, J.-M. (2000). Effects of qualitative and quantitative food restriction treatments on feeding motivational state and general activity level of growing broiler breeders. *Applied Animal Behaviour Science, 69*(2), 135–147. https://doi.org/10.1016/s0168-1591(00)00123-4.

Shrader, H. L. (1952). The chicken-of-tomorrow program; its influence on "meat-type" poultry production. *Poultry Science, 31*(1), 3–10. https://doi.org/10.3382/ps.0310003.

Siegel, P. B., & Dunnington, E. A. (1985). Reproductive complications associated with selection for broiler growth. Poultry genetics and breeding. *Proceedings of the 18th Poultry Science Symposium, in Association with the 25th British Poultry Breeders Round Table, 1983.* 59–72 ref. 144.

Spradley, J. M., Freeman, M. E., Wilson, J. L., & Davis, A. J. (2008). The influence of a twice-a-day feeding regimen after photostimulation on the reproductive performance of broiler breeder hens. *Poultry Science, 87*(3), 561–568. https://doi.org/10.3382/ps.2007-00327.

Tahamtani, F. M., & Riber, A. B. (2020). The effect of qualitative feed restriction in broiler breeder pullets on fear and motivation to explore. *Applied Animal Behaviour Science, 228*, 105009. https://doi.org/10.1016/j.applanim.2020.105009.

Tahamtani, F. M., Moradi, H., & Riber, A. B. (2020). Effect of qualitative feed restriction in broiler breeder pullets on stress and clinical welfare indicators. *Frontiers in Veterinary Science, 7*, 316. https://doi.org/10.3389/fvets.2020.00316.

Taherkhani, R., Zaghari, M., Shivazad, M., & Shahneh, A. Z. (2010). A twice-a-day feeding regimen optimizes performance in broiler breeder hens. *Poultry Science, 89*(8), 1692–1702. https://doi.org/10.3382/ps.2009-00488.

Tallentire, C. W., Leinonen, I., & Kyriazakis, I. (2016). Breeding for efficiency in the broiler chicken: A review. *Agronomy for Sustainable Development, 36*(4), 66. https://doi.org/10.1007/s13593-016-0398-2.

Taylor, P. S., Forder, R., & Morgan, N. (2025). Meat chicken breeder stress: Causes, effects, and mitigation strategies. *World's Poultry Science Journal, 81*(1), 35–64. https://doi.org/10.1080/00439339.2024.2437674.

Tolkamp, B. J., Sandilands, V., & Kyriazakis, I. (2005). Effects of qualitative feed restriction during rearing on the performance of broiler breeders during rearing and lay. *Poultry Science, 84*(8), 1286–1293. https://doi.org/10.1093/ps/84.8.1286.

van Emous, R. (2022). State of the art pullet and breeder nutrition. *Proceedings of the Arkansas Nutrition Conference, 2022*(1), 11.

van Emous, R. A., Kwakkel, R. P., van Krimpen, M. M., van den Brand, H., & Hendriks, W. H. (2015). Effects of growth patterns and dietary protein levels during rearing of broiler breeders on fertility, hatchability, embryonic mortality, and offspring performance. *Poultry Science, 94*(4), 681–691. https://doi.org/10.3382/ps/pev024.

van Emous, R. A., Mens, A. J. W., & Winkel, A. (2021). Effects of diet density and feeding frequency during the rearing period on broiler breeder performance. *British Poultry Science, 62*(5), 686–694. https://doi.org/10.1080/00071668.2021.1918634.

Van Emous, R. A., Kwakkel, R., Van Krimpen, M., & Hendriks, W. (2014). Effects of growth pattern and dietary protein level during rearing on feed intake, eating time, eating rate, behavior, plasma corticosterone concentration, and feather cover in broiler breeder females during the rearing and laying period. *Applied Animal Behaviour Science, 150*, 44–54. https://doi.org/10.1016/j.applanim.2013.10.005.

Van Krimpen, M. M., & De Jong, I. C. (2014). Impact of nutrition on welfare aspects of broiler breeder flocks. *World's Poultry Science Journal, 70*(1), 139–150. https://doi.org/10.1017/s0043933914000129.

Vasdal, G., Gebhardt-Henrich, S. G., Tahamtani, F., & Kittelsen, K. E. (2022a). Effect of perch access on perching, health and production outcomes in commercial broiler breeder flocks. *Poultry Science, 101*(11), 102160. https://doi.org/10.1016/j.psj.2022.102160.

Vasdal, G., Gebhardt-Henrich, S. G., Tahamtani, F., & Kittelsen, K. E. (2022b). Perch use in commercial broiler breeders – Preference for perch material and effect of age. *Applied Animal Behaviour Science, 253*, 105680. https://doi.org/10.1016/j.applanim.2022.105680.

Voitle, R. A., Wilson, H. R., & Harms, R. H. (1970). Effect of body weight restriction of broiler breeder pullets on subsequent performance. *Poultry Science, 49*(5), 1448.

Weaver, E. A., & Ramachandran, R. (2023). Metformin improves ovarian function and increases egg production in broiler breeder hens. *Reproduction, 165*(3), 289–300. https://doi.org/10.1530/REP-22-0256.

Wolanski, N. J., Renema, R. A., Robinson, F. E., & Wilson, J. L. (2004). End-of-season carcass and reproductive traits in original and replacement male broiler breeders. *Journal of Applied Poultry Research, 13*(3), 451–460. https://doi.org/10.1093/japr/13.3.451.

Wurtz, K. E., Thodberg, K., & Riber, A. B. (2025). Impact of feeding a high fiber diet and roughage on stress and clinical welfare indicators in fast-growing broiler breeder pullets. *Poultry Science, 104*(12), 105960. https://doi.org/10.1016/j.psj.2025.105960.

Zuidhof, M. J., Holm, D. E., Renema, R. A., Jalal, M. A., & Robinson, F. E. (2015). Effects of broiler breeder management on pullet body weight and carcass uniformity. *Poultry Science, 94*(6), 1389–1397. https://doi.org/10.3382/ps/pev064.

Zukiwsky, N. M., Afrouziyeh, M., Robinson, F. E., & Zuidhof, M. J. (2021). Feeding, feed-seeking behavior, and reproductive performance of broiler breeders under conditions of relaxed feed restriction. *Poultry Science, 100*(1), 119–128. https://doi.org/10.1016/j.psj.2020.09.081.

Chapter 5

Ducks

Gregory S. Fraley and Jenna M. Schober
Department of Animal Sciences, Purdue University, West Lafayette, IN, United States

Introduction

The most common ducks used for commercial meat production globally include Pekin, Muscovy, and Mule (a hybrid between Pekin and Muscovy) ducks, with varying prevalence depending on the global region (Babington & Campbell, 2022; Cherry & Morris, 2008). Commercial duck production is a globally significant industry. Pekin ducks (*Anas platyrhynchos domesticus*) are the predominant species used for meat production, particularly in the United States and the European Union, where an estimated 200 million ducks are processed yearly for their meat (Makagon & Riber, 2022). Globally, ducks are also raised commercially for eggs, feathers, offal, and foie gras (Babington & Campbell, 2022). Despite the scale of the global duck industry, there is a notable lack of research-based guidance regarding the welfare, sustainability, and optimal housing design for ducks (Bergman et al., 2024; Makagon & Riber, 2022). Commercial duck housing design must balance multiple competing demands: maximizing production efficiency, maintaining animal health and welfare, managing environmental conditions, and meeting regulatory requirements. These demands are addressed differently across regions, resulting in a diverse landscape of housing systems worldwide. The purpose of this chapter is to synthesize available scientific literature on commercial duck housing design, environmental management systems, flooring and litter management, water provision, stocking density, lighting, and biosecurity considerations in relation to welfare and sustainability. Regional differences and emerging technological innovations are also discussed.

Regional production systems

Commercial duck production systems vary considerably across the world. In the United States and the European Union, large-scale indoor production systems predominate, where birds are typically housed in enclosed or semienclosed barns with controlled environments (Cherry & Morris, 2008;

Integrating Welfare and Sustainability in Poultry Systems. DOI: https://doi.org/10.1016/B978-0-443-40594-5.00004-3

Makagon & Riber, 2022). In contrast, in Southeast Asia, particularly in countries such as Thailand, Vietnam, and Cambodia, traditional duck production systems often involve outdoor access, including integration with aquaculture systems such as fish ponds (Cherry & Morris, 2008; Leibler et al., 2009). These integrated duck-fish pond systems, common in Southeast Asia, represent a fundamentally different approach to duck housing compared to the intensive indoor systems prevalent in Western countries (Leibler et al., 2009). Across regions, grower ducks are housed in different types of floor-based systems with a combination of indoor and outdoor areas, or exclusively indoors with enclosed or open-air ventilation (Babington & Campbell, 2022; Chen et al., 2021). In China, large-scale commercial duck production has increasingly adopted fully enclosed, mechanically ventilated housing systems (Hu et al., 2024). The shift toward fully enclosed housing in China reflects broader trends in the intensification of poultry production globally (Jalaludeen & Churchil, 2022; Leibler et al., 2009). The intensification of duck production globally mirrors broader trends in poultry and livestock agriculture. Industrial food animal production systems are characterized by large numbers of animals raised in confinement with high throughput and rapid turnover (Babington & Campbell, 2022; Leibler et al., 2009; Makagon & Riber, 2022). For example, commercial duck operations in the United Kingdom have been documented to house approximately 26000 ducks across two houses (James et al., 2023), while duck houses in China have been documented with capacities of approximately 9500 ducks in a single house of approximately 850 m^2 (Qu et al., 2024), or up to 50,000 ducks in a cage system (GSF, personal observations). In the USA, duck houses can vary among 4000–15,000 ducks depending upon the purpose of the house (meat vs breeder; Cherry & Morris, 2008).

Ventilation systems

Ventilation is one of the most critical aspects of commercial duck house design, affecting air quality, temperature, humidity, and the health and welfare of the birds (Hu et al., 2024; Makagon & Riber, 2022). Two primary ventilation approaches are used in commercial duck production: natural ventilation and mechanical (negative pressure) ventilation. Natural ventilation relies on wind-driven airflow through curtain-sided barns and is the predominant ventilation type globally. Naturally ventilated duck houses, such as those documented in the United Kingdom, allow airflow through the building without mechanical assistance (James et al., 2023). While natural ventilation is simpler and less energy-intensive, it provides less precise environmental control and may be inadequate in extreme weather conditions (James et al., 2023). Although this type provides "natural" ventilation, it does expose the ducks to whatever temperature and humidity exist in the environment at the time, which could have an impact on duck welfare. Mechanical ventilation uses fans to create negative pressure within the building, drawing fresh air in through inlets and exhausting stale air. Fully enclosed duck houses in China use negative-pressure ventilation combined with water curtain filtration to control temperature and humidity (Hu et al., 2024). This approach allows for precise control of the indoor environment but requires significant energy input and careful management to avoid uneven distribution of environmental conditions within the house (Hu et al., 2024).

A significant challenge with longitudinal ventilation systems, such as the Longitudinal Water Curtain Cooling System used in Chinese duck houses, is the creation of environmental gradients within the building. Research has demonstrated that from the water curtain end to the fan cooling end of a duck house, the hourly average temperature decreases by 0.1°C–0.5°C at varying points throughout the barn, while the daily and hourly average relative humidity decreases from the water curtain end to the fan cooling end by 7% to 8%, respectively (Hu et al., 2024). These environmental gradients can affect duck production performance, immune function, and welfare, with differences in dressing weight, skin fat rate, and percentage of abdominal fat observed among ducks housed at different positions within the same house (Hu et al., 2024). The design and operational requirements of large-scale duck houses produce inherent challenges for ventilation and biosecurity. Ventilation of animal houses poses a major, and often unrecognized, gap in biosecurity as the confinement of thousands of animals requires controls to reduce heat and regulate humidity (Leibler et al., 2009). Inadequate ventilation can lead to the accumulation of harmful gases such as ammonia, elevated particulate matter concentrations, and increased disease risk and decreased welfare (Hu et al., 2024; Jalaludeen & Churchil, 2022; Qu et al., 2024).

Flooring systems

Flooring is a critical component of duck house design, affecting litter quality, foot and leg health, and overall welfare. Within indoor housing, birds are typically provided litter or raised plastic/wire flooring (Chen et al., 2021; Fraley et al., 2013; Karcher et al., 2013). The choice of flooring system has significant implications for duck welfare and production performance. Litter-based flooring is the most common system in commercial duck production globally. Litter typically consists of a bedding material such as wood shavings, rice husks, or straw, which absorbs moisture from droppings and spilled water (Makagon & Riber, 2022). The quality of litter is a major determinant of duck welfare, as wet litter (moisture percentage above 25%) can alter microbial activity, temperature, humidity, and ammonia levels (Bergman et al., 2025). Wet litter has been shown to reduce welfare by increasing foot pad dermatitis in ducks (Cherry & Morris, 2008; Jalaludeen & Churchil, 2022). Raised plastic or wire flooring is increasingly used in some commercial duck production systems as a more sustainable substitute for litter. Research in China has documented the use of closed net-flat breeding methods in duck houses, where ducks are raised on raised mesh flooring (Qu et al., 2024). This approach can improve air quality by allowing droppings to fall through the floor, but may have implications for foot and leg health, and could possibly lead to increased lameness; however, that may strictly depend upon the materials used for raised plastic or wire flooring (Chen et al., 2021; Fraley et al., 2013; Karcher et al., 2013). Types of litter are very regionally dictated based upon what is available: pine shavings (sawdust; USA), chopped straw (UK & EU), or rice hulls (Asia), among others (Cherry & Morris, 2008).

Litter quality is closely linked to the design and management of the duck house, particularly water provision and ventilation. The provision of open water sources such as troughs has been correlated with decreased feather and foot pad quality as well as increased bacterial contamination and mortality in commercial duck barns (Jones & Dawkins, 2010; Rodenburg et al., 2005; Schenk et al., 2016; Waitt et al., 2009). This creates a fundamental tension in duck house design between the behavioral needs of ducks as semiaquatic waterfowl and the practical requirements of maintaining litter quality (Bergman et al., 2024; Makagon & Riber, 2022; Schenk et al., 2016; Schober et al., 2023). Research has shown that ducks alter their behaviors and physical condition based on their environment and types of enrichments at a commercial level (Bergman et al., 2025; Jalaludeen & Churchil, 2022). The management of litter quality therefore requires careful attention to the design of water provision systems, ventilation, and stocking density (Bergman et al., 2025; Chen et al., 2021; Cherry & Morris, 2008; Jalaludeen & Churchil, 2022; Makagon & Riber, 2022).

The challenge of water provision

Water provision is one of the most challenging and controversial aspects of commercial duck house design. Domesticated ducks are semiaquatic waterfowl, and a major challenge within the industry is how to safely provide a water source for ducks to express "natural" preening behaviors (Bergman et al., 2025; Chen et al., 2021; Cherry & Morris, 2008; Jalaludeen & Churchil, 2022; Makagon & Riber, 2022). When water is available, domesticated ducks engage in water-related behaviors such as swimming and wet preening (Schenk et al., 2016; Schober et al., 2023; Waitt et al., 2009). It should be noted that "wet preening" is a specific behavior stated in poultry welfare research, but is not a specific behavior noted by naturalists, and cannot be found in any ornithological literature, as all birds use water to preen, but in different frequencies. Preening behaviors are important for feather maintenance and are considered a key component of duck welfare (Makagon & Riber, 2022; Waitt et al., 2009). However, providing open water in commercial duck houses is problematic because it can reduce litter and air quality, and lead to higher duck mortality or illness (Rodenburg et al., 2005; Schenk et al., 2016;

Schober et al., 2023). The tension between the behavioral needs of ducks and the practical requirements of commercial production has led to a range of different approaches to water provision across regions and production systems.

Nipple drinkers are the most common water provision system in commercial duck houses globally, especially in the United States and Asia, and provide water for drinking while minimizing spillage and litter wetting (Cherry & Morris, 2008; Jalaludeen & Churchil, 2022; Makagon & Riber, 2022). However, nipple drinkers do not allow ducks to engage in water-related behaviors such as head dipping, bill washing, or "wet preening" (Makagon & Riber, 2022; Waitt et al., 2009 and link to reference in references section). However, more recent studies have demonstrated that ducks are capable of wet preening with nipple lines and that ducks do, in fact, participate in social behaviors associated with nipple lines (Rice et al., 2014; Schober et al., 2023). Preening cups are semiopen water sources used in some commercial duck barns, particularly in the United States, as an environmental enrichment to increase welfare and improve affective state (Bergman et al., 2025). Currently, semiopen water sources are recommended for commercial duck barns for American Humane welfare certification and are typically placed at approximately 1000–2000 ducks per preening cup (Bergman et al., 2024; Schober et al., 2023). Research on preening cups has found no differences in production values between ducks housed in pens with or without preening cups in an academic setting, but preening cups substantially increase bacterial contamination of the entire barn environment (Schober et al., 2025). Water troughs provide a larger water surface and allow more water-related behaviors, but have been associated with decreased feather and foot pad quality as well as increased bacterial contamination and mortality (Rodenburg et al., 2005; Schenk et al., 2016). Misting systems represent an innovative approach to water provision that attempts to provide some of the benefits of surface water while minimizing litter wetting. Research in Australia found that overhead misting affected duck behavior without compromising welfare, but noted that further research with larger water droplet sizes or extended durations of misting may lead to greater effects (Campbell et al., 2022). Misting systems are an effective way to provide cooling and enrichment; however, their effectiveness is limited to areas around the world with low humidity. The use of misters in areas with higher humidity may provide additional challenges.

In Asia, traditional duck production systems involving integration with fish ponds represent a distinct biosecurity challenge. Duck houses with exits onto fish ponds, as documented in Thailand, and ducks on fish ponds in Vietnam, represent production systems where ducks have direct contact with water bodies that may also be frequented by wild birds (Leibler et al., 2009). These systems have been identified as potential risk factors for the emergence and spread of zoonotic pathogens, including avian influenza (Leibler et al., 2009). The design of duck housing in these systems must therefore consider the biosecurity implications of outdoor access and contact with wild birds (Leibler et al., 2009).

Effects of stocking density on production and welfare

Stocking density is a key variable in commercial duck house design, with significant implications for both production performance and animal welfare. There is abundant evidence of a negative correlation between stocking density and growth in Pekin ducks, irrespective of housing system (Makagon & Riber, 2022). This relationship has been demonstrated across multiple studies, lending support to the importance of appropriate stocking density management in commercial duck production (Makagon & Riber, 2022).

Research has shown that when keeping Pekin ducks at a high stocking density (11 ducks/m^2), administration of dietary tryptophan improved growth and meat quality, suggesting that nutritional interventions may partially mitigate the negative effects of high stocking density (Makagon & Riber, 2022; Wang et al., 2006). However, the broader welfare implications of high stocking density extend beyond production measures, and there is a major research gap with regard to the influence of stocking density on welfare indicators beyond production measures (Makagon & Riber, 2022).

Air quality management

Air quality is a critical concern in commercial duck houses, with implications for both animal health and human occupational health. Research conducted at commercial duck farms in China has documented particulate matter (PM) concentrations inside duck houses that exceed the short-term (24 hours) guideline of the global air quality guidelines for both PM2.5 and PM10 (Qu et al., 2024). Particulate matter concentrations in duck houses are influenced by temperature, humidity, ventilation, and season, with PM concentrations in summer typically being lower than in winter due to better ventilation (Qu et al., 2024). The microbial composition of PM in duck houses is dominated by bacteria, with a total of 1316 bacterial genera and 110 fungal genera identified in PM samples from a single duck house in winter (Qu et al., 2024). Relatively high levels of pathogenic bacteria and conditionally pathogenic bacteria have been identified in duck house PM, raising concerns about public health risks to staff and surrounding areas (Qu et al., 2024). The diversity and relative abundance of microorganisms were found to be similar in the duck house and at downwind points, indicating that microorganisms in the house environment have a significant influence on the air environment around the downwind point (Qu et al., 2024). Ammonia is a major air quality concern in commercial duck houses, arising from the decomposition of nitrogenous organic matter in droppings and litter. Ammonia concentrations in duck houses are influenced by ventilation, temperature, litter moisture content, and management practices (Hu et al., 2024; Makagon & Riber, 2022). Suboptimal housing conditions, including poor ventilation and wet litter, can lead to elevated ammonia concentrations that impair duck welfare and health (Hu et al., 2024; Makagon & Riber, 2022). Research on modern ventilation in duck houses has documented ammonia concentrations within the house, with ventilation management playing a key role in controlling ammonia levels (Hu et al., 2024). The design of ventilation systems must therefore balance the need to remove ammonia and other harmful gases with the need to maintain appropriate temperature and humidity conditions (Cherry & Morris, 2008; Hu et al., 2024; Jalaludeen & Churchil, 2022).

Lighting in commercial duck houses

Lighting is an important component of commercial duck house design, affecting behavior, production performance, and welfare. Research on Pekin ducks has shown that blue and green light can reduce fearfulness (Nelson et al., 2020), suggesting that the light spectrum may be an important consideration in duck house design. While much of the research on lighting in poultry houses has focused on broiler chickens, the principles are likely applicable to duck production as well. A lighting program providing 18 hours of light (photophase) with 6 hours of dark (scotophase) is typical, but waterfowl are adaptable to a range of diurnal photo- and scotoperiods. Waterfowl can be housed under LED, incandescent, or fluorescent lighting or even kerosene lanterns in commercial settings around the world. No adverse effects have been reported for any of the light sources. The output of light (i.e., wavelength and photonic energy) is more important than is the particular bulb type/source (e.g., LED, fluorescent, or incandescent).

White light provides the best environmental conditions for waterfowl at any age. The effects of red or blue lighting have been studied in ducks. Research suggested that in grow-out ducks, red light may have some advantages in terms of reduced activity and reduced feather picking (automutilation) and/or feather pecking (conspecific directed). However, the reduced activity under red light conditions is not associated with improved growth rates or carcass quality (Campbell et al., 2015). Interestingly, unlike chickens, housing ducks under blue light may have detrimental effects in grow-out ducks in terms of behavior, growth rate, welfare, and carcass quality (Campbell et al., 2015). Further, blue light may pose negative impacts on fertility, production, welfare, and behavior of breeder ducks (Haas et al., 2017; Tonissen et al., 2025) as well as with white Roman breeder geese (Chang et al., 2016). Low white-light intensity has also been shown to reduce fertility in drakes, but not hens, for breeder Pekin ducks (Porter et al., 2018). However, more complete studies on light intensity and on relative lengths of photophase versus scotophase have yet to be conducted and specific photoperiods and light intensity may ultimately have an important impact on duck health and welfare.

Nest box design for ducks

The design of nest boxes is an important component of breeder duck house design. Limited, but important, research has shown that singly housed Pekin ducks raised with different combinations of nest box designs preferentially lay their eggs in the nest box that offers the highest degree of enclosure (Makagon & Mench, 2011). When groups of ducks were offered either minimally enclosed (enclosed only on three sides) nest boxes and ones with a roof, both were used to the same extent, possibly because both designs offered more concealment than the floor alone (Makagon & Mench, 2011). These findings suggest that nest box design should prioritize enclosure to meet the behavioral needs of laying ducks (Makagon & Riber, 2022).

Biosecurity considerations in duck house design

Biosecurity is a critical consideration in commercial duck house design, particularly in the context of highly pathogenic avian influenza (HPAI). Research on the 2022–23 clade 2.3.4.4b H5N1 HPAI epizootic in Great Britain has demonstrated that airborne particles harboring infectious HPAI can be translocated short distances (less than 10 m) through the air, while particles containing viral RNA might travel further (less than 50 m; James et al., 2023). Importantly, infectious virus was detected in air samples collected from inside and outside, but in close proximity to, infected duck houses (James et al., 2023). The ventilation system design of duck houses has significant implications for biosecurity. Research has shown that naturally ventilated duck premises yielded greater positivity for HPAI outside the house than mechanically ventilated premises, although many confounding variables exist (James et al., 2023). The design and operational requirements of large-scale duck houses, including ventilation systems, create inherent biosecurity challenges (James et al., 2023; Leibler et al., 2009).

Outdoor access in commercial duck production

While outdoor access is not particularly common in large-scale duck production systems in the United States and the European Union, it is an important component of pasture and organic duck production (Cherry & Morris, 2008; Jalaludeen & Churchil, 2022; Makagon & Riber, 2022). Outdoor access allows ducks to engage in a wider range of natural behaviors, including foraging, swimming, and social interaction, which are important for welfare (Makagon & Riber, 2022). However, outdoor access also introduces biosecurity risks, particularly in the context of avian influenza (James et al., 2023; Leibler et al., 2009). The welfare of ducks raised on farms with outdoor access has been identified as a research need, as this aspect of duck production has been largely overlooked in the scientific literature (Makagon & Riber, 2022).

The design of outdoor areas, including the provision of water features, shade, and shelter, is an important consideration for farms that provide outdoor access.

Environmental monitoring and control

The integration of precision livestock farming (PLF) technologies into commercial duck house design represents an emerging area of innovation. Advanced environmental monitoring systems using sensors and IoT technology can enable real-time monitoring and control of temperature, humidity, ammonia, and other environmental parameters within duck houses (Zhang et al., 2024). Automated monitoring and control systems for livestock and poultry housing environments using IoT and robotics technology have been developed, utilizing IoT robots equipped with sensors to collect environmental data within the animal housing and leveraging WiFi, LoRa wireless communication, and IoT technology to control actuators inside the housing (Zhang et al., 2024). These systems enable data analysis and transmission from sensors to monitoring terminals to smart mobile devices, allowing remote monitoring of the duck house environment (Zhang et al., 2024). The integration of such technologies into commercial duck house design has significant potential to improve environmental control, reduce labor requirements, and enhance animal welfare (Zhang et al., 2024). Artificial intelligence (AI) technologies are increasingly being applied to improve animal welfare and production efficiency in commercial poultry production, including duck production. AI can be used to understand animal behaviors and emotions, providing deeper insights into their well-being and sources of stress (Zhang et al., 2024). In the field of livestock farming, future poultry production could potentially optimize operations through the use of intelligent autonomous systems, remotely managed by human workers and intervening only when necessary (Zhang et al., 2024). Sensor and robot systems for round-the-clock management of poultry houses have been proposed, enabling comprehensive management and control of multiple zones within the housing (Zhang et al., 2024). Automated disinfection robots equipped with air-assisted nozzles capable of independently atomizing and dispersing liquid disinfectant have been demonstrated in poultry houses, navigating autonomously at speeds ranging from 0.1 to 0.5 m/s (Zhang et al., 2024). These technologies have significant potential applications in commercial duck house design, particularly for large-scale operations where manual monitoring and management are impractical (Zhang et al., 2024).

Monitoring using technology can extend to animal behavior as well. Specific to Pekin ducks, the complete repertoire of duck vocalizations has been produced, and the effects of some of these vocalizations on conspecific physiology have been determined (Schober et al., 2024). All birds utilize vocalizations for communication. Vocalizations can be used to determine if an animal is experiencing a specific stimulus, and potentially positive or negative affective states. In the near future, it could be possible to monitor flock vocalizations in order to determine welfare status or identify stressors in real time. However, much more research is required to make this eventuality a reality.

Research gaps

Welfare standards for commercial duck production vary considerably across regions of the planet. In the United States, welfare certification provides guidelines for commercial duck barns, including requirements for semiopen water sources (Bergman et al., 2024; Schober et al., 2023). In the United Kingdom, standards recommend water for head dipping but only if it can be provided in a safe manner (Babington & Campbell, 2022; Heyn et al., 2006; Jalaludeen & Churchil, 2022). Current international duck farming regulations across various regions do not dictate bathing water provision (Babington & Campbell, 2022).

Despite the scale of the global duck industry, there is a significant lack of research-based guidance regarding the welfare and optimal housing design for ducks (Chen et al., 2021; Jalaludeen & Churchil, 2022; Makagon & Riber, 2022). Key research gaps identified in the literature include:

- The influence of stocking density on welfare indicators beyond production measures, particularly at the commercial scale
- The welfare of ducks raised on farms with outdoor access
- The welfare of breeder ducks
- The effects of different water provision systems on duck welfare and litter quality at a commercial scale
- The optimization of ventilation system design to minimize environmental gradients and their associated welfare and health impacts
- The long-term effects of different flooring systems on duck welfare and production performance
- Effects of light cycles and colored lighting systems on grow-out and breeder duck behavior and welfare
- Biosecurity and waste/litter management

The systematic review by Makagon & Riber (2022) identified 63 publications meeting inclusion criteria for a review of Pekin duck welfare research, highlighting the relatively limited evidence base compared to other commercial poultry species. The authors noted that while small-scale studies allow for housing conditions to be standardized, commercial-scale studies can reveal the combined effects of multiple aspects of housing and management, including aspects unique to large farm settings (Makagon & Riber, 2022). The global landscape of commercial duck housing design reflects a complex interplay of production efficiency, animal welfare, biosecurity, and regulatory requirements. The diversity of housing systems across regions—from fully enclosed, mechanically ventilated houses in China (Hu et al., 2024) to open-sided sheds in Australia (Campbell *et al.*, 2022) and integrated duck-fish pond systems in Southeast Asia (Leibler et al., 2009)—highlights the importance of context-specific approaches to duck house design.

Several key themes emerge from the synthesis of available evidence. First, ventilation system design is a critical determinant of environmental quality within duck houses, with implications for duck health, welfare, and production performance (Hu et al., 2024). The challenge of maintaining appropriate temperature, humidity, and air quality across the full length of large duck houses, particularly those using longitudinal ventilation systems, requires careful attention to system design and management. Second, water provision remains one of the most challenging and controversial aspects of commercial duck house design (Babington & Campbell, 2022; Chen et al., 2021; Jalaludeen & Churchil, 2022). The tension between the behavioral needs of ducks as semiaquatic waterfowl and the practical requirements of maintaining litter quality and animal health has not been fully resolved, and current regulatory frameworks provide limited guidance (Babington & Campbell, 2022). Innovative approaches such as misting systems and preening cups represent partial solutions, but further research is needed to develop evidence-based recommendations (Bergman et al., 2025; Campbell et al., 2022; Makagon & Riber, 2022; Schober et al., 2023). Third, the integration of PLF technologies, including IoT-based environmental monitoring and AI-driven management systems, offers significant potential to improve the design and management of commercial duck houses (Zhang et al., 2024). These technologies can enable more precise environmental control, early detection of health and welfare problems, and more efficient use of resources (Zhang et al., 2024). Fourth, biosecurity considerations are increasingly important in commercial duck house design, particularly in the context of HPAI (James et al., 2023; Leibler et al., 2009). The design of ventilation systems, the management of outdoor access, and the separation of duck houses from wild bird habitats are all important biosecurity considerations (James et al., 2023; Leibler et al., 2009).

Welfare considerations

Physical welfare

A duck's physical welfare is influenced by flooring type, stocking density, air quality, and water access. Ducks are often housed indoors on an open floor that is usually litter and/or plastic slatted flooring (Chen et al., 2021; Cherry & Morris, 2008). The plastic slatted flooring can lead to leg problems due to difficulty balancing on slats and injuries from slipping and falling (Raud & Faure, 1994; Rodenburg et al., 2005).

More broadly, ducks are at risk of developing foot and leg problems when housed under suboptimal conditions, such as inappropriate flooring, inadequate ventilation, or poor temperature regulation (i.e., elevated ammonia and humidity levels). In contrast, litter-based systems introduce a different set of concerns, particularly related to dust generation, which may negatively affect respiratory health as well as eye and nostril condition. Evidence comparing these systems remains inconsistent. For instance, Karcher et al. (2013) reported no differences in eye and nostril health between ducks raised on slatted versus solid floors with wood shavings, whereas Abdelfattah et al. (2020) found poorer eye and nostril condition in ducks housed exclusively on slatted flooring. Notably, neither study quantified dust levels, which may partially explain these conflicting findings.

In addition to housing conditions, feather condition represents another critical indicator of physical welfare. Ducks rely on both dry and wet preening to maintain plumage integrity. Wet preening has been described as the act of using the bill to apply water directly to the feathers while nibbling or stroking, or after tossing water over the body (Jones & Dawkins, 2010; Waitt et al., 2009). Ducks use their uropygial gland, also referred to as their preening gland, which is a holocrine gland located at the base of their tail, to preen. The oil secreted from the gland helps maintain and waterproof their feathers, while also providing defense against ectoparasites (Moreno-Rueda, 2017). Lack of enrichment can lead to feather pecking and picking, which may escalate into injurious behavior and even mortality (Blokhuis & Wiepkema, 1998; Savory, 1995).

Given the importance of preening and plumage maintenance, access to water is a central component of duck welfare. As waterfowl, ducks naturally rely on open water sources for feeding, hygiene, and predator avoidance. However, the provision of open water in commercial systems remains a topic of considerable debate. Research findings are mixed; some studies report improved welfare with troughs or baths, while others document poorer hygiene and increased mortality in large-scale systems (Liste et al., 2012; Liste et al., 2013; Lowman et al., 2016; O'Driscoll & Broom, 2011; Schenk et al., 2016). Consequently, several welfare certification programs (e.g., American Humane Society, Global Animal Partnership, A Greener World) now require open water access, despite a lack of consensus in the scientific literature.

Importantly, relatively little research has examined the consequences of completely restricting open water access. For example, Mi et al. (2020) demonstrated that the absence of a water pool significantly inhibited the development of the preening gland in Sanshui white ducks, as evidenced by reduced gland size, weight, and oil secretion. Building on this, more recent work has explored how water design influences duck behavior and welfare outcomes. Schober et al. (2023); Schober et al. (2025) evaluated semiopen water sources (preening cups and Pekinos) and found increased feather pecking and no differences in uropygial gland size. These findings suggest that, particularly in small group settings, limited access points may promote competition or resource guarding. Thus not only the presence but also the design and availability of water resources must be carefully considered when scaling to commercial systems.

Behavioral welfare

While physical conditions set the foundation for welfare, they are closely linked to behavioral outcomes. Ducks have strong motivation for bathing, dabbling, flocking, and synchronized activity, behaviors that are central to their welfare. Water-related behaviors, including wet preening, head-dunking, nostril clearing, and swimming, play essential roles in plumage maintenance, communication, and social signaling.

In addition to water-associated behaviors, ducks are highly social birds. Environments that support coordinated resting, foraging, and preening support social cohesion and reduce stress and abnormal behaviors (Waitt et al., 2009). Conversely, systems that restrict space or limit environmental complexity may disrupt these patterns. Although domesticated ducks do not migrate, they may retain elements of migratory motivation, which can manifest as increased locomotion or exploratory behavior. When such motivations occur, abnormal behaviors such as pacing or displacement activities may emerge.

Overall, behavioral welfare depends on providing environments that allow the expression of natural positive behaviors, social cohesion, exploratory activity, and natural maintenance behaviors. Restrictions on ducks being able to perform these behaviors not only reduce welfare but can also exacerbate physical health problems, underscoring the need for integrated behavioral and environmental management strategies.

Ethical issues

Ethical concerns in duck production encompass housing, handling, transport, slaughter, and specific practices such as force-feeding. Among these, foie gras production remains the most contentious due to the use of gavage. Foie gras is produced from the livers of ducks or geese (see Chapter 6) that are fattened through a controversial procedure known as gavage, during which birds are given fatty, corn-based feed through a tube inserted into their throats. This procedure induces rapid hepatic steatosis, causing the liver to enlarge dramatically and placing substantial physiological strain on the animal (Atallah et al., 2024).

Scientific evaluations consistently highlight the welfare risks associated with foie gras production. During the force-feeding period, the normal function of the liver declines and then ceases as the liver expands and changes histologically, and the birds are slaughtered just before they die of liver failure. These findings have contributed to bans or severe restrictions on foie gras production in several countries, including the United Kingdom, Germany, Italy, and parts of the United States (Scientific Committee on Animal Health and Animal Welfare (SCAHAW), 1998).

Despite these concerns, foie gras production is often defended on cultural and historical grounds. While it is true that waterfowl naturally accumulate fat prior to migration, modern production systems induce a far more extreme and pathological condition. Empirical evidence indicates that force-fed ducks exhibit both behavioral and physiological signs of distress, distinguishing this practice from natural fattening processes (Rochlitz & Broom, 2017). However, see Chapter 6 for a discussion on foie gras production in geese.

Beyond foie gras, several additional ethical issues arise in commercial duck production. One being stocking density, where a higher stocking density (>4–5 birds/m^2) negatively affects average daily weight gain, final body weight, feed intake, feed conversion ratio, dressing percentage, and walking ability (Ahaotu & Agbasu, 2015). Other research further demonstrates that increasing stocking density (e.g., 7–9 vs 5 birds/m^2) reduces body weight, carcass yield, and antioxidant capacity while increasing physiological stress indicators and mortality risk (Nasr et al., 2022). These effects are likely driven by increased competition for resources, reduced resting space, and poorer litter and air quality, all of which are recognized as key welfare concerns in Pekin duck production systems (Makagon & Riber, 2022).

In addition to on-farm conditions, transportation represents another critical welfare challenge. Ducks are particularly susceptible to thermal stress due to their feather insulation and limited capacity for heat dissipation. During transport, birds are exposed to handling stress, crating, feed and water withdrawal, and environmental extremes, all of which can increase injury rates and mortality (Bergman et al., 2025; Fazio & Ferlazzo, 2003). Catching and crating practices, in particular, have been identified as significant sources of stress and physical trauma in commercial duck systems (Duncan, 2010).

Management practices such as bill trimming, implemented in some systems to reduce injurious pecking, raise additional ethical concerns. Studies with mule and Muscovy ducks demonstrate that trimming methods can alter behavior, reduce activity, and cause acute pain, with potential longer-term welfare consequences depending on technique and severity (Ahmed, 2022). Although less prevalent than in laying hens, its continued use in ducks warrants careful consideration.

Finally, slaughter practices remain a significant welfare issue. Inadequate stunning procedures in ducks, due in part to species-specific anatomical differences, can result in ineffective stunning, leading to birds experiencing pain during neck cutting or regaining consciousness prior to death (Gerritzen et al., 2006; Raj & Tserveni-Gousi, 2000). Collectively, these ethical concerns underscore the need for continued refinement of husbandry practices, improved oversight, and the development of production systems that better align with ducks' behavioral and physiological needs. As consumer expectations for transparency and welfare-friendly production continue to rise, addressing these ethical issues is increasingly central to the long-term sustainability and social acceptability of the duck industry.

Sustainability in duck production

Global poultry production has experienced remarkable growth in recent decades, with poultry meat representing approximately 37% of global meat production (Costantini et al., 2021). Within this broader context, duck production constitutes a significant and often underappreciated component of the global livestock sector. The majority of ducks are reared in Asia (89.7%), followed by Europe (6.5%), the Americas (2.3%), Africa (1.4%), and Oceania (0.1%), with China alone housing approximately 712.4 million ducks—nearly two-thirds of the world's duck population (Jalaludeen & Churchil, 2022). This geographic concentration reflects deep cultural, economic, and ecological factors that shape how duck housing systems are designed, managed, and evaluated for sustainability. Sustainability in commercial duck housing must be understood across three interconnected dimensions: environmental, economic, and social (Mostert et al., 2022). Environmental sustainability encompasses greenhouse gas (GHG) emissions, land use, water consumption, and biodiversity impacts. Economic sustainability relates to profitability, resource efficiency, and market viability. Social sustainability includes animal welfare, worker health and safety, food safety, and community impacts (Leinonen & Kyriazakis, 2016). These dimensions are frequently in tension with one another, and progress in one area may generate trade-offs in others (Bakrie et al., 2023). The urgency of addressing sustainability in duck housing is underscored by several converging pressures: (1) rising global demand for animal protein driven by population growth, and (2) increasing wealth in developing regions (Jalaludeen & Churchil, 2022). The growing threat of HPAI has caused devastating outbreaks in commercial duck establishments worldwide (Alexakis et al., 2025), increasing regulatory scrutiny of animal welfare and environmental performance (Oke et al., 2024), and consumer demand for more ethically produced animal products (Leinonen & Kyriazakis, 2016). This chapter section provides a review of these issues and the environmental, economic, and social sustainability considerations associated with duck production across major producing regions, including Europe, Asia, Africa, and the United States.

Life cycle assessment as a framework

Life cycle assessment (LCA) has emerged as the primary methodological tool for evaluating the environmental sustainability of poultry production systems, including duck housing (Costantini et al., 2021; Mostert et al., 2022). LCA is a holistic approach that evaluates environmental impacts across the entire life cycle of a product or process, from raw material extraction through production, processing, and end of life (Costantini et al., 2021). In the context of duck housing, LCA encompasses feed production, housing construction and operation, manure management, processing, and transportation (Mostert et al., 2022). Chapter 15 provides additional discussion of LCA applied to poultry.

A global assessment of GHG emissions from broiler meat production—directly applicable to duck production by analogy—found that the majority of GHG emissions originate from feed production (57%) and land use change associated with feed production (21%; Mostert et al., 2022). This finding underscores the centrality of feed systems to the environmental footprint of any poultry housing operation, including duck production.

Duck production contributes to global food systems, rural livelihoods, and agricultural diversification. Ducks are highly adaptable and can thrive in intensive, semiintensive, and extensive systems, making them an important species for sustainable livestock development. Their natural foraging behavior allows them to utilize local feed resources, enabling ducks to potentially play a unique role in balancing productivity with resource efficiency across diverse production environments. As global demand for poultry products increases, understanding the sustainability dimensions of duck production has become increasingly important.

Greenhouse gas emissions

GHG emissions from commercial duck housing arise from multiple sources: enteric fermentation, manure decomposition (producing methane and nitrous oxide), energy use for heating, ventilation, and lighting, and the upstream emissions associated with feed production (Costantini et al., 2021; Mostert et al., 2022). Feed composition significantly affects the environmental impact of birds by altering the nitrogen and methane content of poultry manure (Leinonen & Kyriazakis, 2016). Optimizing feed formulation is therefore a critical lever for reducing the GHG footprint of duck housing systems. The relationship between housing system intensity and GHG emissions is complex. More intensive systems tend to produce lower GHG emissions per unit of product due to higher feed conversion efficiency, but this advantage may be offset by higher energy consumption and the environmental costs of feed production at scale (Mostert et al., 2022). In the Netherlands, for example, LCA studies of broiler production systems found that higher-welfare systems with lower stocking densities and longer production periods generated higher GHG emissions per kilogram of meat compared to conventional systems, primarily due to increased feed intake and energy use (Mostert et al., 2022). While these findings relate to broilers, the underlying dynamics are directly relevant to commercial duck housing.

Land use and water consumption

Land use is another critical environmental dimension of duck housing sustainability. Feed production for commercial duck operations requires large areas of agricultural land, with soy and grain cultivation being particularly land-intensive (Jalaludeen & Churchil, 2022). The expansion of feed crop production has been linked to deforestation and land use change, particularly in South America, with significant consequences for biodiversity and carbon sequestration (Bakrie et al., 2023).

Water use is especially pertinent to duck production given the species' behavioral need for water access. Ducks require water not only for drinking but possibly also for bathing and preening, which are essential for feather maintenance and welfare (Jalaludeen & Churchil, 2022). Commercial housing systems that restrict water access to nipple drinkers or troughs—common in intensive operations—may reduce water consumption and associated environmental impacts but at the cost of compromised animal welfare, although these studies are contradictory and there is no clear answer as to whether or not commercial ducks require open water systems to optimize health or welfare (Oke et al., 2024; Schenk et al., 2016; Schober et al., 2025; Waitt, Jones & Dawkins, 2009). Balancing water provision with environmental efficiency is therefore a distinctive sustainability challenge in duck housing that does not arise to the same degree in chicken or turkey production.

Manure management

Manure management is a major environmental hotspot in commercial duck housing. Duck manure has a high moisture content compared to chicken manure, making it more challenging to handle and process (Jalaludeen & Churchil, 2022). Poorly managed duck manure can contribute to water pollution through nutrient runoff (nitrogen and phosphorus), ammonia emissions that contribute to air quality problems and ecosystem acidification, and methane emissions from anaerobic decomposition (Mostert et al., 2022). Circular economy approaches offer promising pathways for improving the sustainability of manure management in duck housing. By treating duck manure as a resource rather than a waste product—for example, through composting, anaerobic digestion for biogas production, or direct application as organic fertilizer—producers can reduce environmental impacts while generating economic value (Leinonen & Kyriazakis, 2016).

Energy use in housing systems

Commercial duck housing requires significant energy inputs for heating, ventilation, lighting, and feed delivery systems. In colder climates, heating costs are particularly high, and the energy source used has a major bearing on the GHG footprint of the operation (Mostert et al., 2022). Precision livestock technology, which enables continuous monitoring and control of environmental parameters such as temperature and relative humidity within housing facilities, offers opportunities to optimize energy use and reduce waste (Leinonen & Kyriazakis, 2016). The adoption of renewable energy sources—solar panels, biogas from manure—can further reduce the carbon footprint of duck housing operations but has yet to be fully explored in duck production systems.

Environmental sustainability

Environmental sustainability in duck production is shaped by feed efficiency, land and water use, nutrient loading, and GHG emissions. Ducks generally exhibit strong feed conversion efficiency and can supplement their diets through natural foraging, consuming aquatic weeds, algae, green legumes, fungi, earthworms, maggots, snails, etc. (Jalaludeen & Churchil, 2022). This behavior reduces feed costs and provides ecological services such as mosquito-larvae control, which may help reduce vector-borne diseases. Despite these advantages, intensive systems generate substantial manure and wastewater, requiring effective waste management to prevent nutrient loading in surface waters. Therefore recent assessments highlight the need for improved waste handling and optimized feed strategies to reduce emissions in intensive systems (Priyanto et al., 2025).

Regionally, environmental sustainability varies considerably depending on regulatory frameworks, production intensity, and system integration. Europe's duck industry is shaped by strict environmental regulations, advanced technology, and high animal welfare standards. Ducks are typically raised in indoor or semi-indoor housing, which reduces land use but concentrates waste, necessitating robust manure-handling infrastructure. Feed formulations are optimized for efficiency, lowering emissions per kilogram of meat (*National Research Council, 1994*). Water access is tightly regulated, with open-water systems often restricted to minimize contamination risks.

In contrast, Asia produces over 80% of the world's ducks, with China as the dominant global producer (Jalaludeen & Churchil, 2022; Mottet & Tempio, 2017). Production systems range from backyard flocks to large industrial farms. A defining feature of Asian duck production is rice–duck integrated farming, widely practiced in China, Vietnam, and Indonesia (Cui et al., 2023; Du et al., 2023). Within these systems, ducks significantly reduce the pests and diseases of rice, but increase soil nematodes, which enable greater growth parameters, resulting in higher rice yield (Teng et al., 2016). Smallholders often recycle manure effectively, but large industrial farms may struggle with waste volume and wastewater pollution (Zhang et al., 2015). Overall, Asia's strengths include low land use due to multi-use landscapes and efficient feed use incorporating local byproducts. However, challenges include high water use in some systems, emissions from poorly managed manure, and environmental regulations that sometimes lag behind rapid industrialization (Chen et al., 2024).

Similarly, but at a different stage of development, duck production in Africa remains relatively small but is expanding, particularly in East and West Africa. Production systems are predominantly extensive or semiintensive, relying on scavenging or locally available crop residues; for a review, see Kadurumba et al. (2022). As a result, these low-input systems use minimal synthetic chemicals and maintain low stocking densities, resulting in low per-farm carbon footprints and high compatibility with local biodiversity. Ducks contribute to natural nutrient cycling in mixed crop–livestock systems. Nevertheless, lower productivity can lead to higher emissions per kilogram of meat, and many regions face limited access to clean water (Osisiogu et al., 2024). Looking forward, as commercial duck farming grows in Africa, the lack of waste-management infrastructure and vulnerability to climate variability, such as heat stress, drought, and flooding, pose significant environmental challenges.

In comparison to both highly regulated European systems and integrated Asian systems, the U.S. duck industry is relatively small but highly industrialized, characterized by large-scale, vertically integrated farms. Advanced feed formulations support rapid growth and low emissions, while strict water regulations, especially in states that have a high number of poultry, limit environmental contamination. Dry-litter housing reduces water use and simplifies manure handling. Consequently, strengths include high feed-conversion efficiency, strong manure-management systems, and robust biosecurity. However, concentrated animal feeding operations can create localized pollution risks. Moreover, compared to Asia, U.S. systems show limited integration with crop production, reducing opportunities for circular nutrient flows.

Economic sustainability

Economic sustainability in duck production varies widely across regions due to differences in market structure, production intensity, and cultural demand. In general, it is defined by the balance between rising input costs (feed and energy) and the ability to access high-value or resilient markets. Ducks' rapid growth and adaptability make them economically valuable across both industrial and smallholder contexts. Within this global framework, the EU duck sector is technologically advanced and highly regulated.

By comparison, in Asia, integrated rice–duck systems reduce feed and fertilizer costs, improving profitability (Huo et al., 2021; Yang et al., 2024). In China, which accounts for over 90% of regional consumption, economic sustainability is driven by massive domestic demand and a dominant share of the global down and feather export trade (*International Down and Feather Bureau*, 2023). Additionally, in Southeast Asia, peri-urban duck farming has emerged as a high-margin agribusiness; for instance, medium-scale farmers in Indonesia report a return-to-capital (R/C) ratio of 1.81, significantly outperforming larger industrial operations due to better cost-efficiency and adaptability to local markets (Bakrie et al., 2023). However, increasing industrialization also raises dependence on purchased feed and energy, thereby increasing production costs (Chen et al., 2024).

In contrast to Asia's large-scale and market-driven systems, duck production in Africa is a critical tool for household economic resilience. About 2% of global duck meat production comes from Africa, with Egypt and Madagascar as major producers. Duck production faces risks from fluctuating grain prices and a lack of access to formal credit; lack of credit could limit the ability of smallholders to scale their operations and adopt technologies (Balana & Oyeyemi, 2022).

Finally, duck production in the United States is small and highly specialized, with a few vertically integrated companies that control the breeding, processing, and distribution (IndexBox, 2026). The concentration of duck production in the United States occurs in regions with strong feed grain availability and processing infrastructure, the majority being in the Midwest and the East Coast. The primary cost of production is the feed, and the price volatility in the crops directly impacts profitability. The United States also participates in the global down and feather trade, though at a smaller scale than Asia or the EU. Supermarket sales of duck products have shown growth rates of up to 10% in recent years, driven by consumer interest in specialty proteins (Technavio, 2025). As a result, economic sustainability in the United States is increasingly tied to technological innovation, including the emergence of cellular agriculture and the adoption of AI for procurement and distribution management (Technavio, 2025). Nevertheless, the industry remains sensitive to trade tariff fluctuations and the high costs of maintaining rigorous biosecurity standards (Research & Markets, 2026).

Emerging issues: alternative protein sources

As feed costs rise globally, alternative protein sources such as black soldier fly larvae are gaining attention (see Chapter 14 for a discussion on sustainable feeding). These insects can be produced using organic waste streams and may offer a cheaper, more sustainable protein source for duck diets (Bukchin-Peles et al., 2025). However, future sustainability assessments will need to consider the welfare of insects, the environmental footprint of insect-rearing systems, and how different insect diets influence duck health and product quality (Cattaneo et al., 2025). These considerations reflect a broader shift toward circular, resource-efficient feed systems in poultry production.

Small-scale versus large-scale production

The economic sustainability of duck housing varies considerably between small-scale and large-scale operations. Small-scale backyard duck production, common in much of Asia and Africa, offers lower capital costs and greater flexibility but typically achieves lower production efficiency and may struggle to meet food safety and biosecurity standards (Jalaludeen & Churchil, 2022). Aggregating small-scale farms into larger industrial operations increases efficiency but often negatively affects animal and worker welfare (Leinonen & Kyriazakis, 2016). Finding appropriate scales of production that balance efficiency with welfare and sustainability is a key challenge for duck housing policy and practice worldwide.

Integrating welfare into sustainability assessments

A significant gap in current sustainability assessments of duck housing is the limited integration of animal welfare into LCA frameworks (see Chapter 15). As noted in reviews of poultry LCA literature, social aspects of sustainability, including the macro-topic of animal welfare, are almost totally neglected by current LCA studies (Costantini et al., 2021). Methodological approaches for inserting animal welfare indicators into integrated social-LCA frameworks are being developed, with particular attention to European broiler production (Costantini et al., 2021), but their application to duck production remains nascent. The concept of "One Welfare" (Pinillos et al., 2016) that recognizes that human and animal welfare are intrinsically linked provides a useful framework to understand why animal welfare must be integrated into sustainability assessments of duck housing (Britten & Mahendran, 2025). Improved animal welfare not only benefits the birds themselves but also supports worker well-being, food safety, and consumer trust, all of which are dimensions of social sustainability (Britten & Mahendran, 2025; Leinonen & Kyriazakis, 2016).

Summary and conclusion

There is increased demand from consumers and policymakers for improved welfare and housing systems for all of our poultry species, and the duck is not an exception. Ducks are not simply "chickens that wear a raincoat" but are a unique species (*Anas platyrhynchos domesticus*) from a unique Order of Aves (Anseriformes) compared to chickens, turkeys, and quail (all Galliformes). As such, ducks have their own unique needs for optimized production, behavior, and welfare. Further research is needed on flooring, lighting, ventilation, biosecurity, waste handling, nest boxes, and water systems to maximize production while optimizing the health and welfare of both the ducks and the people who work with the ducks. What we do not know about duck housing needs, behavior, and welfare far exceeds what we do know. Thus research on commercial ducks in both academic and industry settings is not only highly rewarding but also necessary and valuable.

AI disclosure

During the preparation of this work, the author(s) used Microsoft Copilot in order to revise grammar. After using this tool/service, the author(s) reviewed and edited the content as needed and take(s) full responsibility for the content of the publication.

References

Abdelfattah, E., Vezzoli, G., & Makagon, M. M. (2020). On-farm welfare assessment of commercial Pekin ducks: A comparison of methods. *Poultry Science, 99*(2), 689–697. https://doi.org/10.1016/j.psj.2019.10.006.

Ahaotu, E. O., & Agbasu, C. A. (2015). Evaluation of the stocking rate on growth performance, carcass traits and meat quality of male Pekin ducks. *Scientific Journal of Biological Sciences, 4*(3), 23–29. https://europub.co.uk/articles/-A-1118.

Ahmed, F. A. M. (2022). Method of bill trimming can affect the behaviour, performance, and welfare of mule duck. *PSM Veterinary Research, 7*(1), 11–19. https://pubs.thepsm.net/issues/method-of-bill-trimming-can-affect-the-behaviour-performance-and-welfare-of-mule-duck/.

Alexakis, L., Buczkowski, H., Ducatez, M., Fusaro, A., Gonzales, J. L., Kuiken, T., Ståhl, K., Staubach, C., Svartström, O., Terregino, C., Willgert, K., Melo, M., & Kohnle, L. (2025). Avian influenza overview December 2024–March 2025. *EFSA Journal, 23*(4), e9352. https://doi.org/10.2903/j.efsa.2025.9352.

Atallah, E., Trehiou, S., Alquier-Bacquie, V., Lasserre, F., Arroyo, J., Molette, C., & Remignon, H. (2024). Development of hepatic steatosis in male and female mule ducks after respective force-feeding programs. *Frontiers in Physiology, 15*, 1392968. https://doi.org/10.3389/fphys.2024.1392968.

Babington, S., & Campbell, D. L. M. (2022). Water for domestic ducks: The benefits and challenges in commercial production. *Frontiers in Animal Science, 3*. https://doi.org/10.3389/fanim.2022.782507.

Bakrie, B., Sirnawati, E., Rohaeni, E. S., Hutahaean, L., Santoso, A. D., Wasito, W., & Indrawanto, C. (2023). Sustainability analysis of intensive duck farming system in sliyeg district, Indonesia: MDS and MICMAC analysis approach. *International Journal of Sustainable Development and Planning, 18*(10), 3181–3193. https://doi.org/10.18280/ijsdp.181019.

Balana, B. B., & Oyeyemi, M. A. (2022). Agricultural credit constraints in smallholder farming in developing countries: Evidence from Nigeria. *World Development Sustainability, 1*, 100012. https://doi.org/10.1016/j.wds.2022.100012.

Bergman, M. M., Schober, J. M., Novak, R., Grief, A., Plue, C., Frey, D., Parnin, H., & Fraley, G. S. (2025). Differences in central dopamine, but not serotonin, activity and welfare associated with age but not with preening cup use in commercial grow-out Pekin duck barns. *Poultry Science, 104*(1), 104542. https://doi.org/10.1016/j.psj.2024.104542.

Bergman, M. M., Schober, J. M., Oluwagbenga, E. M., Frey, D., Parnin, H., & Fraley, G. S. (2024). Preening cups in duck housing are associated with an increase in central dopamine activity that suggests a negative affective state. *Poultry Science, 103*(5), 103588. https://doi.org/10.1016/j.psj.2024.103588.

Blokhuis, H. J., & Wiepkema, P. R. (1998). Studies of feather pecking in poultry. *Veterinary Quarterly, 20*(1), 6–9. https://doi.org/10.1080/01652176.1998.9694825.

Britten, N., & Mahendran, S. (2025). Environmental sustainability and ruminant production: A UK veterinary perspective. *Veterinary Record, 196*(1), e4703. https://doi.org/10.1002/vetr.4703.

Bukchin-Peles, S., Baker Lozneva, K., Tomberlin, J. K., & Zilberman, D. (2025). From waste management to protein innovation: Black soldier fly as an embodiment of the circular bioeconomy. *Future Foods, 11*. https://doi.org/10.1016/j.fufo.2025.100592.

Campbell, D. L. M., Belson, S., Erasmus, M. A., & Lea, J. M. (2022). Behavior and welfare impacts of water provision via misting in commercial Pekin ducks. *Journal of Animal Science, 100*(12), skac341. https://doi.org/10.1093/jas/skac341.

Campbell, C. L., Colton, S., Haas, R., Rice, M., Porter, A., Schenk, A., Meelker, A., Fraley, S. M., & Fraley, G. S. (2015). Effects of different wavelengths of light on the biology, behavior, and production of grow-out Pekin ducks. *Poultry Science, 94*(8), 1751–1757. https://doi.org/10.3382/ps/pev166.

Cattaneo, A., Belperio, S., Sardi, L., Martelli, G., Nannoni, E., Dabbou, S., & Meneguz, M. (2025). Black soldier fly larvae's optimal feed intake and rearing density: A welfare perspective (Part II). *Insects, 16*(1), 5. https://doi.org/10.3390/insects16010005.

Chang, S.-C., Chiang, H.-I., Lin, M.-J., Jea, Y.-S., Chen, L.-R., Fan, Y.-K., & Lee, T.-T. (2016). Effects of short light regimes and lower dietary protein content on the reproductive performance of White Roman geese in an environment-controlled house. *Animal Reproduction Science, 170*, 141–148. https://doi.org/10.1016/j.anireprosci.2016.05.003.

Chen, X., Shafer, D., Sifri, M., Lilburn, M., Karcher, D., Cherry, P., Wakenell, P., Fraley, S., Turk, M., & Fraley, G. S. (2021). Centennial review: History and husbandry recommendations for raising Pekin ducks in research or commercial production. *Poultry Science, 100*(8), 101241. https://doi.org/10.1016/j.psj.2021.101241.

Chen, Y., Qi, L., & Hussain, H. A. (2024). Greenhouse gas emissions from Chinese livestock sector can be decreased by one third in 2030 by the improvement in management. *Carbon Research, 3*(1), 66. https://doi.org/10.1007/s44246-024-00147-8.

Cherry, P., & Morris, T. (2008). *Domestic duck production: science and practice*. Wallingford, Oxfordshire, UK: CABI. Cambridge, MA. CABI International. https://www.cabidigitallibrary.org/doi/book/10.1079/9780851990545.0000.

Costantini, M., Ferrante, V., Guarino, M., & Bacenetti, J. (2021). Environmental sustainability assessment of poultry productions through life cycle approaches: A critical review. *Trends in Food Science and Technology, 110*, 201–212. https://doi.org/10.1016/j.tifs.2021.01.086.

Cui, J., Liu, H., Wang, H., Wu, S., Bashir, M. A., Reis, S., Sun, Q., Xu, J., & Gu, B. (2023). Rice-animal co-culture systems benefit global sustainable intensification. *Earth's Future, 11*(2), e2022EF002984. https://doi.org/10.1029/2022EF002984.

Du, C., Hu, L., Yuan, S., Xu, L., Wang, W., Cui, K., Peng, S., & Huang, J. (2023). Ratoon rice-duck co-culture maintains rice grain yield and decreases greenhouse gas emissions in central China. *European Journal of Agronomy, 149*, 126911. https://doi.org/10.1016/j.eja.2023.126911.

Duncan, I. J. H. (2010). Animal welfare issues in the poultry industry: Is there a lesson to be learned? *Journal of Applied Animal Welfare Science, 4*(3), 207–221. https://doi.org/10.1207/S15327604JAWS0403_04.

Fazio, E., & Ferlazzo, A. (2003). Evaluation of stress during transport. *Veterinary Research Communications, 27*(1), 519–524. https://doi.org/10.1023/B:VERC.0000014211.87613.d9.

Fraley, S. M., Fraley, G. S., Karcher, D. M., Makagon, M. M., & Lilburn, M. S. (2013). Influence of plastic slatted floors compared with pine shaving litter on Pekin duck condition during the summer months. *Poultry Science, 92*(7), 1706–1711. https://doi.org/10.3382/ps.2012-02992.

Gerritzen, M. A., Lambooij, E., Reimert, H. G., Spruijt, B. M., & Stegeman, J. A. (2006). Susceptibility of duck and turkey to severe hypercapnic hypoxia. *Poultry Science, 85*(6), 1055–1061. https://doi.org/10.1093/ps/85.6.1055.

Haas, R., Alenciks, E., Meddle, S., & Fraley, G. S. (2017). Expression of deep brain photoreceptors in the Pekin drake: A possible role in the maintenance of testicular function. *Poultry Science, 96*(8), 2908–2919. https://doi.org/10.3382/ps/pex037.

Heyn, E., Damme, K., Manz, M., Remy, F., & Erhard, M. H. (2006). Water supply for Peking ducks–possible alternatives for bathing. *DTW. Deutsche Tierarztliche Wochenschrift, 113*(3), 90–93.

Hu, Q., Zhang, T., He, H., Pu, F., Zhang, R., Li, L., Hu, J., Bai, L., Han, C., Wang, J., & Liu, H. (2024). Impacts of longitudinal water curtain cooling system on transcriptome-related immunity in ducks. *BMC Genomics, 25*(1), 333. https://doi.org/10.1186/s12864-024-10179-4.

Huo, W., Weng, K., Gu, T., Luo, X., Zhang, Y., Zhang, Y., Xu, Q., & Chen, G. (2021). Effects of integrated rice-duck farming system on duck carcass traits, meat quality, amino acid, and fatty acid composition. *Poultry Science, 100*(6), 101107. https://doi.org/10.1016/j.psj.2021.101107.

IndexBox. (2026). World - duck and goose meat - market analysis, forecast, size, trends and insights. Report ID: 029384. IndexBox Inc., Wilmington, DE. https://www.indexbox.io/store/world-duck-and-goose-meat-market-analysis-forecast-size-trends-and-insights/.

International Down and Feather Bureau. (2023). https://idfb.net/fileadmin/user_upload/reports/The_Sustainable_and_Ethical_Practices_of_the_Down_and_Feather_Industry_2023.pdf.

Jalaludeen, A., Churchil, R. R. (2022). Duck Production: An Overview. In: Jalaludeen, A., Churchil, R. R., Baéza, E. (eds) Duck Production and Management Strategies. Springer, Singapore. https://doi.org/10.1007/978-981-16-6100-6_1.

James, J., Warren, C. J., De Silva, D., Lewis, T., Grace, K., Reid, S. M., Falchieri, M., Brown, I. H., & Banyard, A. C. (2023). The role of airborne particles in the epidemiology of clade 2.3.4.4b H5N1 high pathogenicity avian influenza virus in commercial poultry production units. Viruses, 15(4), 1002. https://doi.org/10.3390/v15041002

Jones, T. A., & Dawkins, M. S. (2010). Effect of environment on Pekin duck behaviour and its correlation with body condition on commercial farms in the UK. *British Poultry Science, 51*(3), 319–325. https://doi.org/10.1080/00071668.2010.499143.

Kadurumba, O. E., Iloeje, M. U., Ogundu, U. E., Okoli, I. C., Okoro, V. M. O., & Kadurumba, C. (2022). Characteristics of indigenous ducks and their production system in Nigeria: A review. *Agricultural Reviews, 43*(4), 443–449. https://doi.org/10.18805/ag.rf-238.

Karcher, D. M., Makagon, M. M., Fraley, G. S., Fraley, S. M., & Lilburn, M. S. (2013). Influence of raised plastic floors compared with pine shaving litter on environment and Pekin duck condition. *Poultry Science, 92*(3), 583–590. https://doi.org/10.3382/ps.2012-02215.

Leibler, J. H., Otte, J., Roland-Holst, D., Pfeiffer, D. U., Soares Magalhaes, R., Rushton, J., Graham, J. P., & Silbergeld, E. K. (2009). Industrial food animal production and global health risks: Exploring the ecosystems and economics of avian influenza. *EcoHealth, 6*(1), 58–70. https://doi.org/10.1007/s10393-009-0226-0.

Leinonen, I., & Kyriazakis, I. (2016). How can we improve the environmental sustainability of poultry production? *The Proceedings of the Nutrition Society, 75*(3), 265–273. https://doi.org/10.1017/S0029665116000094.

Liste, G., Kirkden, R. D., & Broom, D. M. (2012). Effect of water depth on pool choice and bathing behaviour in commercial Pekin ducks. *Applied Animal Behaviour Science, 139*(1-2), 123–133. https://doi.org/10.1016/j.applanim.2012.03.005.

Liste, G., Kirkden, R. D., & Broom, D. M. (2013). A commercial trial evaluating three open water sources for farmed ducks: Effects on water usage and water quality. *British Poultry Science, 54*(1), 24–32. https://doi.org/10.1080/00071668.2013.763900.

Lowman, Z. S., Parkhurst, C. R., & Romano, J. (2016). Effect of nipple lines vs. Water trough on Pekin duck breeder performance and well-being. *International Journal of Poultry Science, 15*(2), 52–56. https://doi.org/10.3923/ijps.2016.52.56.

Makagon, B. B., & Mench, J. A. (2011). Floor laying by Pekin ducks: Effects of nest box ratio and design. *Poultry Science, 90*(6), 1179–1184. https://doi.org/10.3382/ps.2010-01287.

Makagon, M. M., & Riber, A. B. (2022). Setting research driven duck-welfare standards: A systematic review of Pekin duck welfare research. *Poultry Science, 101*(3), 101614. https://doi.org/10.1016/j.psj.2021.101614.

Mi, J., Wang, H., Chen, X., Hartcher, K., Wang, Y., Wu, Y., & Liao, X. (2020). Lack of access to an open water source for bathing inhibited the development of the preen gland and preening behavior in Sanshui White ducks. *Poultry Science, 99*(11), 5214–5221. https://doi.org/10.1016/j.psj.2020.08.018, https://www.journals.elsevier.com/poultry-science/.

Moreno-Rueda, G. (2017). Preen oil and bird fitness: A critical review of the evidence. *Biological Reviews, 92*(4), 2131–2143. https://doi.org/10.1111/brv.12324.

Mostert, P. F., Bos, A. P., van Harn, J., van Horne, P., & de Jong, I. C. (2022). *Environmental impacts of broiler production systems in the Netherlands* (Wageningen Livestock Research Report No. 1397). Wageningen Livestock Research, Wageningen, the Netherlands. https://doi.org/10.18174/580961. https://research.wur.nl/en/publications/environmental-impacts-of-broiler-production-systems-in-the-nether/

Mottet, A., & Tempio, G. (2017). Global poultry production: Current state and future outlook and challenges. *World's Poultry Science Journal, 73*(2), 245–256. https://doi.org/10.1017/s0043933917000071.

Nasr, M. A. F., Alkhedaide, A. Q., Radwan, M. M. E., Hafez, A. E. S. E., Hussein, M. A., & El Bayomi, R. M. (2022). Growth, carcass parameters, biochemical and oxidative stress indices, and meat traits of duck breeds under different stocking densities. *Poultry Science, 101*(9), 101992. https://doi.org/10.1016/j.psj.2022.101992.

National Research Council (1994). *Nutrient requirements of poultry. 9th rev. ed.* National Academies Press. https://www.nationalacademies.org/publications/2114.

Nelson, J. R., Bray, J. L., Delabbio, J., & Archer, G. S. (2020). Light emitting diode (LED) color and broiler growth: Effect of supplementing blue/green LED to white LED light on broiler growth, stress, and welfare. *Poultry Science, 99*(7), 3519–3524. https://doi.org/10.1016/j.psj.2020.04.020.

O'Driscoll, K. K. M., & Broom, D. M. (2011). Does access to open water affect the health of Pekin ducks (*Anas platyrhynchos*)? *Poultry Science, 90*(2), 299–307. https://doi.org/10.3382/ps.2010-00883.

Oke, O. E., Akosile, O. A., Uyanga, V. A., Oke, F. O., Oni, A. I., Tona, K., & Onagbesan, O. M. (2024). Climate change and broiler production. *Veterinary Medicine and Science, 10*(3), e1416. https://doi.org/10.1002/vms3.1416.

Osisiogu, E. U., Akinrotoye, K. P., Banini, A. E., & Amemo, R. E. (2024). A systematic scoping review of access to safe drinking water in sub-Saharan Africa: Mapping literature on determinants, interventions, and policy implications over the past decade and the path forward. *Ethiopian Journal of Health Sciences, 34*(5), 413–420. https://doi.org/10.4314/ejhs.v34i5.10.

Pinillos, R. G., Appleby, M. C., Manteca, X., Scott-Park, F., Smith, C., & Velarde, A. (2016). One Welfare – A platform for improving human and animal welfare. *Veterinary Record, 179*(16), 412–413. https://doi.org/10.1136/vr.i5470.

Porter, L., Porter, A., Potter, H., Alenciks, E., Fraley, S. M., & Fraley, G. S. (2018). Low light intensity in Pekin duck breeder barns has a greater impact on the fertility of drakes than hens. *Poultry Science, 97*(12), 4262–4271. https://doi.org/10.3382/ps/pey289.

Priyanto, D., Tan, S. S., Diana, E., Tarigan, H., Bakrie, B., Rohaeni, E. S., Wahyuwardani, S., Ariningsih, E., Karsani Apriliyadi, E., Agustine, R., Suryani, S., & Slameto, S. (2025). Developing a sustainable duck farming business model in peri-urban areas of Jakarta, Indonesia. *Cogent Food and Agriculture, 11*(1), 2566898. https://doi.org/10.1080/23311932.2025.2566898.

Qu, Z., Wang, H., Li, N., Guo, Z., Li, J., Lv, X., Cui, Y., Chai, T., & Wang, F. (2024). Detection and high-throughput microbial analysis of particulate matter in houses and downwind areas of duck farms. *Indoor Air, 2024*(1), 7774679. https://doi.org/10.1155/2024/7774679.

Raj, M., & Tserveni-Gousi, A. (2000). Stunning methods for poultry. *World's Poultry Science Journal, 56*(4), 291–304. https://doi.org/10.1079/WPS20000021.

Raud, H., & Faure, J. M. (1994). Welfare of ducks in intensive units. *Revue Scientifique et Technique (International Office of Epizootics), 13*(1), 119–129.

Research and Markets. (2026). Duck Meat Market – Global Forecast 2026-2032. Dublin, Ireland. https://www.researchandmarkets.com/report/duck?srsltid=AfmBOooQBrx2yHQB8-bOsrTYbTgDq4fNUlGvJ0we-V2G-pFujmsDMY3P.

Rice, M., Meelker, A., Fraley, S. M., & Fraley, G. S. (2014). Characterization of Pekin duck drinking and preening behaviors and comparison when housed on raised plastic versus pine litter flooring. *Journal of Applied Poultry Research, 23*(4), 735–741. https://doi.org/10.3382/japr.2014-01037.

Rochlitz, I., & Broom, D. M. (2017). The welfare of ducks during foie gras production. *Animal Welfare, 26*(2), 135–149. https://doi.org/10.7120/09627286.26.2.135.

Rodenburg, T. B., Bracke, M. B. M., Berk, J., Cooper, J., Faure, J. M., Guémené, D., Guy, G., Harlander, A., Jones, T., Knierim, U., Kuhnt, K., Pingel, H., Reiter, K., Servière, J., & Ruis, M. A. W. (2005). Welfare of ducks in European duck husbandry systems. *World's Poultry Science Journal, 61*(4), 633–646. https://doi.org/10.1079/WPS200575.

Savory, C. J. (1995). Feather pecking and cannibalism. *World's Poultry Science Journal, 51*(2), 215–219. https://doi.org/10.1079/wps19950016.

Schenk, A., Porter, A. L., Alenciks, E., Frazier, K., Best, A. A., Fraley, S. M., & Fraley, G. S. (2016). Increased water contamination and grow-out Pekin duck mortality when raised with water troughs compared to pin-metered water lines using a United States management system. *Poultry Science, 95*(4), 736–748. https://doi.org/10.3382/ps/pev381.

Schober, J. M., Merritt, J., Swanson, M., Tetel, V., Oluwagbenga, E., Frey, D., Parnin, H., Erasmus, M., & Fraley, G. S. (2023). Preening cups increase apparent wet preening behaviors, but have no impact on other behaviors, body condition, growth, or body morphometrics of grow-out Pekin ducks. *Poultry Science, 102*(12), 103145. https://doi.org/10.1016/j.psj.2023.103145.

Schober, J. M., Merritt, J., Ulrey, M., Yap, T. Y., Lucas, J. R., & Fraley, G. S. (2024). Vocalizations of the Pekin duck (*Anas platyrhynchos* domesticus): how stimuli, sex, and social groups affect their vocal repertoire. *Poultry Science, 103*(7), 103738. https://doi.org/10.1016/j.psj.2024.103738.

Schober, J. M., Wilson, N., Seyoum, M. M., Lyte, J. M., Bergman, M. M., Oluwagbenga, E. M., & Fraley, G. S. (2025). Effects of semi-open water and non-water enrichment on welfare, production, behavior, and microbial exposure of grow-out Pekin ducks. *Poultry Science, 104*(9), 105477. https://doi.org/10.1016/j.psj.2025.105477.

Scientific Committee on Animal Health and Animal Welfare (SCAHAW). (1998). Welfare aspects of the production of foie gras in ducks and geese. For the European Commission. https://food.ec.europa.eu/system/files/2020-12/sci-com_scah_out17_en.pdf.

Technavio. (2025). Duck meat market in the US by product and distribution channel - Forecast and analysis 2025-2029. Technavio Research, Elmhurst, IL. https://www.technavio.com/report/duck-meat-market-industry-analysis.

Teng, Q., Hu, X. F., Cheng, C., Luo, Z., Luo, F., Xue, Y., Jiang, Y., Mu, Z., Liu, L., & Yang, M. (2016). Ecological effects of rice-duck integrated farming on soil fertility and weed and pest control. *Journal of Soils and Sediments, 16*(10), 2395–2407. https://doi.org/10.1007/s11368-016-1455-9.

Tonissen, S., Emmert, B. J., Schober, J. M., Oluwagbenga, E. M., Karcher, D. M., & Fraley, G. S. (2025). Pulsed alternating wavelength system lighting does not negatively impact production or welfare but reduces dopamine activity and may improve bone growth in grow-out Pekin ducks: Effects of PAWS lighting on meat ducks. *Poultry Science, 104*(3), 104853. https://doi.org/10.1016/j.psj.2025.104853.

Waitt, C., Jones, T., & Dawkins, M. S. (2009). Behaviour, synchrony and welfare of Pekin ducks in relation to water use. *Applied Animal Behaviour Science, 121*(3-4), 184–189. https://doi.org/10.1016/j.applanim.2009.09.009.

Wang, Y. S., Hou, S. S., Huang, W., Zhao, L., Fan, H. P., Xie, M., & Wang, L. H. (2006). Lysine, methionine and tryptophan requirements of beijing ducklings of 0-2 weeks of age. *Agricultural Sciences in China, 5*(3), 228–233. https://doi.org/10.1016/S1671-2927(06)60043-6.

Yang, C., Han, N., Liu, M., Wei, C., Mao, R., & Chen, C. (2024). Effects of long-term different-scale rice–duck farming on the growth and yield of paddy rice. *Journal of the Science of Food and Agriculture, 104*(6), 3729–3735. https://doi.org/10.1002/jsfa.13257.

Zhang, L., Guo, W., Lv, C., Guo, M., Yang, M., Fu, Q., & Liu, X. (2024). Advancements in artificial intelligence technology for improving animal welfare: Current applications and research progress. *Animal Research and One Health, 2*(1), 93–109. https://doi.org/10.1002/aro2.44.

Zhang, T., Yang, Y., & Xie, D. (2015). Insights into the production potential and trends of China's rural biogas. *International Journal of Energy Research, 39*(8), 1068–1082. https://doi.org/10.1002/er.3311.

Chapter 6

Geese

Attila Salamon
Department of Ethology, ELTE Eötvös Loránd University, Budapest, Hungary

Overview of goose production systems

Brief global and regional trends

Geese are raised worldwide for their meat, eggs, foie gras, and feathers (Kozák, 2021). They are currently the fourth most produced poultry species globally, following chickens, turkeys, and ducks (Krunt & Zita, 2026). Approximately 90% of the world's goose population is found in Asia—primarily in China and Myanmar—while about 7% is located in Africa, mainly in Mozambique, Egypt, and Madagascar (El Sabry & Almasri, 2023). China overwhelmingly dominates global production and slaughter, accounting for 95%–97% of output. In Europe, where traditional production practices remain common, the leading producers are France, Poland, and Hungary (Krunt & Zita, 2026). France and Hungary are also the principal producers of fatty liver (Kozák, 2021), although notable foie gras production occurs in Canada, Germany, and Poland as well (Karadağ, 2025). France is the largest consumer of foie gras, with Spain, Switzerland, and Belgium also ranking among the major consumer markets (Karadağ, 2025). Feather and down production is similarly dominated by China (Chen, 2022). Asian countries are the primary consumers of goose eggs (Kozák, 2021), though global data on goose egg production remain unavailable (El Sabry & Almasri, 2023).

Husbandry systems

Domestic geese are either kept in intensive indoor systems, in semi-intensive systems with outdoor access, or in traditional integrated systems (Guémené et al., 2012; Nielsen et al., 2023). In intensive indoor systems, the buildings usually have a concrete floor with litter, slats, or a mixture of litter and slats. In the starting period, the goslings are exclusively kept on a littered floor, while in the growing and finishing periods, all floor types may be used. The buildings can be equipped with automatic lighting and ventilation systems. The houses may have an automated feeder system or portable feeders, and hayracks with

Integrating Welfare and Sustainability in Poultry Systems. DOI: https://doi.org/10.1016/B978-0-443-40594-5.00002-X

nipples as the most common source of drinking water. In the case of breeding geese, the housing contains a nesting area with wooden nest boxes and straw for nest building. In semi-intensive systems, the buildings can be similarly equipped, but the birds are provided with outdoor access. This outdoor area can vary in size and available resources from a concrete yard to a large field with vegetation and/or an open water source (Guémené et al., 2012; Nielsen et al., 2023). In the case of foie gras production, for the duration of the force-feeding period, the geese are moved to a purpose-built building with floor pens or elevated pens with wire mesh or a slatted floor (Nielsen et al., 2023). In the traditional integrated system, the housing consists of a barn, a yard, and a pond with fish. The housing is usually lightproof and automatically ventilated, with the floor being littered or slatted. The pond is used for fish production, but is also the site for mating and preening, as well as providing drinking water for the geese (Guémené et al., 2012).

Production cycle characteristics

The production cycle of geese has three distinct stages comprising the starting, growing, and finishing periods. The starting period lasts up to 4 weeks of age, when the goslings are kept on straw litter in a heated building. The growing period, beginning from week 5, can differ in length depending on the intensity of the production and what type of goose is produced. It can last up to 8–9 weeks in intensive production when geese can be slaughtered before their first molt. The geese are fed with concentrated feed (protein content and amount depending on the final product) and corn and have unlimited access to drinking water via nipple drinkers or water troughs. In the finishing period, a heavier goose is produced with a higher meat yield, when slaughtered before the second molt, in which case the extended fattening period lasts up to 16 weeks with intensive grazing, or after the second molt, around 23 weeks via extensive grazing (Kozák, 2021; Nielsen et al., 2023). There may be an intensive fattening period (not force-feeding) in the last 3–4 weeks of the finishing period (in intensive or extensive grazing) in order to get an extra weight boost by fat deposition (Bogenfürst, 2017; Nielsen et al., 2023).

Based on their weight (which can be achieved by the end of the production), domestic goose breeds can be categorized into heavy-, middle-, and lightweight geese. Heavyweight geese, such as the Embden, Pomeranian, and Toulouse, can weigh 7–10 kg (females 7–9 kg, males 8–10 kg). Middleweight geese, such as the Hungarian, Italian, Landes, and Diepholz, can weigh 5.5–7.5 kg (females 5.5–6.5 kg, males 6.5–7.5 kg). Lightweight geese, such as the Chinese, Czech, and Alsatian, can weigh 4–5.5 kg (females 4–4.5 kg, males 4.5–5.5 kg) (Bogenfürst, 2017; Salamon, 2020b). Domestic goose hybrids can be sorted in a similar manner. Domestic geese can be distinguished based on the production type as well. Egg-type geese, such as the Italian, Chinese, and Czech, have been selected for reproductive performance.

Meat-type geese, such as the Pomeranian and Embden, have exceptional growth rates and meat quality, but have medium or weak reproductive performance. Liver-type geese, such as the Toulouse and Landes, have great fattening ability, but have medium or weak reproductive performance (Bogenfürst, 2017; Buckland & Guy, 2002; Salamon, 2020b). Selective breeding and crossing of various breeds and/or strains provide great improvement in production performance of geese and have been practiced by a few large breeding companies, such as ORVIA (Nielsen et al., 2023).

Welfare considerations

Physical welfare

The physical welfare of the geese may be at risk as a consequence of several conditions, such as damaging environmental features, inadequate or inappropriate equipment or systems in the building, or negative social interactions (Nielsen et al., 2023). Foot-pad dermatitis is a major foot disease that affects poultry health and welfare. Symptoms thereof include hyperkeratosis, scaly lesions, abrasion, ulcers, swelling, cracking, bleeding, inflammation, and tissue necrosis in the footpads that affect the natural behavior of the birds and are likely to induce pain (Liao et al., 2021; Shepherd & Fairchild, 2010). The quality and moisture content of the litter are key factors affecting foot-pad dermatitis, which is most likely to occur in the case of poor litter quality with high moisture content from excreta or from spilled water (Nielsen et al., 2023). If the birds are housed on poor-quality perforated floors, then injury to the feet or toes may be caused by cuts from sharp edges or by trapping of the toes in void areas of inappropriate size (Nielsen et al., 2023). Feather condition is an important indicator of bird health and welfare. Since geese do a considerable amount of pecking, it can be redirected towards the plumage of other birds (Gillette, 1977). However, such feather pecking is rare in geese, and it is generally not injurious (Boz et al., 2021; Gillette, 1977). Access to water is a more important factor affecting feather condition, as it is necessary for preening to maintain clean and tidy plumage. Eye and nostril cleanliness also depends on access to appropriate open water sources (Liao et al., 2022; Nielsen et al., 2023).

Behavioral welfare

Domestic geese are highly social, intelligent animals (Bogenfürst, 2017). They tend to form large flocks naturally, therefore group housing is ideal for them, which enables the formation of complex social structures and reduces stress (Bogenfürst, 2017; Nielsen et al., 2023). Frequent disruption of the flock structure can lead to chronic stress, therefore welfare-friendly systems should facilitate stable flock sizes and minimize reorganization. Access to open water is vital for the expression of natural behavior. It is important not only for the maintenance of plumage and the regulation of temperature (Nielsen et al., 2023), but also to express natural mating behaviors (Salamon, 2025), even though geese can mate on land (Gillette, 1977; Guémené et al., 2012). The European Food Safety Authority (EFSA) (Nielsen et al., 2023) listed the various types of water provision and the behaviors that can be expressed using the given type of equipment. Based on their recommendations, troughs allow the most behaviors to be expressed. Geese can complete the entire bathing sequence and maybe even swimming in these, while showers and deep bell drinkers allow only parts of the bathing sequence, head dipping, wet preening, and nostril cleaning. Natural water sources, such as lakes, ponds, rivers, and streams, allow water bathing (immersion of the whole body in the water) while also enabling swimming and play behavior that are integral components of comfort behavior in domestic geese (Abd-El-Gawad et al., 1993). Water bathing is ideal to get rid of excess heat (Farghly et al., 2017) and to reduce the effects of heat stress (Ewuola et al., 2022). Outdoor access, whether it is continuous or restricted to certain hours of the day, is another key factor for the expression of natural behavior in geese. It should be provided after 5–6 weeks of age to allow geese to choose between different temperatures, light conditions, and substrate quality, and to promote foraging, exploratory, and comfort behaviors. The ground should be mainly covered by vegetation, but areas around water facilities should have concrete or gravel floors. If circumstances, such as disease risk, preclude outdoor access, covered verandas should be provided (Nielsen et al., 2023).

Ethical issues

Overfeeding geese is an ancient tradition dating back to the Egyptians. Overfeeding with large amounts of corn induces liver steatosis, resulting in the fatty liver, or foie gras, but this state of the liver is reversible in geese (Kozák, 2011a). Geese physiologically respond well to overfeeding due to their natural ability to overconsume food to store energy for long-distance migration (Odum, 1960). However, overfeeding is dividing society and is an increasingly questionable practice. The force-feeding procedure, including intubation for corn delivery, is claimed to be stressful and/or painful (Rochlitz & Broom, 2017).

The utilization of feathers and down for bedding was developed as a Celtic-Germanic tradition (Kozák, 2009). Feathers were typically removed from the geese by hand during their natural feather-changing period. The procedure was called feather plucking (Kozák, 2009), which was misinterpreted as the removal of unripe feathers from live geese, which is painful and causes injury. Therefore, new terminology was introduced by the EFSA (EFSA AHAW Panel, 2010). Feather plucking became the removal of unripe feathers (a procedure that is not allowed), while feather harvesting is the procedure of collecting ripe feathers during natural molt (EFSA AHAW Panel, 2010; Kozák, 2011b; Kozák et al., 2010). It was shown by measurements of corticosteroid hormone responses that the capture of the geese is the most stressful part of the entire process (Tóth et al., 2012), as handling and feather harvesting did not increase the stress of the geese.

In intensive production systems, domestic geese are kept indoors all the time. Birds may experience stress and/or negative affective states such as pain, fear, discomfort, and/or frustration because they are unable to move freely, or unable to walk, fly, swim, or dive comfortably (e.g., due to overcrowding, unsuitable floors, gates, or barriers), depending on the system and the resources within the system (Nielsen et al., 2023). Therefore, recommendations have been formulated by the EFSA (Nielsen et al., 2023) regarding stocking density and particular aspects of housing that are important for the welfare of domestic geese (Table 6.1). Furthermore, it was recommended by the EFSA (Nielsen et al., 2023) that birds should not be kept in solitary cages (only relevant in pedigree breeding).

Sustainability dimensions

Environmental dimension

Goose production can contribute to environmental sustainability in many ways. Since geese are well adapted to grazing, they should be kept extensively, or at least have access to outdoor pastures. It is not sustainable to provide geese with cut grass in intensive systems, when they are able to graze (Song et al., 2017). They can consume up to 1.2 kg of grass per day from 13 to 18 weeks of age (Leprettre et al., 2000), so geese are suitable to clean out otherwise unusable land, which reduces competition with human food production. Studies indicate that pasture-based goose systems can reduce the carbon footprint of production by minimizing the need for grain inputs, herbicides, and fertilizers (Buckland & Guy, 2002; Rothrock et al., 2019). Goose manure is high in nitrogen and phosphorus (Bielińska & Mocek-Płóciniak, 2015; Dessborn et al., 2016; Hassan et al., 2017), and when managed correctly, it contributes to soil fertility and alters the soil microbiota community (Massaccesi et al., 2019). Pasture rotation is important, as it is necessary for the prevention of overgrazing and also for the prevention of nutrient (and heavy metal) overload in the soil (Bielińska & Mocek-Płóciniak, 2015). Apart from pastures, other alternative production systems, such as agroforestry systems, can benefit from the presence of geese. For more discussion on silvopasture systems, see Chapter 11.

Goose manure can also impose several environmental burdens on water and air quality. When large numbers of geese congregate near freshwater systems, their droppings introduce substantial loads of nitrogen and phosphorus, which can alter water chemistry, enrich sediments, stimulate algal

TABLE 6.1 Recommendations formulated by the European Food Safety Authority that are important for the welfare of domestic geese. Stocking density (based on 6.7 kg goose weight): the bird is able to move freely, and can perform exploratory, foraging, and comfort behaviors from 7776 cm^2. An additional 1166 cm^2 open water is recommended for complete bathing.

Factor	Recommendation	Desired effect
Stocking density	4958 cm^2/goose 7776 cm^2/goose 7776 cm^2 + 1166 cm^2/goose	minimum can perform head dipping can perform complete bathing
Enclosure height	minimum 127 cm	enables standing upright and spreading wings
Floor	solid and littered perforated around water source	prevents leg problems prevents litter quality deterioration
Litter	dry and friable (e.g., chopped straw), free of physical, chemical, microbiological hazards	maintains bird cleanliness and foot health, allows comfort and foraging behavior
Water source	minimum: bell drinkers, maximum: pond	keeps eyes, nostrils, and plumage clean and intact
Water depth	15 cm 40 cm	enables head dipping enables swimming
Nest	3500 cm^2/goose, 70 cm high, manipulable material (e.g., straw, woodshawings)	allows one bird to express nesting behavior

Based on EFSA AHAW Panel (EFSA Panel on Animal Health and Animal Welfare), Nielsen, S.S., Alvarez, J., Bicout, D.J., Calistri, P., Canali, E., Drewe, J.A., Garin-Bastuji, B., Gonzales Rojas, J.L., Schmidt, C.G., Herskin, M., Michel, V., Miranda Chueca, M.Á., Padalino, B., Roberts, H.C., Spoolder, H., Stahl, K., Viltrop, A., Winckler, C., Berg, C.Velarde, A. (2023). Welfare of ducks, geese and quail on farm. EFSA Journal, 21, 5. https://doi.org/10.2903/j.efsa.2023.7992.

blooms, and reduce invertebrate densities—patterns documented in wild roosting areas (Dessborn et al., 2016). These nutrient inputs can accelerate eutrophication in natural water bodies and in extensive production systems where geese have access to ponds or channels. Although integrated goose–fish systems may partially offset this impact—because fish can consume some of the organic components of the droppings (Jiang et al., 2011; Stojević et al., 2017; Yang et al., 2012)—declining water quality still poses risks. In particular, elevated nutrient loads and degraded conditions can impair goose reproduction by reducing egg fertility and hatchability unless preventative measures, such as adding photosynthetic bacteria to improve pond conditions, are implemented (Yang et al., 2012).

Beyond water pollution, goose manure contributes to atmospheric impacts as well. As herbivorous birds, geese generate methane, a potent greenhouse gas. Experimental work suggests that dietary strategies may help reduce these emissions: alfalfa-based feed shows promise in lowering methane output (Chen et al., 2014). While methane production by geese has been confirmed in vivo, current findings warrant caution, and further research is needed to clarify emission magnitudes and mitigation potential (Clauss et al., 2020).

Economic and social dimensions

Goose meat can serve as an important source of nutrition, particularly in regions where waterfowl account for a significant proportion of total meat production. Because geese tolerate hot, humid climates well, they are often easier to raise than chickens in parts of Asia. On small family farms, geese can be reared with minimal inputs and sustainable use of natural resources, whereas large-scale production requires greater investment in housing, feed, and welfare management (Buckland & Guy, 2002; Kozák, 2021).

Consumers value goose meat for its flavor and high nutritional quality; it is considered a safe product due to its high protein content and favorable fatty acid profile (Kozák, 2021; Krunt & Zita, 2026). Goose production is seasonal, with demand peaking during festive periods in Europe—such as on St. Martin's Day (Kozák, 2021) or during Christmas in Ireland and the UK (Salamon, 2015). This seasonality contributes to irregular, occasion-driven consumption and limits consistent year-round demand (Krunt & Zita, 2026). In China, off-season production strategies are being explored to increase output and improve producer income (Guo et al., 2020; Zhu et al., 2017). In Western markets, more than 80% of goose meat is sold as cut-up or further-processed products, while ready-to-cook and ready-to-eat options remain uncommon (Krunt & Zita, 2026). Nevertheless, goose meat can be used in a wide range of dishes, and many specialty goose products have no direct substitutes, underscoring their uniqueness and economic potential (Kozák, 2021).

Fatty liver (foie gras) is a globally recognized delicacy (Kozák, 2021). In Europe, foie gras is a traditional food (Guémené & Guy, 2004) and has been officially recognized as part of France's gastronomy and cultural heritage since 2006 (Kozák, 2021). Goose liver is seldom sold in portioned or pre-prepared formats suitable for quick cooking or direct consumption, although pâtés and preserved products constitute a small share of the market (Krunt & Zita, 2026).

Goose feathers are valued for their lightness and exceptional insulation properties. They have been used for bedding for millennia and are now widely incorporated into winter clothing, camping equipment, and even furniture (Chen, 2022; Kozák, 2021). Goose down and feathers are generally more sought-after than those from ducks (Chen, 2022), and their global market value ranges from 100 to 130 USD per kilogram (Karadağ, 2025), with white down and feathers being the most valuable (Kozák, 2021).

Goose eggs are mostly produced for incubation in Europe, although they can be boiled and fried, as well as used in the baking industry (Kozák, 2021). However, in China and other Asian countries goose eggs are highly valued as table eggs, and used for making salted or alkalized eggs (Kozák, 2021). Further, double-yolked eggs are considered a sign of good luck in Japanese and Chinese folklore (Salamon & Kent, 2020; Salamon, 2020a; Zhang et al., 2011).

Tensions and trade-offs

Environmental dimension versus welfare

Providing outdoor access for geese offers clear welfare benefits but comes with environmental trade-offs. Outdoor areas allow geese to engage in natural behaviors—such as grazing, bathing, exploratory foraging, and social interactions—that contribute to improved physical health and psychological well-being. However, these same features can create environmental pressures. Grazing geese deposit high concentrations of nutrients in localized areas, increasing the risk of nitrogen and phosphorus leaching into soil and waterways (Bielińska & Mocek-Płóciniak, 2015; Dessborn et al., 2016). Outdoor water bodies or wet areas, while enriching for the birds, can become sources of runoff contamination or microbial proliferation if not carefully managed (Jiang et al., 2011). Additionally, exposure to wild birds and pathogens is generally higher outdoors, raising concerns about disease introduction and spread (Bogenfürst, 2017). Thus, the challenge lies in balancing the welfare gains of outdoor access with sustainable land and nutrient management practices —ensuring geese can experience enriched environments while minimizing the ecological footprint.

Economic dimension versus welfare

The design of husbandry systems often involves a balance between providing geese with higher-quality living conditions and managing the associated production costs. Enhanced welfare practices—such as larger housing spaces, access to open water, enriched environments, and lower stocking densities—-generally improve goose health and reduce stress (Nielsen et al., 2023), but they also require more land, labor, and infrastructure investment as well as management. These added expenses can increase the cost of production for farmers and, ultimately, raise market prices (Kjosevski et al., 2021; Vetter et al., 2014). Direct economic vs animal welfare trade-off analyses are scarce. On one hand, reliable animal welfare measures are required as a first step (Tremolada et al., 2020; Wang et al., 2025), while on the other hand economical assessments of sustainable goose production or competitiveness of the goose sector are also needed. For example, a Hungarian study mentioned that stricter animal welfare and hygiene regulations drive the goose production cost up (Molnár, 2016), but did not go into the details of the cost of specific investments in buildings and equipment that improve the welfare of the birds. Producers face an ongoing trade-off: investing in better welfare can yield long-term benefits like improved animal health and consumer trust, but it must be weighed against the short-term economic pressures of maintaining affordable and efficient operations (Kjosevski et al., 2021).

Balancing strategies

System design

Designing husbandry systems that balance animal welfare and sustainability in goose production requires coordinated management of open-water access, litter quality, and stocking strategies. Providing access to open water supports natural behaviors such as preening, head dipping, dabbling, and (partial) bathing, which in turn improve nostril, eye, and plumage condition and reduce welfare problems like phallus pecking observed in intensive systems without water access (Bogenfürst, 2017; Liao et al., 2021, 2022; Salamon, 2020b). Water access can also enhance productivity through higher egg numbers, fertilization, and hatchability (Bielińska & Herbut, 2002). Complementing water access with good-quality litter helps redirect exploratory pecking and dabbling behaviors, reducing the risk of feather pecking when forage is limited and improving overall plumage condition, particularly when compared with fully intensive systems (Boz et al., 2021; Boz, Sarica, et al., 2017a; Gillette, 1977; Nielsen et al., 2023). Sustainable stocking strategies—such as maintaining lower densities or rotating access to outdoor areas—can prevent declines in growth, walking ability, and feather cleanliness observed at higher stocking densities (>5 birds/m^2) (El Sabry & Almasri, 2023; Wang et al., 2019; Yin et al., 2017). While outdoor access can reduce foot health (Boz, Sarıca, et al., 2017b), access to open water partially offsets these effects, suggesting that a balanced system design can optimize both welfare outcomes and sustainable use of housing and land resources.

Feed innovations

Herbivorous birds produce methane (Chen et al., 2014; Clauss et al., 2020). It was shown that the main origin of methane production is the cecum (Chen et al., 2014; Chen et al., 2003) which provides a habitat for microorganisms, which can ferment and convert nutrients into short-chain fatty acids and biogas (Chen et al., 2014). Feeding experiments showed that supplementation with alfalfa reduced methane production, but the studies came to different conclusions regarding the reasons. Chen et al. (Chen et al., 2014) concluded that the saponin content of the alfalfa meal suppressed methanogenic microbes, similar to other studies, while Clauss et al. (Clauss et al., 2020) argued that methane production may not depend on fiber, but protein digestion via microbial digestion of uric acid. Thus while the reduction in methane production is clear, the mechanisms behind it are not.

Certification and assurance schemes

Balancing animal welfare and sustainability in goose production increasingly relies on robust certification and assurance schemes, such as the Red Tractor and RSPCA in the UK (Main & Mullan, 2017) or Label Rouge in France (Westgren, 1999), that translate broad ethical goals into practical farm-level standards. By setting measurable criteria—such as access to outdoor areas, enriched environments, responsible feed sourcing, limits on slaughter age, and stocking density—certification programs help ensure that welfare improvements go hand-in-hand with reduced environmental impacts. Some schemes also require monitoring of resource use, manure management, biodiversity protection, or even methods of slaughter, creating incentives for farmers to adopt practices that are both humane and ecologically efficient. Transparent auditing and clear labeling then, may allow consumers and buyers to reward producers who meet these dual objectives, helping align market demand with farming practices that support long-term sustainability without compromising goose welfare. This could work among consumers concerned about animal welfare and sustainable production (Alonso et al., 2020; Gorton et al., 2023), but amongst price-sensitive consumers, willingness to pay extra for an animal welfare label was negligible (Gorton et al., 2023; Lagerkvist & Hess, 2011).

Case studies

Example 1: European foie gras debates and regulatory shifts

Ethical concerns about foie gras production came to the forefront in France when the Scientific Committee on Animal Health and Animal Welfare reported to the European Commission that "force-feeding is detrimental to the welfare of the birds" (Le Neindre et al., 1998). Opponents of foie gras production argued that birds experience fear and stress during catching, handling, and force-feeding, and that intubation with the corn-delivery tube may cause pain. However, these claims were based largely on personal impressions and observations rather than on conclusions drawn from controlled scientific studies. Results from several experiments have indicated that these concerns are unfounded (Kozák, 2011a).

Although excessive food intake during overfeeding has been suggested to be painful, geese are naturally capable of grazing up to 1.2 kg of grass per day (Leprettre et al., 2000) or voluntarily consuming 3 kg of carrots (Guémené & Guy, 2004). The corn-delivery procedure is relatively simple and does not typically cause pain or injury because the tube can be inserted directly into the esophagus through the open beak; geese lack a soft palate, which facilitates this process (Sótonyi & Lorászkó, 2008). Nonetheless, it has been acknowledged that flexible rubber tubes are preferable to metal pipes for this purpose (Áprily et al., 2009).

Studies measuring stress indicators such as corticosterone levels and heart rate have shown that force-feeding does not increase corticosterone concentrations in geese or ducks, nor does it elevate heart rate in ducks (Guémené & Guy, 2004). Moreover, geese did not exhibit aversion toward the stockperson performing the force-feeding, suggesting that the procedure is not associated with significant pain (Sótonyi & Lorászkó, 2008).

Despite these findings, foie gras production is banned in several EU countries, prompting efforts to develop an alternative method known as spontaneous fattening. Studies have shown that hyperphagia and spontaneous liver steatosis can be experimentally induced by combining a short photoperiod with a sequence of feed restriction followed by a period of ad libitum corn feeding, both in Greylag geese (Guy et al., 2013) and in Landes geese (Fernandez et al., 2016). Livers produced through spontaneous fattening were approximately 30% lighter than those obtained through overfeeding. Significant differences were also observed in the concentrations of certain chemical compounds—particularly triglycerides and fatty acids—and the sensory profiles of the livers differed markedly. Consumers expressed considerably lower preference for livers produced through spontaneous fattening compared with those obtained by traditional overfeeding (Fernandez et al., 2019). Nevertheless, the spontaneous fattening method remains under development.

Example 2: Goose grazing in vineyards/orchards for weed control

Agroforestry systems—such as fruit orchards, olive groves, and vineyards —offer an alternative to semi-intensive production systems with outdoor access, although they are still relatively uncommon goose production systems (Cartoni Mancinelli et al., 2019, 2020, 2022; Massaccesi et al., 2019). The woody vegetation in these environments provides birds with shade and protection from predators, while grasses and weeds offer additional foraging opportunities. One study found that the presence of geese improved biomass efficiency in the upper soil layer of a vineyard (Cartoni Mancinelli et al., 2019). Several production-related effects have been reported in agroforestry systems, including lower final body weight and reduced feed consumption, as well as improved physical and chemical meat traits—such as antioxidant levels and fatty acid composition—compared with indoor-only systems (Cartoni Mancinelli et al., 2019, 2020). Higher copper concentrations in the liver have also been observed (Massaccesi et al., 2019). However, only one study has assessed goose welfare directly by examining their behavior in these systems. In vineyards, geese displayed more resting, roosting, and stretching, and less walking and drinking, compared with those in orchards (Cartoni Mancinelli et al., 2022). Nevertheless, because no comparisons were made with geese kept exclusively indoors, it remains difficult to evaluate the true welfare impact of agroforestry systems.

Future directions

Research needs

While the production parameters of geese raised in various production systems are relatively broadly studied (Boz, Sarica, et al., 2017a; Cartoni Mancinelli et al., 2019, 2020), there is a considerable gap in the goose literature about the various production systems' effects on the welfare of the birds. A few notable exceptions exist (Boz et al., 2021), but other studies just examined welfare parameters of geese in one production system with/without enrichment (Liao et al., 2021, 2022; Wang et al., 2025), or compared two alternative systems without comparing it to a control intensive system (Cartoni Mancinelli et al., 2022). Further, literature is scarce on the economic aspects of animal welfare and sustainability of goose production, which, for example, would quantify the costs of improving bird welfare compared to intensive systems or would assess how much consumers would/should pay for goose products originating from alternative higher welfare production systems. Additionally, studies would be beneficial from the environmental aspect of sustainability, comparing the inputs and outputs of intensive and alternative goose production systems.

Policy

Policies governing goose production should aim to balance three interconnected pillars: consumer expectations, international trade standards, and practical on-farm management. Consumers increasingly demand high levels of animal welfare, traceability, and environmentally responsible production (Alonso et al., 2020; Mayfield et al., 2007), which encourages producers to adopt transparent practices such as improved housing conditions, responsible feed sourcing, and strict hygiene protocols. For example, the EU is trying to keep up with this demand in its regulatory framework (Nielsen et al., 2023), but further assessments are needed. At the same time, global trade standards—covering food safety, veterinary controls, residue limits, and product labeling—require producers to comply with harmonized regulations to access international markets. On-farm policies should therefore integrate these external demands while remaining feasible for farmers, ensuring that welfare enhancements and compliance measures do not compromise production efficiency or economic viability. Effective policy frameworks and quality assurance schemes, such as RSPCA or Label Rouge, need to support farmers through clear guidelines, training, and risk-based regulation, enabling goose production systems that are ethically responsible, competitive in trade, and aligned with consumer trust.

Opportunities for innovation

There are many opportunities for innovation regarding monitoring behavior and welfare that could improve goose production. Currently, there is a lack of information on the general activities and behavior of geese (e.g., daily time budgets, social networking) (Brassó et al., 2024; Salamon, 2025), and animal-based measures to evaluate goose welfare are quite coarse (Tremolada et al., 2020). This highlights an urgent need for more detailed behavioral research to develop finer, species-specific indicators that can support accurate welfare assessments. Despite this knowledge gap, some studies are already developing and trying innovative technologies to examine behavior and/or welfare, such as thermal comfort assessment with thermal imaging (Chuang et al., 2021), and individual recognition with surveillance cameras and artificial intelligence (Zhang et al., 2025), with cameras and identity collars (Guo et al., 2026) or with radio frequency identification (Zhang et al., 2022). This emerging sensor-based research indicates that precision-farming tools are beginning to enter goose production, yet significant opportunities remain to integrate and standardize these technologies and to expand them to include welfare, environmental, and economic indicators for a more comprehensive view of sustainable production.

Conclusion

The relationship between animal welfare and sustainability in goose production is characterized by both important synergies and unavoidable conflicts. On the one hand, systems that promote natural behaviors—such as grazing, access to open water, and outdoor environments—can enhance welfare while reducing reliance on external inputs (e.g., lower use of concentrated feed, lower use of fertilizers and herbicides, multifunctional land use), thereby supporting environmental sustainability. On the other hand, these same practices may introduce challenges, including nutrient runoff, water pollution, and increased exposure to pathogens. Similarly, efforts to improve welfare through lower stocking densities or enriched housing can raise production costs, creating economic tensions. These trade-offs demonstrate that optimizing one dimension of sustainability does not automatically guarantee improvements in others.

Such complexity underscores the importance of context-specific solutions. Intensive and extensive systems each offer distinct advantages and limitations, shaped by factors such as climate, land availability, disease risk, and market demands. Rather than applying uniform standards, effective management must adapt to local conditions, balancing welfare outcomes with environmental and economic realities in each production setting.

Addressing these challenges requires a truly interdisciplinary approach. Integrating insights from animal welfare science, environmental management, economics, and social sciences is essential to developing holistic strategies that align ethical responsibility with practical feasibility. Future progress will depend on collaborative research, innovative system design, and supportive policy frameworks that recognize these interdependencies. By embracing such integrated approaches, goose production can move toward systems that are not only more humane but also environmentally resilient and economically sustainable.

References

Abd-El-Gawad, E., El-Esssawy, G., & Mervat, K. (1993). Some studies on behavioural, physiological and productive aspects in Egyptian goslings kept under two systems of management. *Veterinary Medical Journal, 41*, 77–82. https://doi.org/10.21608/vmjg.1993.377253.

Alonso, M. E., González-Montaña, J. R., & Lomillos, J. M. (2020). Consumers' concerns and perceptions of farm animal welfare. *Animals, 10*(3), 385. https://doi.org/10.3390/ani10030385.

Áprily, S., Szász, S., & Bogenfürst, F. (2009). Effect of genotype, sex and cramming preparation on performance and fatty liver production in geese. *European Poultry Science, 73*(4), 268–274.

Bielińska, E. J., & Mocek-Płóciniak, A. (2015). Biochemical and chemical indices of soil transformations on goose farms in years 1996-2011. *Archives of Environmental Protection, 41*(1), 80–85. https://doi.org/10.1515/aep-2015-0010.

Bielińska, H., & Herbut, E. (2002). Effect of access to swimming water on welfare of White Kołuda geese® *Annals of Animal Science, S1*, 67–70.

Bogenfürst, F. (2017). *Lúdtenyésztők kézikönyve*. Udine: Forum.

Boz, M. A., Sarica, M., & Yamak, U. S. (2017a). Production traits of artificially and naturally hatched geese in intensive and free-range systems: I. Growth traits. *British Poultry Science, 58*(2), 132–138. https://doi.org/10.1080/00071668.2016.1261997.

Boz, M. A., Sarıca, M., & Yamak, U. S. (2017b). Effect of production system on foot pad dermatitis (FPD) and plumage quality of geese. *European Poultry Science, 81*, 1–10. https://doi.org/10.1399/eps.2017.216.

Boz, M. A., Sarıca, M., Yamak, U. S., & Erensoy, K. (2021). Behavioral traits of artificially and naturally hatched geese in intensive and free-range production systems. *Applied Animal Behaviour Science, 236*, 105273. https://doi.org/10.1016/j.applanim.2021.105273.

Brassó, L. D., Komlósi, I., & Barta, Z. (2024). Behaviour frequencies, spatial distribution and social network of Grimaud geese during the laying season. *Applied Animal Behaviour Science, 275*, 106288. https://doi.org/10.1016/j.applanim.2024.106288.

Buckland, R. B., & Guy, G. (2002). *Goose production, vol 154*. Rome: Food and Agriculture Organisation.

Cartoni Mancinelli, A., Mattioli, S., Dal Bosco, A., Piottoli, L., Ranucci, D., Branciari, R., Cotozzolo, E., & Castellini, C. (2019). Rearing Romagnola geese in vineyard: Pasture and antioxidant intake, performance, carcass and meat quality. *Italian Journal of Animal Science, 18*(1), 372–380. https://doi.org/10.1080/1828051X.2018.1530960.

Cartoni Mancinelli, A., Mattioli, S., Menchetti, L., Dal Bosco, A., Chiattelli, D., Angelucci, E., & Castellini, C. (2022). Validation of a behavior observation form for geese reared in agroforestry systems. *Scientific Reports, 12*(1), 15152. https://doi.org/10.1038/s41598-022-18070-6.

Cartoni Mancinelli, A., Mattioli, S., Bosco, A. D., Ciarelli, C., & Castellini, C. (2020). Grass intake and meat oxidative status of geese reared in three different agroforestry systems. *Slovak University of Acta Fytotechnica et Zootechnica, 23*, 308–315. https://doi.org/10.15414/afz.2020.23.mi-fpap.308-315, http://www.acta.fapz.uniag.sk/journal/index.php/on_line/article/view/741.

Chen, Y. H., Lee, S. M., Hsu, J. C., Chang, Y. C., & Wang, S. Y. (2014). Methane generation from the intestine of muscovy ducks, mule ducks and White Roman geese. *Aerosol and Air Quality Research, 14*(1), 323–329. https://doi.org/10.4209/aaqr.2013.05.0180.

Chen, Y. H., Wang, S. Y., & Hsu, J. C. (2003). Effect of caecectomy on body weight gain, intestinal characteristics and enteric gas production in goslings. *Asian-Australasian Journal of Animal Sciences, 16*(7), 1030–1034. https://doi.org/10.5713/ajas.2003.1030.

Chen, Y. K. Y. (2022). Value addition of feather and down. In: A. Jalaludeen, R.R. Churchil, E. Baéza, (eds) *Duck production and management strategies*. Australia: Springer Nature, 531–548. https://doi.org/10.1007/978-981-16-6100-6_13.

Chuang, C. H., Chiang, C. Y., Chen, Y. C., Lin, C. Y., & Tsai, Y. C. (2021). Goose surface temperature monitoring system based on deep learning using visible and infrared thermal image integration. *IEEE Access, 9*, 131203–131213. https://doi.org/10.1109/ACCESS.2021.3113509.

Clauss, M., Frei, S., Hatt, J.-M., & Kreuzer, M. (2020). Methane emissions of geese (*Anser anser*) and turkeys (*Meleagris gallopavo*) fed pelleted lucerne. *Comparative Biochemistry and Physiology Part A: Molecular & Integrative Physiology, 242*, 110651. https://doi.org/10.1016/j.cbpa.2020.110651.

Dessborn, L., Hessel, R., & Elmberg, J. (2016). Geese as vectors of nitrogen and phosphorus to freshwater systems. *Inland Waters, 6*(1), 111–122. https://doi.org/10.5268/IW-6.1.897.

EFSA AHAW Panel. (2010). Scientific Opinion on the welfare aspects of the practice of harvesting feathers from live geese for down production. *EFSA Journal, 8*, 1886. https://doi.org/10.2903/j.efsa.2010.1886.

El Sabry, M. I., & Almasri, O. (2023). Global waterfowl production: Stocking rate is a key factor for improving productivity and well-being—a review. *Tropical Animal Health and Production, 55*, 419. https://doi.org/10.1007/s11250-023-03835-6.

Ewuola, E. O., Akinbola, E. T., Ogundele, A. A., & Oyewale, J. O. (2022). Effects of wallowing, sex and post wallowing time on the blood profile and thermoregulatory parameters of geese during the period of high temperature humidity index. *Thai Journal of Agricultural Science, 55*, 112–123.

Farghly, M. F. A., Mahrose, K. M., Ullah, Z., Rehman, Z. U., & Ding, C. (2017). Influence of swimming time in alleviating the deleterious effects of hot summer on growing Muscovy duck performance. *Poultry Science, 96*(11), 3912–3919. https://doi.org/10.3382/ps/pex207.

Fernandez, X., Guy, G., Laverze, J. B., Bonnefont, C., Knudsen, C., & Fortun-Lamothe, L. (2016). A kinetic study of the natural induction of liver steatosis in Greylag Landaise geese: The role of hyperphagia. *Animal, 10*(8), 1288–1295. https://doi.org/10.1017/S1751731116000161.

Fernandez, X., Lazzarotto, V., Bernadet, M. D., & Manse, H. (2019). Comparison of the composition and sensory characteristics of goose fatty liver obtained by overfeeding and spontaneous fattening. *Poultry Science, 98*(11), 6149–6160. https://doi.org/10.3382/ps/pez289.

Gillette, D. D. (1977). Mating and other behavior of domestic geese. *Applied Animal Ethology, 3*(4), 305–319. https://doi.org/10.1016/0304-3762(77)90056-6.

Gorton, M., Yeh, C.-H., Chatzopoulou, E., White, J., Tocco, B., Hubbard, C., & Hallam, F. (2023). Consumers' willingness to pay for an animal welfare food label. *Ecological Economics, 209*, 107852. https://doi.org/10.1016/j.ecolecon.2023.107852.

Guémené, D., & Guy, G. (2004). The past, present and future of force-feeding and "foie gras" production. *World's Poultry Science Journal, 60*(2), 210–222. https://doi.org/10.1079/WPS200314.

Guémené, D., Shi, Z. D., & Guy, G. (2012). Production systems for waterfowl. In: V. Sandilands and P. Hocking (eds.) *Alternative systems for poultry: health, welfare and productivity*, France: CABI Publishing, 128–154. https://www.cabidigitallibrary.org/doi/10.1079/9781845938246.0128.

Guo, Y., Zhao, Z. H., Huang, R. S., Chen, X. Q., Li, L. X., Balasubramanian, B., Jia, R. M., & Liu, W. C. (2020). Comparative study on reproductive performance of black and yellow-footed Chinese indigenous goose (*Shitou goose*) under off-season production conditions. *Journal of Animal and Plant Sciences, 31*(2), 630–634. https://doi.org/10.36899/JAPS.2021.2.0252.

Guo, D., Xiao, D., Liu, Y., Wang, J., Tang, Z., Peng, S., Lv, S., & Cheng, L. (2026). A real-time visual monitoring system for egg production of breeding geese in small-group natural mating cages. *Biosystems Engineering, 265*, 104442. https://doi.org/10.1016/j.biosystemseng.2026.104442.

Guy, G., Fortun-Lamothe, L., Bénard, G., & Fernandez, X. (2013). Natural induction of spontaneous liver steatosis in Greylag Landaise geese (*Anser anser*). *Journal of Animal Science, 91*(1), 455–464. https://doi.org/10.2527/jas.2012-5492.

Hassan, M., Ding, W., Umar, M., & Rasool, G. (2017). Batch and semi-continuous anaerobic co-digestion of goose manure with alkali solubilized wheat straw: A case of carbon to nitrogen ratio and organic loading rate regression optimization. *Bioresource Technology, 230*, 24–32. https://doi.org/10.1016/j.biortech.2017.01.025.

Jiang, D. L., Liu, L., Wang, C. L., Chen, F., Sun, A. D., & Shi, Z. D. (2011). Raising on water stocking density reduces geese reproductive performances via water bacteria and lipopolysaccharide contaminations in "Geese-Fish" production system. *Agricultural Sciences in China, 10*(9), 1459–1466. https://doi.org/10.1016/S1671-2927(11)60139-9.

Karadağ, S. (2025). Alternative species in poultry farming and their economic, social, and cultural importance. *Journal of Istanbul Veterinary Sciences, 9*(1), 30–37. https://doi.org/10.30704/http-www-jivs-net.1620490.

Kjosevski, M., Nikolovski, M., Ilievska, K., Pendovski, L., & Ilieski, V. (2021). The three pillars of applied farm animal welfare. *Veterinarski Glasnik, 75*(2), 99–111. https://doi.org/10.2298/VETGL210629014K.

Kozák, J., Gara, I., & Kawada, T. (2010). Production and welfare aspects of goose down and feather harvesting. *World's Poultry Science Journal, 66*(4), 767–778. https://doi.org/10.1017/s0043933910000723.

Kozák, J. (2009). Expertise on the feather plucking of live geese. *Animal Welfare, Ethology and Housing Systems, 5*(2), 100–107.

Kozák, J. (2011a). Foie gras production: Pros and cons in the light of animal protection. *Animal Welfare, Ethology and Housing Systems, 7*(3), 200–209.

Kozák, J. (2011b). Gathering feathers: Pros and cons in the light of animal protection. *Animal Welfare, Ethology and Housing Systems, 7*(4), 433–442.

Kozák, J. (2021). Goose production and goose products. *World's Poultry Science Journal, 77*(2), 403–414. https://doi.org/10.1080/00439339.2021.1885002.

Krunt, O., & Zita, L. (2026). Goose meat production and consumption: Current knowledge and market challenges. *World's Poultry Science Journal, 82*(1), 79–91. https://doi.org/10.1080/00439339.2025.2564803.

Lagerkvist, C. J., & Hess, S. (2011). A meta-analysis of consumer willingness to pay for farm animal welfare. *European Review of Agricultural Economics, 38*(1), 55–78. https://doi.org/10.1093/erae/jbq043.

Le Neindre, P., Willeberg, P., Jensen, P., Broom, D., Harting, J., Dantzer, R., Morton, D., Bénard, P., Verga, M., Faure, J., Nicks, B., & Estevez, I. (1998). *Welfare aspects of the production of foie gras in ducks and geese*. Report of the Scientific Committee on Animal Health and Animal Welfare. https://food.ec.europa.eu/system/files/2020-12/sci-com_scah_out17_en.pdf.

Leprettre, S., Auvergne, A., Dubois, J.P., Babilé, R., Lavigne, F. (2000). Système d'élevage des oies et performances de croissance. In: 4èmes Journées de la Recherche sur les Palmipèdes à Foie Gras, 105–108.

Liao, S. C., Lu, P. X., Shen, S. Y., Hsiao, C. C., Lien, C. Y., Wang, S. D., Lin, T. Y., & Tu, P. A. (2021). Effects of different swimming pool conditions and floor types on growth performance and footpad dermatitis in indoor-reared white roman geese. *Animals, 11*(6), 1705. https://doi.org/10.3390/ani11061705.

Liao, S. C., Lyu, P. X., Shen, S. Y., Hsiao, C. C., Lien, C. Y., Wang, S. D., Lin, T. Y., & Tu, P. A. (2022). Effects of swimming pool conditions and floor types on white Roman geese's physical condition scores and behaviors in an indoor rearing system. *Animals, 12*(23), 3273. https://doi.org/10.3390/ani12233273.

Main, D., & Mullan, S. (2017). A new era of UK leadership in farm animal welfare. *Veterinary Record, 181*(2), 49–50. https://doi.org/10.1136/vr.j3273.

Massaccesi, L., Mancinelli, A. C., Mattioli, S., Feudis, M. D., Castellini, C., Bosco, A. D., Marongiu, M. L., & Agnelli, A. (2019). Geese reared in vineyard: Soil, grass and animals interaction. *Animals, 9*(4), 179. https://doi.org/10.3390/ani9040179.

Mayfield, L. E., Bennet, R. M., Tranter, R. B., & Wooldridge, M. J. (2007). Consumption of welfare-friendly food products in Great Britain, Italy and Sweden, and how it may be influenced by consumer attitudes to, and behaviour towards, animal welfare attributes. *International Journal of Sociology of Agriculture and Food, 15*(3), 59–73. https://doi.org/10.48416/ijsaf.v15i3.284.

Molnár, S. (2016). A magyar lúdágazat aktuális problémái és hazai, nemzetközi szintű kihívásai. *Acta Agraria Kaposváriensis, 20*(1), 62–81.

EFSA AHAW Panel (EFSA Panel on Animal Health and Animal Welfare), Nielsen, S. S., Alvarez, J., Bicout, D. J., Calistri, P., Canali, E., Drewe, J. A., Garin-Bastuji, B., Gonzales Rojas, J. L., Schmidt, C. G., Herskin, M., Michel, V., Miranda Chueca, M.Á., Padalino, B., Roberts, H. C., Spoolder, H., Stahl, K., Viltrop, A., Winckler, C., Berg, C., ... Velarde, A. (2023). Welfare of ducks, geese and quail on farm. *EFSA Journal. 21*(5), e07992. https://doi.org/10.2903/j.efsa.2023.7992.

Odum, E. P. (1960). Premigratory hyperphagia in birds. *The American Journal of Clinical Nutrition, 8*(5), 621–629. https://doi.org/10.1093/ajcn/8.5.621.

Rochlitz, I., & Broom, D. M. (2017). The welfare of ducks during foie gras production. *Animal Welfare, 26*(2), 135–149. https://doi.org/10.7120/09627286.26.2.135.

Rothrock, M. J., Gibson, K. E., Micciche, A. C., & Ricke, S. C. (2019). Pastured poultry production in the United States: Strategies to balance system sustainability and environmental impact. *Frontiers in Sustainable Food Systems, 3*, 74. https://doi.org/10.3389/fsufs.2019.00074.

Salamon, A. (2015). Maternal investment – and its constraints – in the egg of domestic waterfowl. PhD Dissertation. Dublin: University College Dublin.

Salamon, A., & Kent, J. P. (2020). The double-yolked egg: from the 'miracle of packaging' to nature's 'mistake' *World's Poultry Science Journal, 76*(1), 18–33. https://doi.org/10.1080/00439339.2020.1729671.

Salamon, A. (2020a). Factors affecting the production of double-yolked eggs. *World's Poultry Science Journal, 76*(4), 815–826. https://doi.org/10.1080/00439339.2020.1830011.

Salamon, A. (2020b). Fertility and hatchability in goose eggs: A review. *International Journal of Poultry Science, 19*(2), 51–65. https://doi.org/10.3923/ijps.2020.51.65.

Salamon, A. (2025). The reproductive behaviour of geese and its implications on production and welfare. *World's Poultry Science Journal, 81*(1), 103–116. https://doi.org/10.1080/00439339.2024.2437173.

Shepherd, & Fairchild (2010). Footpad dermatitis in poultry. *Poultry Science, 89*(10), 2043–2051. https://doi.org/10.3382/ps.2010-00770.

Song, Y., Li, Y., Zheng, S., Dai, W., Shen, X., Zhang, Y., Zhao, W., Chang, G., Xu, Q., & Chen, G. (2017). Effects of forage feeding versus grain feeding on the growth performance and meat quality of Yangzhou geese. *British Poultry Science, 58*(4), 397–401. https://doi.org/10.1080/00071668.2017.1307942.

Sótonyi, P., & Lorászkó, G. (2008). A libatömés nem állatkínzás. *Magyar Mezőgazdaság, 63*(42), 22–32.

Stojević, Z., Vranković, L., Petrinec, Z., Aladrović, J., Župan, B., & Božić, M. (2017). Geese raising on fish ponds. *Croatia Veterinarska Stanica, 48*(1), 31–34. https://hrcak.srce.hr/221506.

Tóth, P., Bódi, L., Maros, K., Szűcs, E., & Janan, J. (2012). Blood corticosterone levels in growing geese around feather gathering. *Acta Veterinaria Hungarica, 60*(4), 477–487. https://doi.org/10.1556/AVet.2012.042.

Tremolada, C., Bielińska, H., Minero, M., Ferrante, V., Canali, E., & Barbieri, S. (2020). Animal-based measures for the on-farm welfare assessment of geese. *Animals, 10*(5), 890. https://doi.org/10.3390/ani10050890.

Vetter, S., Vasa, L., & Ózsvári, L. (2014). Economic aspects of animal welfare. *Acta Polytechnica Hungarica, 11*(7), 119–134.

Wang, C., Liu, Z. L., Xue, J. J., Wang, Y. M., Huang, X. F., & Wang, Q. G. (2019). Effect of stocking density on growth performance, feather quality, carcass traits, and muscle chemical component of geese from 49 to 70 days of age. *Journal of Applied Poultry Research, 28*(4), 1297–1304. https://doi.org/10.3382/japr/pfz097.

Wang, G., Chen, J., Guo, Y., Weng, K., Zhang, Y., Zhang, Y., Chen, G., Xu, Q., & Chen, Y. (2025). Effects of free-range systems on muscle fiber characteristics and welfare indicators in geese. *Animals, 15*(3), 304. https://doi.org/10.3390/ani15030304.

Westgren, R. E. (1999). Delivering food safety, food quality, and sustainable production practices: The Label Rouge poultry system in France. *American Journal of Agricultural Economics, 81*(5), 1107–1111. https://doi.org/10.2307/1244092.

Yang, X. W., Liu, L., Jiang, Dl, Wang, Cl, Sun, A. D., & Shi, Z. D. (2012). Improving geese production performance in " goose-fish" production system by competitive reduction of pathogenic bacteria in pond water. *Journal of Integrative Agriculture, 11*(6), 993–1001. https://doi.org/10.1016/S2095-3119(12)60091-4.

Yin, L. Y., Wang, Z. Y., Yang, H. M., Xu, L., Zhang, J., & Xing, H. (2017). Effects of stocking density on growth performance, feather growth, intestinal development, and serum parameters of geese. *Poultry Science, 96*(9), 3163–3168. https://doi.org/10.3382/ps/pex136.

Zhang, H., Wu, Z., Zhang, T., Lu, C., Zhang, Z., Ye, J., Yang, J., Yang, D., & Fang, C. (2025). Improved YOLO-goose-based method for individual identification of lion-head geese and egg matching: Methods and experimental study. *Agriculture, 15*(13), 1345. https://doi.org/10.3390/agriculture15131345.

Zhang, T., Li, H. F., Chen, K., Zhao, Y., Chang, H., Xuec, M., & Zhang, S. (2011). Analysis of fitness predominance for Gaoyou duck's double yolk egg. *Journal of Animal and Veterinary Advances, 10*(3), 367–371. https://doi.org/10.3923/javaa.2011.367.371.

Zhang, Y., Ge, Y., Yang, T., Guo, Y., Yang, J., Han, J., Gong, D., & Miao, H. (2022). An IoT-based breeding egg identification and coding system for selection of high-quality breeding geese. *Animals, 12*(12), 1545. https://doi.org/10.3390/ani12121545.

Zhu, H., Shao, X., Chen, Z., Wei, C., Lei, M., Ying, S., Yu, J., & Shi, Z. (2017). Induction of out-of-season egg laying by artificial photoperiod in Yangzhou geese and the associated endocrine and molecular regulation mechanisms. *Animal Reproduction Science, 180*, 127–136. https://doi.org/10.1016/j.anireprosci.2017.03.009.

Chapter 7

Turkeys

Karen Schwean-Lardner, Tory Shynkaruk, and Carolin Adler
College of Agriculture and Bioresources, University of Saskatchewan, Saskatoon, SK, Canada

Introduction

Sustainability in turkey production is a balancing act between biology, management, and innovation. Unlike broiler chickens, turkeys present unique challenges to researchers and producers alike. Their impressive size, physical strength, and innate curiosity make them both fascinating and difficult study subjects. Designing experiments that account for these traits is complex, resource-intensive, and often logistically demanding. Consequently, the scientific foundation for turkey-specific management remains comparatively thin, and many practices are still adapted from broiler research (Schwean-Lardner, 2025). However, as the global poultry industry faces increasing pressure to balance productivity, animal welfare, and environmental responsibility, the need for data-driven, turkey-focused sustainability strategies has never been greater.

This chapter examines three pillars of sustainability (environmental, social, and economic) in turkey production; however, the literature in each area is often not available with a turkey focus. Each pillar will be discussed in each section, but only when the literature is available. Regardless, the chapter attempts to explore the interplay between animal welfare, efficiency, and environmental impact across key stages of turkey production. By integrating current research, this chapter aims to highlight pathways toward sustainable turkey systems that meet both the birds' needs and the expectations of a changing world.

Hatchery processes

Commercial hatcheries house the highest bird densities within poultry production systems, making hatchery practices critical to both immediate and long-term poult welfare. While numerous procedures occur at this stage, research specific to turkeys remains limited. Existing studies—primarily conducted in other species—have examined factors such as conveyor belt speed, drop height,

Integrating Welfare and Sustainability in Poultry Systems. DOI: https://doi.org/10.1016/B978-0-443-40594-5.00017-1

beak and toe treatments, euthanasia methods, handling, and shipping box stocking density. However, substantial knowledge gaps persist across nearly all aspects of poult management. Moreover, much of the available literature predates recent technological and operational advancements in hatcheries, with earlier work largely emphasizing economic outcomes rather than animal welfare considerations.

Social sustainability

In commercial turkey hatcheries, social management practices primarily involve beak, toe, and occasionally snood treatment. These interventions can reduce injurious pecking and improve welfare in some respects, but may also limit natural behaviors.

Infrared (IR) beak treatment is now widely used globally. This method exposes the beak tip of day-old poults to high-intensity IR light, which penetrates the rhamphotheca and disrupts keratin-producing cells, causing the treated portion to erode and slough while minimizing bacterial infiltration (Glatz, 2000). The result is a shortened upper beak with an intact lower beak (Struthers et al., 2022a and Struthers et al., 2022b) reported that IR-treated turkeys were heavier at 12 weeks than sham-treated controls, and mortality from aggressive pecking tended to be lower (1.91% vs 6.10%, P=.09). Pecking-related injuries requiring deterrent treatment were significantly reduced in IR-treated birds (8.23% vs 26.17%), consistent with findings by Busatta et al. (2024), who noted that enrichment alone did not mitigate damage. Stress indicators did not differ until 20 days of age, when untreated birds exhibited higher stress, coinciding with increased pecking activity. Additionally, pecking force at 1 week was greater in IR-treated poults, suggesting minimal pain posttreatment (Struthers et al., 2022a and Struthers et al., 2022b).

Claw treatment in turkey poults aims to reduce carcass damage caused by sharp claws, which is particularly important for whole-bird markets. Previously, treatment involved microwave energy (MCP), resulting in tissue necrosis and sloughing within 1–3 weeks (Fournier et al., 2014, 2015). Studies using MCP reported sex-specific effects: treated toms showed reduced activity early in life and altered posture later, suggesting possible imbalance and welfare concerns. However, no reduction in carcass scratching was achieved, leading the authors to suggest that claw treatment was not required for heavy toms (Fournier et al., 2012). Hens exhibited significant reductions in carcass scratches (12% vs 46% untreated), most exceeding 5 cm (Fournier et al., 2014). The authors concluded that the treatment led to clear benefits for hens, reducing pain and infection risk and improving carcass quality, which has economic implications for Grade A markets.

In response to the research mentioned above, the company responded and has instituted improved methodologies for this treatment. Claw treatment now

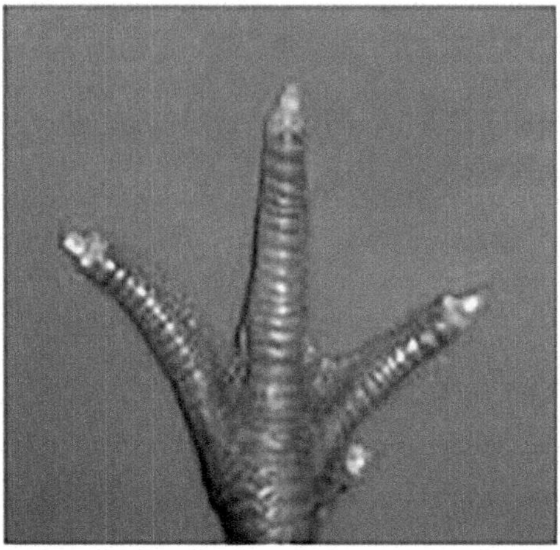

FIGURE 7.1 The area of toe treatment can be seen by the light-colored tissue. *Courtesy Nova-Tech Engineering LLC.*

uses a new application method versus the older microwave wavelength and waveguide technology (https://nteglobal.com/poultry/). The applicator emits electromagnetic energy to treat the tips of the claws (see Fig. 7.1). To date, research has not been reported on this new methodology to the authors' knowledge.

Economic sustainability

Early-life conditions can have lasting effects on turkey health and welfare. Hatchery sanitation is particularly important, as inadequate hygiene has been associated with increased mortality later in life, directly impacting production efficiency and costs (Carver et al., 2000; Carver et al., 1991; Gardner et al., 1980). Infertile eggs, dead embryos, and cull poults represent direct production losses, and egg handling before and during incubation significantly influences these outcomes. Research has long focused on reducing embryonic mortality. Early studies demonstrated that storage beyond 7 days markedly decreases hatchability (Becker, 1963; Kosin, 1954). More recent work suggests that preincubation for 12 hours may mitigate this decline (Fasenko et al., 2001), though findings remain inconsistent (Filho et al., 2020). Other strategies include modifying incubation conditions: supplemental CO_2 during the first 10 days (Gildersleeve & Boeschen, 1983), increased oxygen at high altitudes (Christensen & Bagley, 1988), and short-term elevated temperatures for long-stored eggs (Christensen et al., 2003).

Conclusion

Knowledge of poult welfare at the hatchery remains limited, despite growing evidence that early-life conditions influence welfare throughout production. Ethical challenges also arise during hatchery processing. For example, IR beak treatment is currently the most effective method for reducing the consequences of injurious pecking—a major welfare concern—yet the practice is controversial due to perceptions of the invasiveness thereof.

Transportation of poults

After hatching and processing at the hatchery, poults are transported to the farm.

Social sustainability

The number of poult hatcheries tends to be fewer than for other species, so transportation of poults may occur by ground or by air. Fahd et al. (2024) (2025) studied the effects of air transportation on poult behavior during brooding, comparing a short flight (France to Egypt; 4.3 hours) and a long flight (Tunisia to Egypt with a stopover in Turkey; 30 hours). Long transportation inconsistently affected behavior in the first 3 weeks on farm: foraging, walking, and running increased, while standing, sitting, preening, and wing flapping decreased. Feeding, drinking, dustbathing, wing stretching, and feather pecking were unaffected.

Economic sustainability

Short-distance air transport (4.3 hours) resulted in higher body weights, weight gain (Fahd et al., 2024), and increased feed intake, though feed efficiency remained unchanged (Fahd et al., 2025), compared to long-distance transport (30 hours).

Carver et al. (2002) identified truck temperature and the number of poults dead on arrival (DOA) as risk factors for hen flock mortality within the first 14 days on farm. For toms, additional risk factors included shipping time and weather conditions at placement. Due to their inability to regulate body temperature, poults require warm environments from hatch through brooding. Female poults transported in cooler conditions had increased mortality compared to male poults, suggesting greater cold temperature tolerance in males. Hen flocks with more DOAs had higher mortality odds, whereas DOA level had no impact on mortality within the first 2 weeks on farm in some flocks. Transportation duration did not affect hen mortality but increased tom mortality risk with each additional hour.

Conclusion

Although limited, existing research highlights that transportation duration, temperature, and method influence poult behavior, growth, and mortality. Shorter, warmer transport conditions appear more favorable, especially for female poults. Further studies are needed to address environmental factors and optimize transportation protocols for improved welfare and productivity in different geographical and climatic regions.

Stocking density

Stocking density is a key management factor in turkey production, influencing bird performance, welfare, carcass quality, and economic outcomes. Literature reviews by Erasmus (2017) and Krautwald-Junghanns & Sirovnik (2022) provide insights into the effects of stocking density on commercial turkeys.

Social sustainability

Feather pecking and cannibalism are significant welfare concerns as they result in injury, with birds experiencing pain and distress, and can be severe enough to require culling or result in mortality. Early studies (Denbow et al., 1984; Leighton et al., 1985) found no impact of stocking density on feather cover, but recent research (Beaulac & Schwean-Lardner, 2018; Jhetam et al., 2022a) reported poorer feather cover with increased stocking density. Pecking behavior in hens rose with higher stocking density (Jhetam et al., 2022a; Leighton et al., 1985), though aggressive pecking and the number of culls decreased. In toms, very low density (30 kg/m^2) was associated with more damage due to aggression (Beaulac & Schwean-Lardner, 2018).

As stocking density increases, resting disturbances increase (Beaulac & Schwean-Lardner, 2018; Jhetam et al., 2022a; Martrenchar et al., 1999); however, increased disturbances have also been observed under low density (Beaulac & Schwean-Lardner, 2018). Behavioral shifts may also occur, such as more time at the feeder and resting in hens (Jhetam et al., 2022a), and varied activity patterns in toms across various ages (Beaulac & Schwean-Lardner, 2018).

Additional negative consequences of higher stocking densities in turkeys include:

- Poorer gait scores (Beaulac & Schwean-Lardner, 2018; Martrenchar et al., 1999)
- Increased incidence and severity of footpad dermatitis (Beaulac et al., 2019; Jhetam et al., 2022a; Martrenchar et al., 1999)
- Increased litter moisture (Beaulac et al., 2019; Jhetam et al., 2022a)
- Dirtier feathers (Boussaada et al., 2022; Fidan et al., 2020)
- Increased stress (Beaulac & Schwean-Lardner, 2018; Jhetam et al., 2022a)

Economic sustainability

Lower stocking densities generally yield higher body weights and gains (Bartz et al., 2020; Beaulac et al., 2019; Denbow et al., 1984; Jhetam et al., 2022b; Martrenchar et al., 1999), with Leighton et al. (1985) reported no effect in hens. Feed intake and efficiency declined with higher densities, especially in toms (Bartz et al., 2020; Beaulac et al., 2019; Denbow et al., 1984). No significant effects on flock uniformity, mortality, or carcass grades have been observed (Beaulac et al., 2019; Denbow et al., 1984; Jhetam et al., 2022b; Leighton et al., 1985). However, higher stocking densities result in an increased net income for producers (Jhetam et al., 2022b).

Conclusion

Stocking density significantly influences turkey welfare and performance. Research to date has compared various densities to one another, and relative to lower densities, higher densities may improve economic returns, but they are associated with increased stress, poorer feather and litter condition, and behavioral changes that may compromise welfare. Lower stocking densities increase aggression in toms, likely because of healthier birds that are more able to move and interact with one another. There is a gap in the existing literature regarding the impact of stocking density on the environmental sustainability of turkey production. Future research could focus on factors including land and energy use, feed and water consumption, and manure output. To date, no single optimal stocking density has been identified, partly because the welfare impacts associated with stocking density are also associated with other environmental and management factors that are related to stocking density, such as litter quality, air quality, and access to feed and water. Optimal stocking density strategies should balance productivity with animal welfare.

Lighting

Lighting programs for turkeys have often been based on broiler research due to their shared status as meat-type birds. However, turkeys differ substantially in behavior, temperament, and production traits, making species-specific data essential for commercial application (Schwean-Lardner et al., 2016).

Light duration

Social sustainability

Animal welfare considerations have evolved from focusing solely on biological functioning to incorporating affective states and natural behaviors. Long photophases (e.g., 23 L:1D) are associated with increased skeletal disorders, impaired mobility, and ocular health issues, including cataracts and blindness (Leis et al., 2017; Vermette et al 2016a; Vermette et al 2016b). Intermittent lighting programs can reduce pecking injuries but may introduce welfare concerns such as blindness in some birds (Sherwin & Devereux, 1999; Sherwin et al., 1999a). Programs that include substantial dark periods improve walking ability and reduce metabolic disease, though cannibalism risk may rise, likely because the birds with darkness exposure are healthier and more mobile (Classen et al., 1994). This may mean that new methodologies for controlling this injurious behavior need to be investigated, which remains a significant gap in the turkey literature. Behavioral studies confirm that shorter photophases promote activity and natural rhythms, supporting welfare objectives (Prusik & Lewczuk, 2020; Zawilska et al., 2007).

Environmental sustainability

Lighting programs in turkey production influence energy consumption and environmental impact. Early studies primarily used incandescent lighting with intensities ranging from 2.5 to 108 lux (Hester & Kohl, 1989; Marsden et al., 1966; Siopes et al., 1986). Extended photophases (e.g., 23 L:1D) increase electricity demand compared to diurnal, intermittent, or step-up/step-down programs, which incorporate dark periods and reduce energy use (Classen et al., 1994; Hamilton & Kennie, 1997). Recent research also explored LED systems and natural light supplementation to minimize energy requirements while maintaining production outcomes (Bartz & Grimes, 2021). These strategies align with sustainability goals by reducing resource consumption and supporting lower carbon footprints.

Economic sustainability

Lighting strategies significantly affect growth performance and production efficiency. Extended photophases often reduce body weight and feed efficiency compared to intermittent or step-up/step-down programs (Lilburn et al., 1992; Siopes et al., 1986; Vermette et al 2016a; Vermette et al 2016b). For example, body weight decreased linearly with increasing day length, while mortality and skeletal defects were highest under 23 L:1D (Vermette et al 2016a; Vermette et al 2016b). Intermittent programs have demonstrated improved feed conversion and market weights over extended photophases, but may increase leg abnormalities (Clarke et al., 1993).

Light flickering or intensity

Reduced light intensity can lower bird activity in general, and this is often used to aid in controlling outbreaks of injurious pecking behavior in turkeys. While very little recent work has focused on light intensity for turkeys, more recently, there has been an interest in the effects of light flickering, which is most often caused by changes in currents supplied to the light source.

Light intensity

Social sustainability

Welfare implications of light intensity include behavioral preferences and stress responses. Turkeys generally avoid extremely dim environments (<1 lux) and show age-dependent preferences for brighter areas (Barber et al., 2004; Sherwin, 1998). Very bright conditions can increase injurious pecking and aggression, compromising welfare (Duggan et al., 2014; Moinard et al., 2001). Stress indicators such as cortisol rise under higher intensities (Mohammed et al., 2016), and feather cover deteriorates in brightly lit curtain-sided barns, reducing thermoregulation and skin protection (Duggan et al., 2014).

Environmental sustainability

Light intensity affects energy use and environmental impact. Higher intensities (e.g., 220 lux) require more electricity compared to dim lighting (Siopes et al., 1983, 1989). Recommendations for uniform intensity across barns may not align with bird preferences, suggesting that variable intensity systems could optimize energy efficiency while meeting birds' welfare needs (Barber et al., 2004). Use of fluorescent and LED lighting, as explored in later studies, offers potential for reducing energy consumption compared to traditional incandescent sources (Moinard et al., 2001).

Economic sustainability

Light intensity influences productivity and health outcomes. While early studies reported minimal effects on mortality and feed efficiency (Siopes et al., 1983, 1989), recent work shows that lower intensities (e.g., 5 lux) can improve body weight and reduce stress compared to higher intensities (Mohammed et al., 2016). However, brighter environments may increase pecking-related injuries, leading to higher culling rates and economic losses (Duggan et al., 2014).

Light flicker

Social sustainability

Flicker frequency influences welfare indicators. Severe flicker (30 Hz) negatively affected early body weight and increased mortality compared to a higher frequency of 195 Hz (Hammond et al., 2024a). Eye morphology changes and behavioral alterations, such as reduced wing flapping and exploratory pecking, were observed under low flicker rates, indicating potential sensory stress (Hammond et al., 2024a; 2024b). However, reduced aggressive pecking under 30 Hz suggests complex welfare outcomes (Hammond, Buchynski, Shynkaruk, Brown, et al., 2024). Raabe et al. (2024) reported no significant effects on stress hormones or mortality at higher flicker rates, supporting the use of high-frequency flicker (>165 Hz) for welfare optimization.

Environmental sustainability

LED lighting systems are generally more energy-efficient than incandescent or fluorescent sources, reducing electricity consumption and environmental impact. Studies on flicker frequency (Hammond et al., 2024a; 2024b) confirm that varying flicker rates (30 or 90 Hz as compared to 195 Hz) does not compromise productivity, suggesting that energy-efficient LED systems can be adopted without negative productivity trade-offs.

Economic sustainability

While LED flicker did not affect long-term productivity in Raabe et al. (2024), Hammond et al., 2024a; 2024b found early growth suppression and higher mortality under visible flicker (30 Hz), which would increase production costs. High-frequency flicker (≥195 Hz) maintained performance and reduced mortality, supporting its economic viability. Adoption of LED systems with appropriate flicker rates can balance energy savings with profitability.

Light wavelength and source

Due to birds' exceptional visual acuity, light wavelength has long been a research focus, primarily in chickens rather than turkeys. The introduction of LED technology enabled precise control of spectral output. Research has also examined light source characteristics, which often overlap with wavelength effects. Light sources differ in multiple aspects, including spectrum, beam directionality (modifiable in LEDs), and heat production.

Social sustainability

Due to their ability to see across the UV and visible light spectrum, there is the potential to use light of various wavelengths to improve turkey welfare. However, research to date has assessed the combined effects of light wavelength and environmental enrichment on turkey behavior and welfare, and the effects of wavelength alone are not fully understood. UV supplementation combined with enrichments reduced injurious pecking and culling losses (Lewis et al., 2000). Preference studies (Moinard & Sherwin, 1999) showed turkeys favored chambers with supplemental UV light over chambers lit with fluorescent light only, indicating UV light is not aversive. Sherwin & Devereux (1999) suggested that the lack of visible UV patterns on feathers under standard lighting may contribute to aggression, while enriched environments (Sherwin et al., 1999a) reduced wing and tail injuries. Fluorescent lighting was not found to be aversive (Sherwin & Devereux, 1999), and higher intensities combined with enrichments helped mitigate pecking injuries (Moinard et al., 2001). These findings emphasize lighting and enrichment as key factors influencing turkey welfare.

Hormonal studies (Martyniuk et al., 2020) revealed that light wavelength and intensity affect serotonin and melatonin rhythms, likely through variable activation of the pineal gland that occurs with various wavelengths and light intensities. While the authors did not examine turkey behavior, the impact of light wavelength and intensity on serotonin and melatonin has implications for turkey behavior and welfare outcomes that warrant further investigation.

Environmental sustainability

Studies incorporating UV light (Bartels et al., 2017; Lewis et al., 2000) highlight its potential role in naturalistic (mimicking sun rays) environments, as UV reflection varies by feather color and age. However, these interventions require careful evaluation to balance energy use and ecological footprint. Limited information is available to be able to evaluate the environmental sustainability of light wavelength and source in turkey production.

Conclusion

Lighting strategies influence productivity and cost efficiency. Blue light reduced growth compared to red or white light (Levenick & Leighton, 1988), which could negatively impact profitability; however, the results from the 1988 study may not be directly translatable to modern, commercial turkey genetic lines that are more efficient than genetic lines used in older studies. While UV supplementation did not improve feed efficiency or leg health (Lewis et al., 2000), its role in reducing culling losses due to pecking injuries offers indirect economic advantages. Managing light intensity and enrichment (Moinard et al., 2001) can minimize welfare-related losses, supporting sustainable production systems. Hormonal regulation under different lighting conditions (Martyniuk et al., 2020) may further inform strategies to optimize growth and reduce stress-related inefficiencies and welfare impacts.

Environmental enrichment

Environmental enrichment in turkey production aims to enhance animal welfare by increasing housing complexity and promoting physical and mental stimulation. Common strategies are straw bales, raised platforms, pecking devices, visual barriers, or substrate materials (Enrichment for broilers and turkeys - from theoretical consideration to practical application, 2021). Their application depends on factors such as effectiveness, safety, cost, labor, and biosecurity (Enrichment for broilers and turkeys - from theoretical consideration to practical application, 2021; Jones et al., 2020; Van de Weerd & Day, 2009). When appropriately implemented, enrichments can improve animal welfare and broader sustainability goals.

Social sustainability

Enrichments can improve musculoskeletal health, feather and skin condition, and reduce injurious behaviors in turkeys by promoting activity, providing outlets for natural behaviors, or offering refuge from pecking. Sherwin et al. (1999b) found that items such as straw, boards, hooks, chains, cabbage, plastic conduit, lighting, and AstroTurf increased latency to sit after forced standing in male turkeys, suggesting improved musculoskeletal function. Dong et al. (2023) reported that although enrichments had limited effects on gait, straw bales and tunnels improved feather condition, and straw bales specifically reduced footpad dermatitis under suboptimal litter conditions. In another study, perch design affected leg health: 5 cm round dowels improved gait, likely by requiring greater balance and deliberate movement, whereas 15 cm square perches provided more comfortable resting surfaces that encouraged birds to spend time off moist litter, improving footpad condition (Buchynski et al., 2024). Similarly, raised platforms, straw bales, and hay-filled baskets reduced activity on the structures without changing overall activity, and birds spent more time on elevated areas, likely due to reduced disturbance (Letzguß & Bessei, 2009; Spindler & Hartung, 2009).

Effects of enrichment on stress and immune function are variable. Elevated platforms improved immune measures, including increased MHC class II^+ lymphocytes and reduced thrombocyte counts, likely by allowing birds to avoid aggression (Lindenwald et al., 2021). In contrast, metal objects, straw, and perches had little effect on stress indicators (Martrenchar et al., 2001), and early enrichment with frequently rotated toys negatively affected immune responses (Huff et al., 2003), possibly due to stress from novelty (Erasmus, 2017).

Overall, enrichment strategies can reduce pecking at conspecifics and improve feather and skin condition by diverting the attention of birds that perform injurious pecking or providing opportunities for birds to avoid being pecked (Crowe & Forbes, 1999; Sherwin et al., 1999a; Cottin, 2004; Letzguß & Bessei, 2009; Lindenwald et al., 2021; Martrenchar et al., 2001; Michel & Boilletot, 2003). Providing wheat grain feeders had partial benefits for feather condition and aggression (Berk et al., 2017), whereas colored plastic balls or dyed water bottles were ineffective (Busatta et al., 2024; Duggan et al., 2014), indicating that the type of enrichment strongly influences welfare outcomes.

Environmental sustainability

Research on the environmental impact of enrichment in turkeys is limited. Hinz & Berk (2002) found that elevated platforms, straw bales, and access to an open area did not affect dust, CO_2, NH_3, or endotoxin concentrations or emissions. Enrichments should also be evaluated for resource use, durability, reusability, and waste. Straw can promote welfare (Enrichment for broilers and turkeys - from theoretical consideration to practical application, 2021) but may pose biosecurity risks if contaminated (Lüning et al., 2023), whereas durable platforms or plastic objects reduce material turnover but increase water, energy, and disinfectant use. Further studies under commercial conditions, including lifecycle assessments, are needed to evaluate both in-barn and broader environmental impacts. For more in-depth information on lifecycle assessment, see Chapter 15.

Economic sustainability

Bessei (Enrichment for broilers and turkeys - from theoretical consideration to practical application, 2021) reported unpublished data based on a commercial flock of 4000 male turkeys. Using market prices and various enrichment types, additional labor associated with enrichment use ranged from 0.1 to 0.25 manpower-hours per flock, and total cost per bird ranged from 0.12 € to 0.68 €, with lower costs for straw provided "on demand" and higher costs for continuously supplied pecking blocks and hay.

Metal square perches placed between feeders and drinkers reduced performance in male turkeys under tropical conditions (González-Zapata et al., 2022), while perch design had no effect on performance in turkey hens raised under standard conditions (Buchynski et al., 2024). Cottin (2004) found that elevated platforms, straw bales, a veranda, and free-range access affected performance differently by strain: enriched slow-growing males were slightly heavier than controls, whereas fast-growing birds were lighter. Mortality was lower in enriched groups for both strains. Similarly, Busatta et al. (2024) reported that turkey hens with access to plastic bottles containing dyed water consumed slightly less feed and had better feed conversion ratios in nonbeak-treated birds, with fewer carcass injuries at slaughter, regardless of beak treatment. Dennis et al. (2024) reported that enrichment caused some variation in turkey meat quality traits but did not negatively affect fresh or processed meat, indicating that enrichment can be adopted to enhance welfare without compromising product quality.

Conclusion

Current evidence for the environmental impact of environmental enrichments for turkeys is lacking, and further research on resource use, durability, and biosecurity is needed. In addition, research is needed to establish the optimal ratio of enrichments to turkeys that provides both welfare and economic benefits. Economically, enrichment increases costs and labor, and production benefits are inconsistent, making its viability largely dependent on market premiums for higher welfare products. Overall, enrichment supports social sustainability by meeting public expectations and is increasingly recognized in regulations and certification schemes, but further work is needed to identify strategies that help turkey producers maximize turkey welfare while remaining practical and cost-effective.

Environmental conditions

Environmental conditions such as temperature, humidity, CO_2, and NH_3 affect all aspects of turkey sustainability. Sensors enable continuous monitoring and automated management, supporting welfare, productivity, and environmental outcomes (Aviagen Turkeys, 2024; Hybrid Turkeys, 2024). However, most turkey studies focus mainly on litter or flooring management.

Social, economic, and environmental sustainability

Litter-based strategies demonstrate how substrate choice affects welfare and environmental emissions. Comparisons among clover straw, corn stalk straw, and wheat straw showed largely similar welfare indicators (footpad dermatitis, hock discoloration, breast blisters) and production performance, although corn stalk straw reduced litter moisture and NH_3 concentrations compared with wheat straw (Farghly, 2012). Long straw particles were associated with greater caking and a higher incidence of footpad dermatitis, whereas chopped materials or wood shavings altered the temporal pattern of NH_3 emissions, suggesting that litter structure plays a key role in both environmental emissions and bird health (Mayne et al., 2007; Furo, 2019; Youssef et al., 2010, 2011). Moisture-management strategies such as floor heating or lignocellulose inclusion reduced footpad dermatitis despite high litter moisture, likely due to improved water-binding capacity (Youssef et al., 2010; Abd El-Wahab et al., 2011), whereas miscanthus grass increased litter moisture and lesion prevalence (Furo, 2019). Alternative substrates may also influence economic outcomes, as chopped paper-based litter (GroChips) improved body weight in toms and hens (Grimes et al., 2007).

Similar to litter-based management, flooring systems influence all pillars of sustainability by shaping barn environment, bird welfare, and production efficiency. Chen et al. (1991) examined partially slatted flooring systems using fiberglass, concrete, wood, or PVC. Their results indicated reduced litter moisture without negative effects on growth, mortality, or feed efficiency, while welfare indicators such as footpad dermatitis and leg problems remained largely unchanged. However, wood slats increased breast blisters, suggesting that material properties and slat design may influence bird comfort. Similarly, Farghly et al. (2018) compared littered flooring with cement, wire net, rubber mat, plastic slats, and wooden slatted floors under hot-climate conditions. They concluded that plastic and wooden slats have beneficial effects on production performance and welfare, as they lower body temperature and stress compared to other flooring materials. Airborne dust was lower on wire net, rubber mats, plastic slats, and wooden slats, while NH_3 concentrations were reduced mainly on wire net and plastic slats than on littered floors (Farghly et al., 2018). However, earlier work reported increased dust despite lower NH_3 on partially slatted floors, possibly because drier litter generates more particulate matter

(Noll et al., 1997). These contrasting outcomes emphasize trade-offs between environmental emissions, welfare indicators, and production value.

Conclusion

Proper litter or flooring management can reduce health issues, improve production efficiency, and minimize environmental impact, benefiting birds, farmers, and the broader environment. However, variability in outcomes and limited research in turkeys underscore the need for continued evaluation and on-farm strategies. Further research with current genetic lines of turkeys that are more efficient at converting feed into meat, and that grow at rates different from birds in older studies, is needed to assess the sustainability and welfare impacts of environmental conditions.

Individual bird euthanasia

Euthanasia of sick or injured birds is a necessary and routine practice on farms, aimed at preventing unnecessary suffering. Performing euthanasia promptly and humanely is the producer's responsibility. Several effective methods are available, each with varying impacts on bird welfare.

Social sustainability

Cervical dislocation is a common method of euthanasia, involving the severing of the spinal cord and separation of the vertebrae. It can be performed manually or mechanically. Hernandez et al. (2019) found that manual cervical dislocation resulted in brain death within approximately 120 seconds, compared to 300 seconds for mechanical dislocation using the Koechner euthanizing device (KED). Stiewert et al. (2021) reported that manual dislocation caused less vertebral damage and achieved more accurate separation, while mechanical methods, including the KED and broomstick technique, led to longer latencies in the cessation of brainstem reflexes.

Blunt force trauma using captive bolt devices is another effective method. Woolcott et al. (2018) compared the Zephyr-EXL and Turkey euthanasia device (TED), finding the Zephyr-EXL had a higher success rate (98% vs 89%) and induced more significant subdural hemorrhage, indicating effective brain trauma. Stiewert et al. (2021) evaluated four devices—TED, Zephyr-EXL, Jarvis stunner, and an experimental crossbow—and concluded that captive bolt devices resulted in immediate loss of brainstem reflexes, suggesting they may be a more humane euthanasia method compared to cervical dislocation methods. This is in agreement with Erasmus et al. (2010), who reported that cervical dislocation (manual and mechanical) did not result in immediate insensibility, whereas the Zephyr and blunt trauma (metal bar/bat) did cause immediate insensibility in turkeys of three different sizes (4, 11, and 13 kg). The authors noted that the location where the Zephyr was placed was important; the application needed to be directly over the cerebral cortex, and therefore suggested the Zephyr be applied twice in rapid succession to more consistently achieve insensibility. It was also reported that the size of the bird and the training received by stock people can influence the effectiveness of the euthanasia method and should be considered.

Carbon dioxide and electrical euthanasia methods have also been studied. Boyal et al. (2020) compared four methods—mechanical cervical dislocation, captive bolt, carbon dioxide, and electrical euthanasia—and found all were effective. For smaller turkeys, success rates ranged from 93% (KED) to 100% (electrical), while for larger birds, rates ranged from 92% (TED) to 100% (KED and carbon dioxide). Jacobs et al. (2021) reported that mechanical cervical dislocation resulted in longer durations of reflexes and more frequent unsuccessful kills compared to other methods. Electrical euthanasia was deemed the most humane due to its rapid and consistent cessation of reflexes and minimal physical trauma.

Conclusion

While multiple euthanasia methods are available and effective, their impacts on bird welfare vary. Manual cervical dislocation is quicker and less damaging than mechanical methods, but captive bolt devices offer faster insensibility. Carbon dioxide and electrical methods are highly effective, with electrical euthanasia emerging as the most humane method due to its consistency and minimal distress. Producers should consider bird age, size, and welfare outcomes when selecting euthanasia techniques to ensure ethical and efficient management practices. It should also be noted that unique challenges are associated with implementing different euthanasia methods on farms, including physical strength requirements, operator training and safety, equipment maintenance and cost, which have direct impacts on the economic sustainability of the method.

Transportation of market age turkeys

At the end of the production cycle, turkeys face multiple concurrent stressors, including feed/water withdrawal, handling, transport, lairage, shackling, and euthanasia/slaughter. Understanding how these factors impact birds is essential for improving welfare and economic outcomes/slaughter.

Social sustainability

Focus groups with US live haul crews identified animal (sex, age, weight, mobility), farm (lighting, litter, equipment, temperature), and human (training, physical demands) factors as key influences on handling efficiency (Hernandez et al., 2025).

Manual loading increased stress in turkeys compared to systems that allowed birds to walk into modules, which was associated with a lower heart rate (Prescott et al., 2000). Crate height affects behavior, with taller crates promoting standing and activity but also harmful behaviors such as wing flapping and mounting (Wichman et al., 2010, 2012; Di Martino, Capello, Stefani, et al., 2017). No clear welfare benefit from increased crate height has been reported (Wichman et al., 2012).

Curvy and busy roads elevate stress behaviors (Di Martino, Capello, Stefani, et al., 2017). Microclimate within trucks varies by location, with thermal cores located in the middle/rear bottom regions (Carvalho et al., 2017, 2018). Misting birds pretransport lowers temperatures but raises humidity. Cold transport (−18°C) lowers core temperature and glucose, with toms coping better than hens (Henrikson et al., 2018; Munson et al., 2024b), and hot transport (35°C) causes distress, especially in toms (Vermette et al., 2017). Both extremes impair behavior, blood gases, and stress markers (Munson et al., 2024b), with longer transport durations further compromising welfare. Catching, transport, and lairage increase corticosterone and heterophil to lymphocyte ratios, indicating physiological stress (Scanes et al., 2020).

Economic sustainability

Dead-on arrival (DOA) rates range from 0.14%–0.38% and are influenced by temperature, transport duration, sex, genetic line, and truck conditions (Di Martino et al., 2017; Gickel and Visscher, 2025; Machovcova et al., 2017; Petracci et al., 2006; Voslářová et al., 2007). Farm management plays a significant role in carcass downgrading (McEwen & Barbut, 1992). Crate design also impacts carcass damage—walking into modules reduces injury (Prescott et al., 2000); taller crates increase wing damage, shorter crates increase rump bruising (Wichman et al., 2012).

Transported toms had higher pH and darker meat compared to birds that were not transported (Owens & Sams, 2000). Temperature and duration of transport influence pH and color (Carvalho et al., 2017; Carvalho et al., 2018; Vermette et al., 2017; Munson et al., 2024a). High temperatures increase redness and yellowness in hens (Munson et al., 2024a). Water-holding capacity decreases in transported birds, especially toms under high heat and humidity (Munson et al., 2024a).

Conclusion

Transportation of market-age turkeys involves multiple stressors that can impact bird welfare and meat quality. Factors such as handling methods, crate design, transport conditions, and ambient temperature play critical roles. While some systems and practices can mitigate stress and injury, extremes in temperature and prolonged transport durations consistently compromise welfare and product quality. Optimizing transport protocols is essential to enhance all pillars of sustainability.

Conclusion

The scientific research focusing on turkeys is certainly lacking for many reasons, including the difficulty in the work itself, the cost, the time it takes to conduct the work, and potentially the lower consumption worldwide. As a result, conclusions often drawn from broiler research are used in turkey production. However, as explained by Schwean-Lardner et al. (2016), turkeys do not always respond similarly to broilers with respect to photoperiods, which draws into question the validity of using conclusions drawn from any broiler research for turkeys. The research gaps focused on turkey production sustainability and welfare are wide, and include so many factors, including light wavelength and intensity, enrichment, environmental conditions, and almost all factors relating to breeding turkeys.

AI disclosure

During the preparation of this work, the author(s) used CoPilot in order to ensure grammatical accuracy and improve conciseness. After using this tool/service, the author(s) reviewed and edited the content as needed and take(s) full responsibility for the content of the publication.

References

Abd El-Wahab, A., Beineke, A., Beyerbach, M., Visscher, C. F., & Kamphues, J. (2011). Effects of floor heating and litter quality on the development and severity of foot pad dermatitis in young turkeys. *Avian Diseases, 55*(3), 429–434. https://doi.org/10.1637/9684-021011-reg.1.

Aviagen Turkeys Management Guidelines. (2024). https://aviagenturkeys.us/documents/?_english.

Barber, C., Prescott, N., Wathes, C., Le Sueur, C., & Perry, G. (2004). Preferences of growing ducklings and turkey poults for illuminance. *Animal Welfare, 13*(2), 211–224. https://doi.org/10.1017/S0962728600026956.

Bartels, T., Lütgeharm, J.-H., Wähner, M., & Berk, J. (2017). UV reflection properties of plumage and skin of domesticated turkeys (*Meleagris gallopavo* f. dom.) as revealed by UV photography. *Poultry Science, 96*(12), 4134–4139. https://doi.org/10.3382/ps/pex256.

Bartz, B. M., Anderson, K. A., Oviedo-Rondón, E. O., Livingtson, K., & Grimes, J. L. (2020). Effects of stocking density on large white, commercial tom turkeys reared to 20 weeks of age: 1. Growth and performance. *Poultry Science, 99*(11), 5582–5586. https://doi.org/10.1016/j.psj.2020.08.024.

Bartz, B., & Grimes, J. L. (2021). The use of mixed light-emitting diodes and natural light in combination with daylength affects turkey hen performance, eye development, and feather coverage. *German Journal of Veterinary Research, 1*(3), 67–77. https://doi.org/10.51585/gjvr.2021.3.0023.

Beaulac, K., Classen, H.l., Gomis, S., Sakamoto, K. S., Crowe, T. G., & Schwean-Lardner, K. (2019). The effects of stocking density on turkey tom performance and environment to 16 weeks of age. *Poultry Science, 98*(7), 2846–2857. https://doi.org/10.3382/ps/pez087.

Beaulac, K., & Schwean-Lardner, K. (2018). Assessing the effects of stocking density on turkey tom health and welfare to 16 weeks of age. *Frontiers in Veterinary Science, 5*(SEP). https://doi.org/10.3389/fvets.2018.00213.

Becker, W. A. (1963). Length of pre-incubation storage of turkey eggs and its effects on body weight. *Poultry Science, 42*(6), 1356–1359. https://doi.org/10.3382/ps.0421356.

Berk, J., Stehle, E., & Bartels, T. (2017). Enrichment device - A way to reduce injurious pecking in fattening turkeys with intact beaks? Beschäftigungsmaterial - Eine möglichkeit zur reduktion von "beschädigungspicken" bei mastputen mit unkupierten schnäbeln? *Praktische Tierarzt, 99*(2), 190–201. https://doi.org/10.2376/0005-9366-16033.

Boussaada, T., Lakhdari, K., Benatallah, S. A., & Meradi, S. (2022). Effects of common litter types and their physicochemical properties on the welfare of broilers. *Veterinary World, 15*(6), 1523–1529. https://doi.org/10.14202/vetworld.2022.1523-1529.

Boyal, R. S., Buhr, R. J., Harris, C. E., Jacobs, L., & Bourassa, D. V. (2020). Equipment and methods for poultry euthanasia by a single operator. *Journal of Applied Poultry Research, 29*(4), 1020–1032. https://doi.org/10.1016/j.japr.2020.09.010.

Buchynski, K., Jhetam, S., Hargis, B. M., & Schwean-Lardner, K. (2024). Perch use in 11-wk-old turkey hens: Impact on performance, health, and behavior. *Journal of Applied Poultry Research, 33*(3). https://doi.org/10.1016/j.japr.2024.100432.

Busatta, J. G., Foppa, L., Sperandio, J., Martins, D. S., & Titto, C. G. (2024). Effects of environmental enrichment and lack of beak trimming on aspects of health, behaviour and production of commercial turkeys Efectos del enriquecimiento ambiental y la falta de recorte del pico en aspectos de salud, comportamiento y producción de pavos comerciales. *Spanish Journal of Agricultural Research, 22*(4). https://doi.org/10.5424/sjar/2024224-20638.

Carvalho, R. H., Honorato, D. C. B., Guarnieri, P. D., Soares, A. L., Pedrão, M. R., Oba, A., Paião, F. G., Ida, E. I., & Shimokomaki, M. (2017). In-transit development of color abnormalities in turkey breast meat during winter season. *Journal of Animal Science and Technology, 59*. https://doi.org/10.1186/s40781-017-0157-1.

Carvalho, R. H., Honorato, D. C. B., Guarnieri, P. D., Soares, A. L., Pedrão, M. R., Oba, A., Paião, F. G., Ida, E. I., & Shimokomaki, M. (2018). Assessment of turkey vehicle container microclimate on transit during summer season conditions. *International Journal of Biometeorology, 62*(6), 961–970. https://doi.org/10.1007/s00484-018-1498-2.

Carver, D. K., Wages, D. P., & Cowen, P. (1991). Effects of poult source on turkey production parameters. *Avian Diseases, 35*(1), 231–234. https://doi.org/10.2307/1591321.

Carver, D. K., Fetrow, J., Gerig, T., Correa, M. T., Krueger, K. K., & Barnes, H. J. (2000). Use of statistical modeling to assess risk for early poult mortality in commercial turkey flocks. *Journal of Applied Poultry Research, 9*(3), 303–318. https://doi.org/10.1093/japr/9.3.303.

Carver, D. K., Fetrow, J., Gerig, T., Krueger, T., & Barnes, H. J. (2002). Hatchery and transportation factors associated with early poult mortality in commercial turkey flocks. *Poultry Science, 81*(12), 1818–1825. https://doi.org/10.1093/ps/81.12.1818.

Chen, F., Noll, S. L., Clanton, C. J., Janni, K. A., & Halvorson, D. A. (1991). Market turkey performance affected by floor type and brooding method. *Applied Engineering in Agriculture, 7*(5), 606–612. 10.13031/2013.26277.

Christensen, Grimes, Wineland, & Davis (2003). Accelerating embryonic growth during incubation following prolonged egg storage 1. Embryonic livability. *Poultry Science, 82*(12), 1863–1868. https://doi.org/10.1093/ps/82.12.1863.

Christensen, & Bagley (1988). Improved hatchability of turkey eggs at high altitudes due to added oxygen and increased incubation temperature. *Poultry Science, 67*(6), 956–960. https://doi.org/10.3382/ps.0670956.

Clarke, J. P., Ferket, P. R., Elkin, R. G., McDaniel, C. D., McMurtry, J. P., Freed, M., Krueger, K. K., Watkins, B. A., & Hester, P. Y. (1993). Early dietary protein restriction and intermittent lighting. 1. Effects on lameness and performance of male turkeys. *Poultry Science, 72*(11), 2131–2143.

Classen, H. L., Riddell, C., Robinson, F. E., Shand, P. J., & McCurdy, A. R. (1994). Effect of lighting treatment on the productivity, health, behavior and sexual maturity of heavy male turkeys. *British Poultry Science, 35*(2), 215–225. https://doi.org/10.1080/00071669408417686.

Cottin, E. (2004). Stiftung Tierärztliche Hochschule Hannover Influence of enriched husbandry environment and origin on performance, behaviour, plumage condition, leg position, running ability and tibial dyschondroplasia in male fattening turkeys. https://d-nb.info/973938447/34 .

Crowe, & Forbes (1999). Effects of four different environmental enrichment treatments on pecking behaviour in turkeys. *British Poultry Science, 40 Suppl*, 11–12. https://doi.org/10.1080/00071669986558.

Denbow, D. M., Leighton, A. T., & Hulet, R. M. (1984). Behavior and growth parameters of large white turkeys as affected by floor space and beak trimming. 1. Males. *Poultry Science, 63*(1), 31–37. https://doi.org/10.3382/ps.0630031.

Dennis, M., Dong, Y., Guo, Y., Kim, Y. H. B., Erasmus, M., & Scramlin Zuelly, S. M. (2024). Environmental enrichment has minimal impact on fresh and processed meat quality of turkeys. *Meat and Muscle Biology, 8*(1). https://doi.org/10.22175/mmb.16759.

Di Martino, G., Capello, K., Russo, E., Mazzucato, M., Mulatti, P., Ferrè, N., Garbo, A., Brichese, M., Marangon, S., & Bonfanti, L. (2017a). Factors associated with pre-slaughter mortality in turkeys and end of lay hens. *Animal, 11*(12), 2295–2300. https://doi.org/10.1017/S1751731117000970.

Di Martino, G., Capello, K., Stefani, A. L., Tripepi, L., Garbo, A., Speri, M., Trolese, M., Brichese, M., Marangon, S., & Bonfanti, L. (2017b). The effect of crate height on the behavior of female turkeys during commercial pre-slaughter transportation. *Animal Science Journal, 88*(10), 1651–1657. https://doi.org/10.1111/asj.12823.

Dong, Y., Fraley, G. S., Siegford, J. M., Zhu, F., & Erasmus, M. A. (2023). Comparing different environmental enrichments for improving the welfare and walking ability of male turkeys. *PLoS One, 18*(5). https://doi.org/10.1371/journal.pone.0285347.

Duggan, G., Widowski, T., Quinton, M., & Torrey, S. (2014). The development of injurious pecking in a commercial turkey facility. *Journal of Applied Poultry Research, 23*(2), 280–290. https://doi.org/10.3382/japr.2013-00860.

Lohmann Breeders. 2021. Enrichment for broilers and turkeys - from theoretical consideration to practical application. Enrichment for broilers and turkeys - from theoretical consideration to practical application. (2021). https://lohmann-breeders.com/lohmanninfo/enrichment-for-broilers-and-turkeys-from-theoretical-consideration-to-practical-application/.

Erasmus, M. A., Lawlis, P., Duncan, I. J. H, & Widowski, T. M. (2010). Using time to insensibility and estimated time of death to evaluate a non penetrating captive bolt, cervical islocation, and blunt trauma for on-farm killing of turkeys. *Poultry Science, 89*(7), 1345–1354. https://doi.org/10.3382/ps.2009-00445.

Erasmus, M. A. (2017). A review of the effects of stocking density on turkey behavior, welfare, and productivity. *Poultry Science, 96*(8), 2540–2545. https://doi.org/10.3382/ps/pex075.

Fahd, A. M., Yousef, M., Yousef, I., Saleem, A., Fakhry, A., & Donia, G. (2025). Effect of transportation on maintenance, behavior, and performance in turkey poults during the brooding period. *Zagazig Veterinary Journal, 53*(1), 82–93. https://doi.org/10.21608/zvjz.2025.344675.1272.

Fahd, A. M. M., Saleem, A.-S. K., Ibrahim, M. Y., Fattah, A. F. A., & Donia, G. R. (2024). Impact of transportation stress on growth performance, behaviour and welfare of turkey pullets. *Advances in Animal and Veterinary Sciences, 12*(Special issue1), 353–360. https://doi.org/10.17582/journal.aavs/2024/12.s1.353.360.

Farghly, M. (2012). Evaluation of clover and corn stalks straw as alternative litter materials to wheat straw for raising local turkey. *Egyptian Journal of Animal Production, 49*(2), 161–172. https://doi.org/10.21608/ejap.2012.94332.

Fasenko, G. M., Christensen, V. L., Wineland, M. J., & Petitte, J. N. (2001). Examining the effects of prestorage incubation of turkey breeder eggs on embryonic development and hatchability of eggs stored for four or fourteen days. *Poultry Science, 80*(2), 132–138. https://doi.org/10.1093/ps/80.2.132.

Fidan, E. D., Kaya, Nazligul, & Türkyilmaz (2020). The effects of perch cooling on behavior, welfare criteria, performance, and litter quality of broilers reared at high temperatures with different litter thicknesses. *Revista Brasileira de Ciencia Avicola / Brazilian Journal of Poultry Science, 22*(3), 1–12. https://doi.org/10.1590/1806-9061-2019-1083.

Filho, J. K., Zampar, A., Boiago, M. M., & Da Silva, A. S. (2020). Effect of heat treatment on pre-stocking of turkey eggs and its impact on incubation and hatching. *Journal of Thermal Biology, 88*. https://doi.org/10.1016/j.jtherbio.2019.102501.

Fournier, J., Schwean-Lardner, K., Knezacek, T. D., Gomis, S., & Classen, H. L. (2014). The effect of toe trimming on production characteristics of heavy turkey toms. *Poultry Science, 93*(9), 2370–2374. https://doi.org/10.3382/ps.2014-04044.

Fournier, J., Schwean-Lardner, K., Knezacek Tennille, D.G., Susantha, C., Henry, L. (2012). Impact of toe trimming on productivity, carcass grading, and welfare of turkey hens raised to 15 weeks of age. Proceedings of the 24th World's Poultry Congress Salvador, Brazil .

Fournier, J., Schwean-Lardner, K., Knezacek, T. D., Gomis, S., & Classen, H. L. (2015). The effect of toe trimming on behavior, mobility, toe length and other indicators of welfare in tom turkeys. *Poultry Science, 94*(7), 1446–1453. https://doi.org/10.3382/ps/pev112.

Farghly, M F A, Mahrose, K M, Cooper, R G, Ullah, Z, Rehman, Z, & Ding, C (2018). Sustainable floor type for managing turkey production in a hot climate. *Poultry Science, 97*(11), 3884–3890. https://doi.org/10.3382/ps/pey280.

Furo, G. (2019). Bedding material and stocking density influence the performance, and the occurrence of footpad dermatitis in turkey hens. *Retrieved from the University Digital Conservancy* https://hdl.handle.net/11299/206184.

Gardner, F. A., Golan, F. A., & Denton, J. H. (1980). Microbiological parameters associated with hatching turkeys. *Poultry Science, 59*(1), 966–970. https://doi.org/10.3382/ps.0590966.

Gickel, J., Visscher, C., Kemper, N., & Spindler, B. (2025). Factors determining the dead-on-arrival (DOA) rate of turkeys transported for slaughter in Germany. *Poultry Science, 104*(4). https://doi.org/10.1016/j.psj.2025.104951.

Gildersleeve, R. P., & Boeschen, D. P. (1983). The effects of incubator carbon dioxide level on turkey hatchability. *Poultry Science, 62*(5), 779–784. https://doi.org/10.3382/ps.0620779.

Glatz, P. C. (2000). Beak trimming methods - Review. *Asian-Australasian Journal of Animal Sciences, 13*(11), 1619–1637. https://doi.org/10.5713/ajas.2000.1619.

González-Zapata, F. A., Sanginés-García, J. R., Piñero-Vázquez, A. T., Velázquez-Madrazo, P. A., Itzá-Ortíz, M. F., Bello-Pérez, E. V., Chay-Canul, A. J., & Aguilar-Urquizo, E. (2022). Performance of turkeys in enrichment environment with perches and outdoor access under tropical conditions. *Revista Brasileira de Ciencia Avicola / Brazilian Journal of Poultry Science, 24*(2). https://doi.org/10.1590/1806-9061-2021-1553.

Grimes, J. L., Carter, T. A., Gernat, A. E., & Godwin, J. L. (2007). A novel bedding material made from cotton waste, gypsum, and old newsprint for rearing turkeys. *Journal of Applied Poultry Research, 16*(4), 598–604. https://doi.org/10.3382/japr.2007-00034.

Hamilton, R. M. G., & Kennie, J. (1997). The effects of lighting program, ingredient particle size and feed form on the performance of broiler turkeys. *Canadian Journal of Animal, 77*(3), 503–508. https://doi.org/10.4141/A97-028.

Hammond, A., Buchynski, K., Shynkaruk, T., Crowe, T., & Schwean-Lardner, K. (2024a). Are turkey hens affected by light flicker? Effects on performance and health. *Poultry Science, 103*(6). https://doi.org/10.1016/j.psj.2024.103747.

Hammond, A., Buchynski, K., Shynkaruk, T., Brown, J., Crowe, T., & Schwean-Lardner, K. (2024b). Do flickering lights impact turkey hen behavior, stress, and fear? *Poultry Science, 103*(6). https://doi.org/10.1016/j.psj.2024.103699.

Henrikson, Z. A., Vermette, C. J., Schwean-Lardner, K., & Crowe, T. G. (2018). Effects of cold exposure on physiology, meat quality, and behavior of Turkey hens and toms crated at transport density. *Poultry Science, 97*(2), 347–357. https://doi.org/10.3382/ps/pex227.

Hernandez, E., James, F., Torrey, S., Widowski, T., Schwean-Lardner, K., Monteith, G., & Turner, P. V. (2019). Electroencephalographic, physiologic and behavioural responses during cervical dislocation euthanasia in turkeys. *BMC Veterinary Research, 15*(1). https://doi.org/10.1186/s12917-019-1885-x.

Hernandez, L. A., Merenda, V. R., Pairis-Garcia, M. D., Coady, M., Amaya, A., Grimes, J. L., & Pullin, A. N. (2025). Handling and loading market-ready turkeys – Focus groups exploring the perspectives of live haul crew workers in the turkey industry. *Journal of Applied Poultry Research, 34*(3). https://doi.org/10.1016/j.japr.2025.100547.

Hester, P. Y., & Kohl, H. (1989). Effect of intermittent lighting and time of hatch on large broad-breasted white turkeys 1. *Poultry Science, 68*(4), 528–538. https://doi.org/10.3382/ps.0680528.

Hinz, T., & Berk, J. (2002). Airborne contaminants in tom turkey production under enriched husbandry conditions. *Annals of Animal Science, 2*, 39–41. https://www.openagrar.de/receive/timport_mods_00041525.

Huff, G. R., Huff, W. E., Balog, J. M., & Rath, N. C. (2003). The effects of behavior and environmental enrichment on disease resistance of turkeys. *Brain, Behavior, and Immunity, 17*(5), 339–349. https://doi.org/10.1016/S0889-1591(03)00035-7.

Hybrid Turkeys. (2024). Commercial Management Guide. https://www.hybridturkeys.com/en/resources/commercial-management/.

Jacobs, L., Bourassa, D. V., Boyal, R. S., Harris, C. E., Bartenfeld Josselson, L. N., Campbell, A., Anderson, G., & Buhr, R. J. (2021). Animal welfare assessment of on-farm euthanasia methods for individual, heavy turkeys. *Poultry Science, 100*(3). https://doi.org/10.1016/j.psj.2020.11.001.

Jhetam, S., Buchynski, K., Shynkaruk, T., & Schwean-Lardner, K. (2022a). Evaluating the effects of stocking density on the behavior, health, and welfare of turkey hens to 11 weeks of age. *Poultry Science, 101*(7). https://doi.org/10.1016/j.psj.2022.101956.

Jhetam, S., Buchynski, K., Shynkaruk, T., & Schwean-Lardner, K. (2022b). Stocking density effects on turkey hen performance to 11 weeks of age. *Poultry Science, 101*(6). https://doi.org/10.1016/j.psj.2022.101874.

Jones, P. J., Tahamtani, F. M., Pedersen, I. J., Niemi, J. K., & Riber, A. B. (2020). The productivity and financial impacts of eight types of environmental enrichment for broiler chickens. *Animals, 10*(3). https://doi.org/10.3390/ani10030378.

Kosin, I. L. (1954). Length of pre-incubation storage and heredity as factors affecting the hatchability of turkey eggs. *Poultry Science, 33*(1), 24–28. https://doi.org/10.3382/ps.0330024.

Krautwald-Junghanns, M.-E., & Sirovnik, J. (2022). The influence of stocking density on behaviour, health, and production in commercial fattening turkeys - A review. *British Poultry Science, 63*(4), 434–444. https://doi.org/10.1080/00071668.2022.2050673.

Leighton, A. T., Denbow, D. M., & Hulet, R. M. (1985). Behavior and growth parameters of large white turkeys as affected by floor space and beak trimming. II. Females. *Poultry Science, 64*(3), 440–446. https://doi.org/10.3382/ps.0640440.

Leis, M. L., Dodd, M.-M. U., Starrak, G., Vermette, C. J., Gomis, S., Bauer, B. S., Sandmeyer, L.S., Schwean-Lardner, K., Classen, H. L., & Grahn, B. H. (2017). Effect of prolonged photoperiod on ocular tissues of domestic turkeys. *Veterinary Ophthalmology, 20*(3), 232–241. https://doi.org/10.1111/vop.12395.

Letzguß, H., & Bessei, W. (2009). Welfare of meat producing birds Poultry Welfare Symposium Cervia, Italy. Effects of environmental enrichment on the locomotor activity of turkeys. https://www.cabi.org/Uploads/animal-science/worlds-poultry-science-association/WPSA-italy-2009/39_welfare2009_letzguss.pdf.

Levenick, C. K., & Leighton, A. T. (1988). Effects of photoperiod and filtered light on growth, reproduction, and mating behavior of turkeys. 1. Growth performance of two lines of males and females. *Poultry Science, 67*(11), 1505–1513. https://doi.org/10.3382/ps.0671505.

Lewis, P. D., Perry, G. C., Sherwin, C. M., & Moinard, C. (2000). Effect of ultraviolet radiation on the performance of intact male turkeys. *Poultry Science, 79*(6), 850–855. https://doi.org/10.1093/ps/79.6.850.

Lilburn, M. S., Renner, P. A., & Anthony, N. B. (1992). Interaction between step-up versus step-down lighting from four to sixteen weeks on growth and development in turkey hens from two commercial strains. *Poultry Science, 71*(3), 419–426. https://doi.org/10.3382/ps.0710419.

Lindenwald, Schuberth, Spindler, & Rautenschlein (2021). Influence of environmental enrichment on circulating white blood cell counts and behavior of female turkeys. *Poultry Science, 100*(9). https://doi.org/10.1016/j.psj.2021.101360.

Lüning, J., Wunderl, D., Rautenschlein, S., & Campe, A. (2023). Histomonosis in German turkey flocks: Possible ways of pathogen introduction. *Avian Pathology Journal of the W.V.P.A, 52*(3), 199–208. https://doi.org/10.1080/03079457.2023.2191833.

Machovcova, Z., Vecerek, V., Voslarova, E., Malena, M., Conte, F., Bedanova, I., & Vecerkova, L. (2017). Pre-slaughter mortality among turkeys related to their transport. *Animal science journal=Nihon chikusan Gakkaiho, 88*(4), 705–711. https://doi.org/10.1111/asj.12692.

Marsden, S. J., Lucas, L. M., & Wilson, S. P. (1966). The influence of daylength and environment on reproduction, broodiness, and mortality of turkeys. *Poultry Science, 45*(4), 668–675. https://doi.org/10.1080/00071669987403.

Martrenchar, A., Huonnic, D., Cotte, J. P., Boilletot, E., & Morisse, J. P. (1999). Influence of stocking density on behavioural, health and productivity traits of turkeys in large flocks. *British Poultry Science, 40*(3), 323–331. https://doi.org/10.1080/00071669987403.

Martrenchar, A., Huonnig, D., & Cotte, J. P. (2001). Influence of environmental enrichment on injurious pecking and perching behaviour in young turkeys. *British Poultry Science, 42*(2), 161–170. https://doi.org/10.1080/00071660120048393.

Martyniuk, K., Hanuszewska, M., & Lewczuk, B. (2020). Metabolism of melatonin synthesis-related indoles in the turkey pineal organ and its modification by monochromatic light. *International Journal of Molecular Sciences, 21*(24), 1–21. https://doi.org/10.3390/ijms21249750.

Mayne, R. K., Else, R. W., & Hocking, P. M. (2007). High litter moisture alone is sufficient to cause footpad dermatitis in growing turkeys. *British Poultry Science, 48*(5), 538–545. https://doi.org/10.1080/00071660701573045.

McEwen, S. A., & Barbut, S. (1992). Survey of turkey downgrading at slaughter: Carcass defects and associations with transport, toenail trimming, and type of bird. *Poultry Science, 71*(7), 1107–1115. https://doi.org/10.3382/ps.0711107.

Michel, V., & Boilletot, E. (2003). Influence of environmental enrichment on the behaviour, body state and zootechnical performances of turkey broilers. *Sciences & Techniques, 3*, 18–23.

Mohammed, H., Ibrahim, M., & Saleem, A.-S. (2016). Effect of different light intensities on performance, welfare and behavior of Turkey poults. *Journal of Advanced Veterinary and Animal Research, 3*(1), 18–23. https://doi.org/10.5455/javar.2016.c126.

Moinard, Lewis, Perry, & Sherwin (2001). The effects of light intensity and light source on injuries due to pecking of male domestic Turkeys (*Meleagris gallopavo*). *Animal Welfare, 10*(2), 131–139. https://doi.org/10.1017/S0962728600023800.

Moinard, & Sherwin (1999). Turkeys prefer fluorescent light with supplementary ultraviolet radiation. *Applied Animal Behaviour Science, 64*(4), 261–267. https://doi.org/10.1016/S0168-1591(99)00043-X.

Munson, B. R., Buchynski, K., Shynkaruk, T., Adler, C., Crowe, T., & Schwean-Lardner, K. (2024a). The effects of simulated transportation conditions on 12-week-old turkey hen and 16-week-old turkey tom meat quality. Poultry Science Association Annual Meeting, Louisville, Kentucky, USA.

Munson, B. R., Buchynski, K., Shynkaruk, T., Adler, C., Crowe, T., & Schwean-Lardner, K. (2024b). The effects of simulated transportation conditions on 12-week-old turkey hen and 16-week-old turkey tom physiology and behavior. Poultry Science Association Annual Meeting, Louisville, Kentucky, USA.

Noll, S. L., Janni, K. A., Halvorson, D. A., & Clanton, C. J. (1997). Market turkey performance, air quality, and energy consumption affected by partial slotted flooring. *Poultry Science, 76*(2), 271–279. https://doi.org/10.1093/ps/76.2.271.

Owens, C. M., & Sams, A. R. (2000). The influence of transportation on turkey meat quality. *Poultry Science, 79*(8), 1204–1207. https://doi.org/10.1093/ps/79.8.1204.

Petracci, M., Bianchi, M., Cavani, C., Gaspari, P., & Lavazza, A. (2006). Preslaughter mortality in broiler chickens, turkeys, and spent hens under commercial slaughtering. *Poultry Science, 85*(9), 1660–1664. https://doi.org/10.1093/ps/85.9.1660.

Prescott, Berry, Haslam, & Tinker (2000). Catching and crating turkeys: Effects on carcass damage, heart rate, and other welfare parameters. *Journal of Applied Poultry Research, 9*(3), 424–432. https://doi.org/10.1093/japr/9.3.424.

Prusik, M., & Lewczuk, B. (2020). Diurnal rhythm of plasma melatonin concentration in the domestic turkey and its regulation by light and endogenous oscillators. *Animals, 10*(4). https://doi.org/10.3390/ani10040678.

Raabe, J., Raveendran, G., Otten, W., Homeyer, K., & Bartels, T. (2024). Research note: Irritating flashing light or poultry-friendly lighting-are flicker frequencies of LED luminaires a potential stress factor in the husbandry of male fattening turkeys? *Poultry Science, 103*(1). https://doi.org/10.1016/j.psj.2023.103214.

Scanes, C. G., Hurst, K., Thaxton, Y., Archer, G. S., & Johnson, A. (2020). Effect of transportation and shackling on plasma concentrations of corticosterone and heterophil to lymphocyte ratios in market weight male turkeys in a commercial operation. *Poultry Science, 99*(1), 546–554. https://doi.org/10.3382/ps/pez485.

Schwean-Lardner, K., Vermette, C., Leis, M., & Classen, H. L. (2016). Basing Turkey lighting programs on broiler research: A good idea? A comparison of 18 daylength effects on broiler and Turkey welfare. *Animals, 6*(5). https://doi.org/10.3390/ani6050027.

Schwean-Lardner, K. (2025). A study of the behavioural output differences in turkeys vs broilers response to photoperiod. Poultry Science Association Annual Meeting Raleigh, North Carolina, USA., p 149. https://higherlogicdownload.s3.amazonaws.com/POULTRYSCIENCE/d925b046-3667-43aa-aeb6-5a36c497c07a/UploadedImages/2025_PSA_Annual_Meeting_Abstract_Book_FINAL.pdf.

Sherwin, C. M. (1998). Light intensity preferences of domestic male turkeys. *Applied Animal Behaviour Science, 58*(1-2), 121–130.

Sherwin, C. M., & Devereux, C. L. (1999). Preliminary investigations of ultraviolet-induced markings on domestic turkey chicks and a possible role in injurious pecking. *British Poultry Science, 40*(4), 429–433. https://doi.org/10.1080/00071669987151.

Sherwin, C. M., Lewis, P. D., & Perry, G. C. (1999a). Effects of environmental enrichment, fluorescent and intermittent lighting on injurious pecking amongst male turkey poults. *British Poultry Science, 40*(5), 592–598. https://doi.org/10.1080/00071669986954.

Sherwin, C. M., Lewis, P. D., & Perry, G. C. (1999b). The effects of environmental enrichment and intermittent lighting on the behaviour and welfare of male domestic turkeys. *Applied Animal Behaviour Science, 62*(4), 319–333. https://doi.org/10.1016/S0168-1591(98)00215-9.

Siopes, T. D., Timmons, M. B., Baughman, G. R., & Parkhurst, C. R. (1983). The effect of light intensity on the growth performance of male turkeys. *Poultry Science, 62*(12), 2336–2342. https://doi.org/10.3382/ps.0622336.

Siopes, T. D., Baughman, G. R., Parkhurst, C. R., & Timmons, M. B. (1989). Relationship between duration and intensity of environmental light on the growth performance of male turkeys. *Poultry Science, 68*(11), 1428–1435. https://doi.org/10.3382/ps.0681428.

Siopes, T. D., Parkhurst, C. R., & Baughman, G. R. (1986). Intermittent light and growth performance of male turkeys from 2 to 22 weeks of age. *Poultry Science, 65*(12), 2221–2225. https://doi.org/10.3382/ps.0652221.

Spindler, B., & Hartung, J. (2009). Influence of environmental enrichment on the behaviour of female Big 6 turkeys reared on an ecological farm. Vechta, Germany Proceedings of the 14th International Congress of the International Society for Animal Hygiene (ISAH).

Stiewert, A. M., Wooming, B., & Archer, G. S. (2021). Comparing various euthanasia devices and methods on 8 and 12-week-old turkey hens. *Poultry Science, 100*(5). https://doi.org/10.1016/j.psj.2021.101053.

Struthers, S., Fiss, T., Classen, H. L., Gomis, S., Dickinson, T., Crowe, T., Herwig, E., & Schwean-Lardner, K. (2022a). The effect of infrared beak treatment on the welfare of turkeys reared to 12 weeks of age. *Poultry Science, 101*(4). https://doi.org/10.1016/j.psj.2022.101728.

Struthers, S., Fiss, T., Classen, H. L., Gomis, S., Herwig, E., & Schwean-Lardner, K. (2022b). The impact of infrared beak treatment on turkey tom and hen beak length and performance to 12 weeks of age. *Poultry Science, 101*(4). https://doi.org/10.1016/j.psj.2022.101737.

Van de Weerd, H. A., & Day, J. E. L. (2009). A review of environmental enrichment for pigs housed in intensive housing systems. *Applied Animal Behaviour Science, 116*(1), 1–20. https://doi.org/10.1016/j.applanim.2008.08.001.

Vermette, Schwean-Lardner, Gomis, Grahn, Crowe, & Classen (2016b). The impact of graded levels of day length on turkey health and behavior to 18 weeks of age. *Poultry Science, 95*(6), 1223–1237. https://doi.org/10.3382/ps/pew078.

Vermette, Schwean-Lardner, Gomis, Crowe, & Classen (2016a). The impact of graded levels of daylength on turkey productivity to eighteen weeks of age. *Poultry Science, 95*(5), 985–996. https://doi.org/10.3382/ps/pew060.

Vermette, C. J., Henrikson, Z. A., Schwean-Lardner, K. V., & Crowe, T. G. (2017). Influence of hot exposure on 12-week-old turkey hen physiology, welfare, and meat quality and 16-week-old turkey tom core body temperature when crated at transport density. *Poultry Science, 96*(11), 3836–3843. https://doi.org/10.3382/ps/pex220.

Voslářová, E., Janáčková, Rubešová, Kozák, Bedáňová, Steinhauser, & Večerek (2007). Mortality rates in poultry species and categories during transport for slaughter. *Acta Veterinaria Brno, 76*(SUPPL. 8). https://doi.org/10.2754/avb200776S8S101.

Wichman, A., Norring, M., Pastell, M., Algers, B., Pösö, R., Valros, A., Saloniemi, H., & Hänninen, L. (2010). Effect of crate height during short-term confinement on the welfare and behaviour of turkeys. *Applied Animal Behaviour Science, 126*(3-4), 134–139. https://doi.org/10.1016/j.applanim.2010.06.003.

Wichman, A., Norring, M., Voutila, L., Pastell, M., Valros, A., Algers, B., & Hänninen, L. (2012). Influence of crate height during slaughter transport on the welfare of male turkeys. *British Poultry Science, 53*(4), 414–420. https://doi.org/10.1080/00071668.2012.711465.

Woolcott, C. R., Torrey, S., Turner, P. V., Serpa, L., Schwean-Lardner, K., & Widowski, T. M. (2018). Evaluation of two models of non-penetrating captive bolt devices for on-farm euthanasia of Turkeys. *Animals, 8*(3). https://doi.org/10.3390/ani8030042.

Youssef, I. M. I., Beineke, A., Rohn, K., & Kamphues, J. (2010). Experimental Study on effects of litter material and its quality on foot pad dermatitis in growing turkeys. *International Journal of Poultry Science, 9*(12), 1125–1135. https://doi.org/10.3923/ijps.2010.1125.1135.

Youssef, I. M. I., Beineke, A., Rohn, K., & Kamphues, J (2011). Effects of litter quality (moisture, ammonia, uric acid) on development and severity of foot pad dermatitis in growing turkeys. *Avian Diseases, 55*(1), 51–58. https://doi.org/10.1637/9495-081010-reg.1.

Zawilska, J. B., Lorenc, A., Berezínska, M., Vivien-Roels, B., Pévet, P., & Skene, D. J. (2007). Photoperiod-dependent changes in melatonin synthesis in the turkey pineal gland and retina. *Poultry Science, 86*(7), 1397–1405. https://doi.org/10.1093/ps/86.7.1397.

Chapter 8

Dual-purpose and specialty poultry breeds

Inga Tiemann[1] and Stefano Paolo Marelli[2]

[1]*Precision Livestock Farming, Faculty of Agricultural Sciences and Landscape Architecture, Osnabrueck University of Applied Sciences, Osnabrueck, Germany,* [2]*Department of Veterinary Medicine and Animal Science DIVAS, University of Milan, Lodi, Italy*

Introduction

Poultry is considered one of the most sustainable forms of animal-based food production. This sustainability is, in part, due to commercial poultry breeding and production, in which genetically selected domesticated lines convert feed into food (meat and eggs) with high efficiency (Vaarst et al., 2015). Before commercial poultry farming arose, poultry production was based on the typical use of roosters and hens. Increasingly in today's world, where ethical, moral, sustainable, and social attitudes toward food significantly influence consumer behavior, specialized breeds and dual-purpose breeds are being rediscovered and gaining interest (Escobedo del Bosque et al., 2021). These so-called alternative (compared to commercial lines) genetics may be attributed with characteristics that take on a whole new meaning in the context of resilience and robustness, in terms of animal welfare and climate adaptation. Coupled with desirable diversification, both in the biology of the genetics used and in product variety, these alternative genetics are certainly of interest in the production landscape (Yang & Jiang, 2005). However, if sustainability is translated as feed conversion efficiency, specialized breeds and dual-purpose chickens will have a hard time. They often eat more without laying more eggs or putting on more meat (Giersberg & Kemper, 2018). Although efforts are being made in this context to feed animals in a more appropriate manner, with less energy-dense feed and thus more cost-effectively, they still do not fully achieve the advantage of efficient high performance that is seen in standard commercial lines. If sustainability also means improved animal welfare, greater longevity and lower mortality, climate resilience, and even consumer confidence, then expectations of specialized breeds and dual-purpose breeds are high. The description that these breeds are intended to turn back the clock

Integrating Welfare and Sustainability in Poultry Systems. DOI: https://doi.org/10.1016/B978-0-443-40594-5.00015-8

is only partially accurate. Many of these breeds originated in the days before commercial poultry farming, and most have since been bred little or not at all for performance traits and thus for food production (Becker et al., 2023). But some of them have retained characteristics, such as their genetic diversity, that are taking on new significance in today's changing world, increasing the potential value of these birds (Groeneveld et al., 2010).

The focus of commercial poultry farming on either egg production or meat production has resulted in two different types of use. One type of use, meat production, involves fattening both sexes. Males and females are both ready for slaughter within a few weeks. In the other area of use, egg production, males of the modern laying hen are superfluous because they do not lay eggs. Additionally, their physiology is so one-sided that, like their sisters, they do not have any meat that can be fattened (Giersberg & Kemper, 2018). That is why, for decades, these male animals were killed immediately after hatching (van Niekerk & Workamp, 2022). However, this practice is increasingly becoming a topic of social and ethical debate, resulting in the first countries in Europe banning the killing of day-old chicks by law (Krautwald-Junghanns et al., 2018). Some countries still allow culled chicks to be sold as pet food, to zoos and bird sanctuaries (van Niekerk & Workamp, 2022). The integration of roosters and hens into human food production as dual-purpose chickens that produce both meat and eggs is an example of how changing consumer behavior is leading to the use of alternatives for protein supply.

Alternative genetics might offer other approaches for poultry production, also referred to as heritage, native, traditional, indigenious, or local breeds, hereafter referred to as local breeds. All of these breeds are adapted to a local environment, and some might be endangered as these alternatives are rarely economically relevant at present. There are a few pioneering examples, such as Bresse in France (Lambertz et al., 2018), Three Yellow in China (Yang & Jiang, 2005), Lohmann Dual (LD) in Switzerland (Gangnat et al., 2020), local crossbreeds in the RegioHuhn project (Siebenmorgen et al., 2024), and very active breeding lines in organic farming (Werner et al., 2023). All these origins and approaches share a niche market—a market that works best in direct marketing with the products being sold directly off-farm (Siebenmorgen et al., 2026), in local value chains, or because of a commitment in the food trade. However, together they contribute more to organic and genetic diversity in poultry production than the efficient high-performance lines, of which there are only a handful in two large global companies.

The diversity of local breeds also means that knowledge about animal welfare and sustainability is rarely researched in detail. The focus is more on performance parameters to achieve comparability with high-performance genetics. However, since the arguments for using these local breeds are rarely based on performance, scientists are called upon to integrate animal welfare and sustainability into their research projects. Given the increasing potential value of these birds, there is a growing body of literature on local breeds and

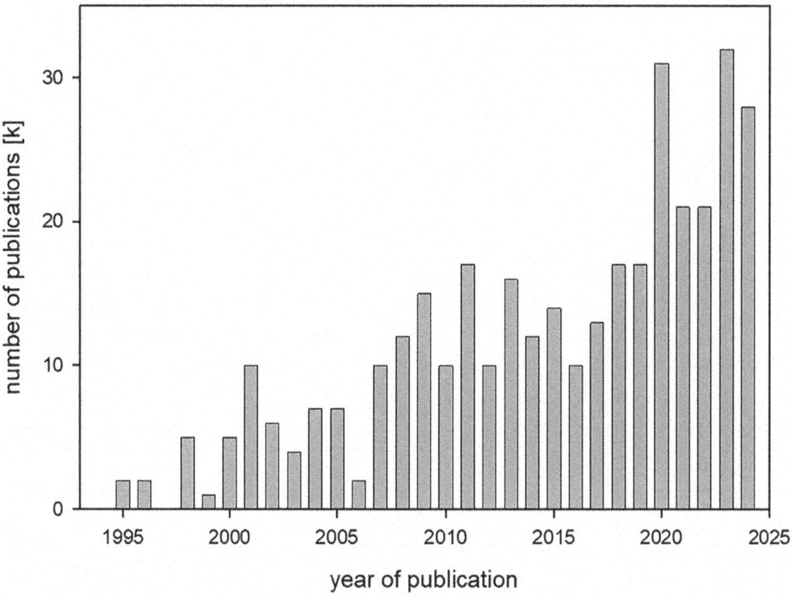

FIGURE 8.1 Number of publications per year.

dual-purpose breeds (TS = domestic* chick* AND breed OR native OR local OR endangered OR indigenous OR special, 05.09.2025 in webofscience.com, Fig. 8.1).

Classification and typology of local poultry breeds

Unlike the large groups of broilers and laying hens, local breeds can only be roughly and partially distinguished from one another in continuous steps, depending on the criteria applied. One of the most common keywords in this context is dual-purpose. However, there is no such thing as *the* dual-purpose chicken. Rather, we distinguish between dual-purpose breeds that are primarily used for laying or fattening and their origin, that is, whether they are specific to a region or commercial hybrids. The global Food and Agriculture Organization (FAO) estimates that there are approximately 1600 different chicken breeds worldwide (FAO Gateway to poultry production & products, 2026), but it is probably even more. These breeds, most of which are locally adapted, have one advantage: their adaptability. In times of climate change and scarce resources, local breeds could make a definitive contribution to resilient agriculture—provided that their production also allows the producing farmer to survive in economic terms. The genetics are bred either privately, coordinated through associations and initiatives, or anchored in industry. However, this origin determines whether the animals are available as an open source or part of an economic chain. Both approaches have advantages and disadvantages and are just as much a part of a selection strategy as adapting the animals to the prevailing ecological niche in the local area. DAD-IS is a section of the Domestic Animal Diversity Information System provided by FAO (Domestic Animal Diversity Information System DAD-IS, 2026); it is a data analysis and visualization tool that deals with domestic animal breeds, productive traits, population consistency, and related data. It reports the extinction risk of every listed breed. Data about production, reproduction, and adaptability are reported, divided by breed within every species. Data about the country of origin and development, production system, and productive attitude are also reported. DAD-IS is an effective tool supporting biodiversity conservation, small breeds management, and choice, and the development of appropriate sustainable animal production strategies.

The categorization into pure breeds, crossbreeds, and commercial origins reflects a division into the main large groups.

Purebred animals are defined by the fact that they can reproduce within their population, and the next generation exhibits a comparable phenotype. These breeds are typically adapted to their local environment and often originated in the region that gave them their name or were bred to "flourish" there. For example, today's laying hybrids originated from Mediterranean chickens, the most prominent of which are called Italian or Leghorn. Even today, this breed is kept in countless color varieties worldwide, but for shows rather than agricultural use. In this context, other breeds appear, for example, the French Bresse Gauloise, the Chinese silkie chickens, the German Ramelsloher, the British Ixworth (Fig. 8.2; Fig. 8.3), the Swiss chicken, the Hungarian naked neck chickens, and the Italian Paduan. Each of these breeds is adapted to its "ecological niche." This also applies to Indian and African chickens, as well as the Horro chicken from Ethiopia (Dana et al., 2011).

The Bresse breed has its roots in local fancy chicken breeds and develops as a premium meat product, marketed as a delicacy with a recognized labeling as Protected Designation of Origin. With blue legs, white plumage, and red combs, Bresse chickens correspond to the French national colors. The history of this breed started with official recognition of the production area in 1936. and the protection of the definition "Volaille de Bresse" traces back to 1957 laws. They are marketed as a dual-purpose breed and have pronounced exploratory behavior, which is reflected in the production landscape by an equally pronounced use of free-range farming. Strict production protocols are applied based on animals' growth status, defining feed formula, birds' density, housing system, and open-air access. Slaughtering and carcass processing must follow specific procedures, too. Meat production requires a slaughter age ranging between 112 and 150 days of age; the minimum required live weight at slaughter is 1.5 kg for male chickens. Minimum carcass weight is defined as 1.2 kg for males, 1.8 kg for females, and 3.0 kg for capons. Bresse Gauloise birds share the same genetic and production criteria as Bresse birds but are

FIGURE 8.2 Hens of the breed British Ixworth. British Ixworth is a purebred dual-purpose breed. *From Senta Becker, Nideggen, Germany.*

raised outside the Bresse region of France (Verrier et al., 2005). Bresse are also known for their comparatively good laying performance, with hens producing up to 240–250 eggs per year and cocks reaching around 2.5 kg at 16 weeks of age (Lambertz et al., 2018; Nolte et al., 2020a, Nolte et al., 2020b). Egg weight ranges from 50 g in the first weeks of lay to 68 g in the last weeks of lay. Studies show that Bresse chickens and their crossbreeds perform well in alternative production systems aiming to increase sustainability by, for example, using regional protein sources like faba beans instead of imported soy, without adverse effects on growth, health, or product quality (Escobedo del Bosque et al., 2022). Bresse chickens also demonstrate unique characteristics, with high carcass yields, good meat quality, and adaptability to free-range and organic systems (Baldinger & Bussemas, 2021). Crossbreeding Bresse with commercial layers can further enhance egg production and bone strength, making them a promising option for ethical and niche market poultry production (Nolte et al., 2020a). Overall, Bresse chickens combine tradition, quality, and sustainability, making them a model breed for modern dual-purpose poultry farming.

Silkie chickens are a popular breed worldwide with a characteristic fluffy plumage, used as broody hens or pets (Fig. 8.4, Fig. 8.5). They are Chinese slow-growing birds with an average live weight at 59 days of around 520 g. The adult weight of Silkie males ranges between 1000 and 1500 g, and that of Silkie females ranges between 500–1750 g. Carcass yield averages around 394 g (76%), and carcass breast yield is 10.7% (Choo et al., 2014). Eggshell color is defined (European breed standard) to be light brown with an average egg weight of 40 g. This breed is prized not only for its ornamental qualities

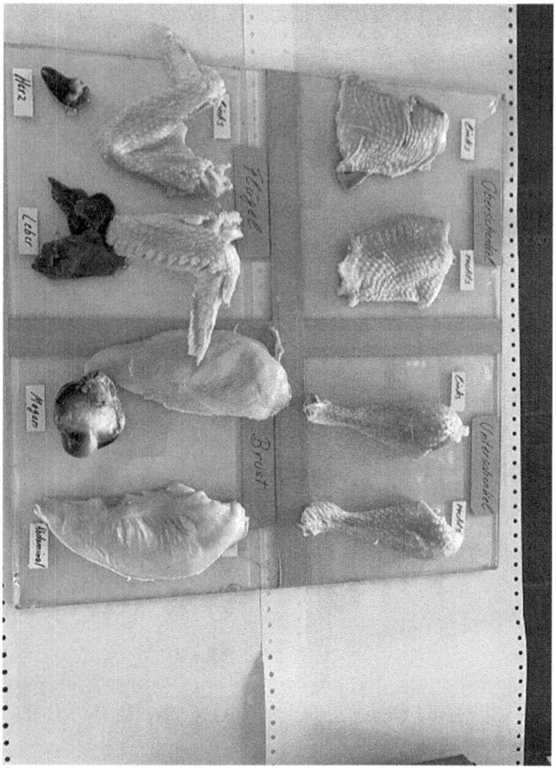

FIGURE 8.3 **Meat quality assessment in British Ixworth.** Valuable parts of the carcass of a British Ixworth cockerel after 10 weeks of life.
From Senta Becker, Nideggen, Germany.

but also for its culinary and medicinal significance. Silkie chickens are consumed as a delicacy in China, with their unique-tasting carcass traditionally used in medicinal cuisine (Zhu et al., 2025). Their skin is black, due to their striking internal pigmentation caused by a genetic mutation (Choo et al., 2014; Han et al., 2015; Prakash et al., 2023). Due to fibromelanosis, a hyperpigmentation from the overexpression of the EDN3 gene, flesh, bones, and connective tissues in this breed are dark colored too. Their meat is rich in amino acids, trace elements, unsaturated fatty acids, and carnosine, contributing to its reputation for health benefits such as immune enhancement and antiaging properties (Liao et al., 2022; Xiong et al., 2024). Reproductive traits in Silkie chickens are characterized by pronounced broodiness and relatively short egg-laying periods, with recent genomic and transcriptomic studies identifying key genes and regulatory networks involved in egg production and brooding behavior (Tan et al., 2024). Overall, Silkie chickens represent a valuable genetic resource with unique biological, nutritional, and cultural

FIGURE 8.4 **Rooster of the Silkie Chicken breed.** Silkie chickens are typically kept as fancy fowl in Europe, but used for meat production in Asia.
From Sabine Gebhardt-Henrich, Switzerland.

attributes, making them a subject of ongoing research in genetics, nutrition, and poultry science (Huang et al., 2024).

More often, however, initiatives and active farmers are attempting to reintroduce local breeds into agricultural use. This applies, for example, to the Italian Paduan and other local breeds. The Padovana breed is a local Italian breed that originated in the region of Venice, Northern Italy. The breed has been present since 1500 CE in the area of Padua, the city that the breed was named after. Padovana is a crested breed characterized by a cerebral hernia with a big tuft (Verdiglione & Rizzi, 2018) and presents with different plumage colors. Meat production has been investigated as an important productive trait where Chamois Padovana males showed an average daily weight gain of 8.13 g/d as recorded at 3 days of age, growing to 15.6 g/d at 15–18 weeks of age. Age at slaughter was determined to be 1744 g at 18 weeks of age. The dressing percentage was calculated to be 74.6% with a breast proportion of 17.6% (Rizzi, 2019). However, Padovana is traditionally mainly oriented to egg production, with a white eggshell color. A study conducted on Chamois Padovana hens using a laying curve between 39 and 50 weeks of age revealed an average whole egg weight of 53.4 g; Golden Padovana hens of the

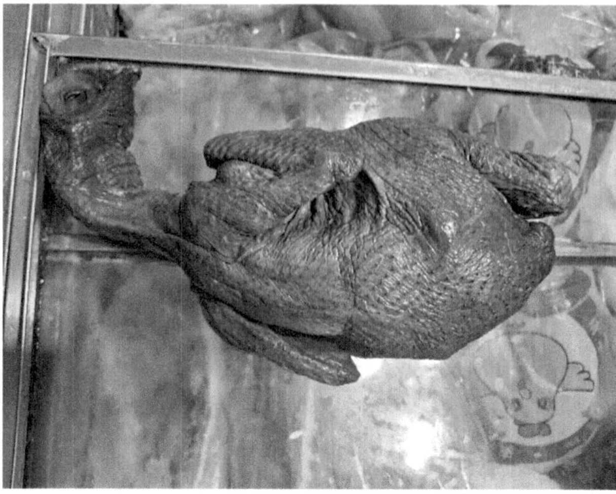

FIGURE 8.5 **Carcass of a Silkie chicken.** Silkie chickens are used for meat production in Hong Kong, China.
From Inga Tiemann, Neuss, Germany.

same age laid a slightly heavier egg (55.8 g). The average albumen ratio was 58%, and the average yolk ratio was 32.6%. Padovana production traits are similar to those described in other traditional heritage breeds (Rizzi, Cendron, et al., 2023). A focal point is the impact of the albumen ratio on the edible part of the egg: hybrid hens have a higher percentage of albumen, which is a trait under genetic selection to increase the egg weight and improve market gain.

These examples of some local breeds illustrate their priceless biological value in animal production, where their genetic uniqueness strongly influences the added economic value of their product. Their high adaptability to open and marginal environments, their behavioral plasticity and resilience, and their link to tradition and culture must be considered exclusive traits to be preserved and promoted. Although quantitative information on these added values is more or less completely absent so far when compared to commercial breeding programs (Cheng, 2010). Furthermore, their genomic landscape can represent a valuable reservoir of unique traits. However, often, the level of performance is so low that ad hoc agricultural use does not appear possible. In such cases, one solution is to cross local breeds with high-performance breeds. Crossbreeding significantly increases performance, and using local breeds as one (often the male) parent integrates them into the value chain.

Crossbreeding, that is, the targeted crossing of commercial parent lines with local chicken breeds, is used worldwide to improve the productivity, adaptability, resilience, and profitability of poultry farming. These crossbreeds can achieve positive heterosis effects (hybrid advantages), which are reflected in increased laying performance, improved growth, and higher egg production compared to the local breed (Alene et al., 2024; Dzungwe et al., 2024). Studies show that crossbred offspring often achieve higher laying persistence and greater egg weight than the local parent breeds and, in some cases, can even outperform commercial lines for certain traits (Mokoena et al., 2025). At the same time, the adaptability to local environmental conditions and the robustness of local breeds are partially preserved, which is particularly advantageous for alternative forms of husbandry and sustainable production (Menchetti et al., 2024). Overall, crossbreeding offers an effective strategy for increasing the genetic performance of local chicken populations while preserving their valuable adaptive traits (Tudorache et al., 2022).

Modern commercial dual-purpose hybrids are a result of the debate surrounding the killing of male day-old chicks from laying lines. Even though there are at least two suppliers of commercial dual-purpose hybrids, they are struggling to gain a foothold in the market (I. Tiemann, Pers. comm.). Profit margins in the food chain are generally too low for farmers to be able to afford to produce an animal that eats more, is ready for slaughter later, and lays fewer eggs. In a closed system, for example, in Switzerland, however, this is possible with the help of food retailers.

LD (Lohmann Dual) is a modern dual-purpose hybrid line developed to avoid the killing of male day-old chicks from laying hybrids. LD combines moderate laying performance of hens with sufficient fattening performance of roosters, thus representing an ethical and animal welfare-oriented alternative to specialized laying and broiler hybrids. Compared to intensive broiler hybrids, LD roosters grow more slowly and reach lower final weights (Damme et al., 2015). In terms of behavior, LD hens show a significantly lower tendency for feather pecking and cannibalism, as well as lower fear responses than conventional laying hybrids, which improves animal welfare (Giersberg et al., 2020; Rieke et al., 2021). The laying performance and persistence of LD hens are lower than that of specialized laying hybrids, but feed conversion and eggshell quality are comparable (Fig. 8.6; Fig. 8.7). Meat characteristics differ from broilers, with LD showing higher leg weights and differences in meat color (Siekmann et al., 2018). The fattening performance of roosters is sufficient to enable economic use, especially if higher market prices can be achieved for products with added animal welfare value (Gangnat et al., 2020; Meuser et al., 2021). Overall, the LD offers a practical solution for greater animal welfare and sustainability in poultry farming, even if it cannot compete with specialized hybrids in terms of performance.

However, there are also settings in which the focus is not on the specific food, but rather on the specific adaptability. In times of climate change, the physiological adaptability of chickens is also of particular interest (Lin et al., 2006). One example of this is naked neck chickens, which are better able to cope with thermal challenges than other breeds. This means that, over the course of their breeding history, they have been optimally adapted to their environment, in this case, the Hungarian Puszta and its hot summers.

FIGURE 8.6 **Lohmann Dual (LD) hens.** LD is a dual-purpose hybrid still under development. *From Animalco, Switzerland.*

Other studies also show that Chinese Lingshan chickens have an advantage over modern broilers in terms of heat tolerance (Xu et al., 2018). Another study confirms that Brazilian Peloco and Caneluda have greater heat tolerance (Cedraz et al., 2017). The challenges in the Global South are quite different. Availability, disease resistance, and hatchability are goals based on modern high-performance genetics, but in a completely different setting (Abdelqader et al., 2007). Smallholders are to be empowered to provide for themselves and their entire village communities. There are special projects here that support women in rural areas, in particular, through chicken breeds.

Smallholder situations that have limited infrastructure need highly adaptable and resilient animals that are able to grow and produce in marginal areas, often characterized by harsh climatic conditions. In these contexts, local breeds with their low body weight, high adaptability, and high broodiness represent the historical choice for family animal production. In addition, dual-purpose breeds could be considered a valuable alternative with higher productive potential in small production contexts. Hybrid vigor and access to genomic selection procedures have improved dual-purpose breeds, which are now spreading in Asia and Africa. Dual-purpose breeds could improve animal protein production and food security at the same time with lower input requirements compared to high-performance hybrids. Family poultry production, mainly based on women's work, can be considered a basic component of improving nutritional and economic conditions where resources and infrastructures are limited (Kebede et al., 2024).

FIGURE 8.7 **Eggs of Lohmann Dual (LD) hens.** Eggs of LD hens are showing an intermediate, slightly variable brown eggshell color.
From Animalco, Switzerland.

Functional traits and performance profiles

Local and dual-purpose breeds will never lay eggs or produce meat as efficiently as high-performance hybrids. However, they have other qualities that may be of interest for agricultural production, in addition to keep these breeds for pleasure only.

The laying performance of dual-purpose chickens and local breeds is moderate compared to specialized laying hybrids, but offers advantages in terms of genetic diversity and possibly also in terms of animal welfare and sustainability. Studies show that local breeds such as the Branca chicken or Italian dual-purpose breeds (e.g., Robusta maculata), as well as crosses between local and higher-performance lines, typically lay between 100 and 250 eggs per year, with performance depending heavily on genotype, husbandry, and feeding (Lordelo et al., 2020; Manyelo et al., 2020). Nevertheless, their eggs may be enough for self-sufficiency, niche markets, and direct marketing (Bamidele et al., 2020; Becker et al., 2023). While commercial hybrids usually produce larger eggs and more eggs, local breeds often achieve comparable egg quality and are better adapted to regional environmental conditions (Lordelo et al., 2020). They also contribute to the conservation of genetic resources and enable more sustainable, animal welfare-oriented production systems.

Across local and dual-purpose breeds, the lifetime performance in terms of egg numbers remains unclear. The hens can live for several years, with a gradual reduction in laying performance and usually an increase in egg size. However, if feed costs and maintenance costs are calculated per egg, farmers face considerable economic challenges. These can only be solved if self-sufficiency is the priority, a special niche is filled (e.g., shell color), or, in most cases, if the products are sold through direct marketing.

The shell color of eggs from dual-purpose chickens and local breeds is genetically and phenotypically very diverse. In local breeds such as Italian landraces, breed-specific genes play an important role in pigment synthesis and thus in the expression of shell color, which is relevant for both breed identification and ecological adaptations (Perini et al., 2024). In Chinese breeding programs, Dongxiang blue eggshell and Jiangshan black-bone lines were crossed to create lines that combine both blue eggshells and dark skin characteristics—an example of the targeted combination of shell color and other breeding goals in dual-purpose chickens (Wang et al., 2021). The shell color ranges from white in local and dual-purpose breeds to chocolate brown in Marans and blue and green in Araucana, with genetic diversity often greater than in commercial hybrids (Lordelo et al., 2020; Zhao et al., 2006). Research shows that shell color is not only an esthetic feature, but also important for marketing, consumer perception, and the conservation of genetic resources. Modern molecular methods increasingly enable targeted selection for specific shell colors, with the conservation of genetic diversity in local breeds continuing to play a central role (Li et al., 2022; Perini et al., 2024; Wang et al., 2023).

The egg quality of dual-purpose chickens and local breeds varies significantly depending on genotype. Local and dual-purpose breeds usually lay fewer eggs than high-performance hybrids, but show greater diversity in egg size, shell color, egg weight, yolk-to-albumen ratio, and shell thickness.

Dual-purpose breeds such as Ermellinata di Rovigo and Robusta Maculata produce medium-sized eggs with stable shell and albumen quality, while local breeds often have a higher yolk content and more variability in egg quality (Rizzi et al., 2023). Eggs from local breeds often contain more yolk but are smaller and have thicker shells. Blood and meat spots occur more frequently in tinted eggs than in white eggs. Overall, local and dual-purpose breeds offer a wide range of egg qualities that are attractive for biodiversity conservation and niche markets (Hammershøj et al., 2021; Rizzi, 2023).

The fattening performance of dual-purpose and local chicken breeds is significantly lower than that of specialized fattening hybrids. Dual-purpose hybrids such as LD reach a live weight of around 2 kg after just 9 weeks, while local breeds such as the Rhinelander chicken need 15 weeks to do so. In addition, dual-purpose hybrids show more favorable feed conversion and production figures than local breeds, which often have unfavorable feed conversion (Nolte, Jansen, Weigend, et al., 2020; Tiemann et al., 2020). Crossbreeds from old local breeds and layer breeds can improve fattening performance and efficiency, with final weights of 2.9–3.1 kg possible after 18 weeks and feed conversion approaching that of established dual-purpose breeds (Werner et al., 2023). However, compared to fast-growing broiler hybrids, the daily weight gain of dual-purpose and local breeds lags significantly behind (20–30 g/day vs 68 g/day for broilers), and the breast meat percentage is also lower (Mueller et al., 2018).

Nevertheless, these breeds might offer advantages in terms of animal welfare, robustness, that is, higher stress resilience, and better adaptability to extensive farming conditions (Stefanetti et al., 2023), making them attractive for niche markets and alternative production systems (Torres et al., 2019). Fattening performance can be further improved through targeted crossbreeding and adapted feeding management (Kreuzer et al., 2020). Various research projects and local networks are attempting to add value to these breeds (Sharma et al., 2023; Siebenmorgen et al., 2024). Be it through the aspect of dual-purpose use and the associated use of male animals instead of killing chicks, or through local relevance, right through to special sensory characteristics.

The meat quality of dual-purpose and local chicken breeds differs significantly in various characteristics from that of intensively bred broiler breeds and is a popular focus of scientific studies (Jayasena et al., 2013). Dual-purpose breeds such as LD, Bresse Gauloise, or Swiss chicken generally show slower growth and lower breast meat content, but provide meat with good sensory properties, firmer textures, and a differentiated aroma that can be positively evaluated by consumers, especially when regional feed such as field beans is used (Escobedo del Bosque et al., 2022; Mueller et al., 2018; Wattanachant et al., 2004). Volatile organic compounds, in particular, contribute to this diversification (Jin et al., 2021). Local and native breeds such as the Portuguese "Branca" or Italian landraces are characterized by a favorable nutrient profile with a high content of minerals and essential fatty acids, which makes the meat appear particularly healthy and high-quality (Jaturasitha et al., 2008; Meira et al., 2022; Pellattiero et al., 2020). Skin color is a particular characteristic that also gives the carcass, especially of Asian breeds, an unusual appearance (Zhang et al., 2015). Even if sensory characteristics lead to a unique selling point, this is usually offset by a longer fattening period and poor feed conversion.

Information on reproduction is also likely to come mainly from research projects. Fertilization, hatch, and mortality rates have so far only been collected to a limited extent in practice (Haunshi et al., 2011). Since the husbandry practices for these breeds vary greatly, ranging from research facilities to small-scale organic farms, these data are only comparable to a limited extent. Several studies show that dual-purpose chickens and local breeds often have lower mortality rates than conventional hybrid chickens. In a study conducted under semicommercial conditions, conventional laying hens (Lohmann Brown Plus) had higher mortality rates than dual-purpose chickens (LD), which was attributed, among other things, to a lower incidence of feather pecking and cannibalism among the dual-purpose chickens (Giersberg et al., 2020). Local breeds have also been found to show better adaptability and lower mortality rates in alternative farming systems, such as free-range farming, than conventional hybrids bred for intensive fattening (Stefanetti et al., 2023). Whether their typically reduced level of meat and egg production also leads to greater robustness is the subject of current research projects.

In terms of production characteristics, egg and meat performance, and feed conversion, local or dual-purpose breeds are rarely an efficient alternative. However, in terms of lifetime performance and longevity, as well as mortality—provided that the husbandry method is adapted to the breed—they represent an alternative and locally adapted form of production, especially compared to high-performance breeds. To illustrate the advantages of local adaptation to higher temperatures, the naked neck chicken features as a prominent example.

Naked neck chickens have a genetic mutation (Na gene) that leads to reduced feathering on the neck and gives them significantly better heat tolerance (Eberhart & Washburn, 1993). Birds presenting this mutation can better dissipate body heat, maintaining a lower body temperature; in addition, their coping ability is higher in hot climates, having lower heterophil/lymphocyte ratios and reduced plasma cholesterol compared to fully feathered birds. This trait makes them particularly valuable for poultry farming in tropical and subtropical regions, as they can dissipate excess body heat more efficiently, thus suffering less from heat stress (Desta, 2021; Fernandes et al., 2023). In hot climates, naked neck chickens often show better growth rates, feed conversion, disease resistance, and improved egg and meat production compared to normally feathered breeds. Pure breeds and hybrids showing the "naked neck phenotype" are found worldwide. The most common local breeds are Transylvanian and Italian naked neck. In countries such as China and Bangladesh, naked neck chickens are considered a valuable genetic resource because they are not only robust and adaptable, but also appreciated by consumers for their taste and meat quality (Islam & Nishibori, 2009; Shafiq et al., 2022).

The targeted breeding and conservation of naked neck breeds is recommended as an important strategy for adapting poultry production to climate change and for sustainable agriculture (Fernandes et al., 2023). Local adaptation strategies may become another unique feature of specialty breeds in the future. Research on physiological adaptation strategies to high altitudes has been conducted, for example, on the Tibetan chicken (Wang et al., 2015; Zhang et al., 2007). Nevertheless, the question of robustness and resilience of speciality breeds, that is, the description of the niche a single breed is adapted to and quantitative data to prove the adaptation, remains the focus of research. Finally, the availability of the animals is likely to be a decisive factor in the use of local breeds. Often, they are only commercially available locally in smaller hatcheries, if at all. However, the number of hatcheries is declining, so it is not uncommon for research institutions to maintain nucleus flocks. Integration into the value chain, which was established for high-performance animals, often fails due to the lack of uniformity of the animals and their products (Xiao et al., 2024), as well as the usually limited hygiene status.

Animal welfare aspects

Robustness and resilience are among the most prominent attributes of dual-purpose chickens and local breeds. However, quantitative, and thus scientifically based, knowledge on this subject is limited. Robustness describes the ability to remain in balance despite changing environmental conditions, while resilience means responding to environmental changes in such a way that balance is quickly restored (at the same level or through adaptation).

Various aspects lead to the assumption that local breeds are more robust and resilient:

1. The breeds in question do not perform as well as high-performance chickens and are therefore somewhat relaxed in terms of egg production or weight gain. This reduced performance may allow physiological capacities to be invested in maintaining homeostasis rather than maintaining performance in response to environmental change.
2. The high degree of inbreeding that has led to the stabilization of performance levels in commercial lines may also have led to a reduction in behavioral diversity. Behavioral diversity enables adaptation to changing environmental conditions and therefore supports the resilience of the individual animal.
3. Local breeds are also locally adapted. This pattern contrasts with that of a globally applicable chicken in poultry production.

4. Injurious or abnormal behavior, such as feather pecking and cannibalism, occurs in almost all chickens, as do keel bone fractures and the like. But at least the prevalence in the population varies, to the advantage of the speciality breeds (Giersberg et al., 2020; Jung et al., 2024), even if there is little data available on this.

However, the assumption that high performance is potentially associated with suffering, pain, and harm to the animal, and that local and dual-purpose breeds have a blanket advantage in this respect, is not scientifically based and will also not be universally valid. Rather, it is the interaction between the (production) niche and the individual animal that should concern us in terms of suitability and thus the promotion of animal welfare. In addition to the physiological characteristics mentioned above, local and dual-purpose breeds also have breed-specific behavioral traits.

The debate about whether behaviors differ between chicken breeds can already be deduced from the different structures of their brains (Mehlhorn & Caspers, 2021; Rehkämper et al., 2003). Various studies focus on anxiety and exploratory behavior due to their central importance and mutual influence on animal welfare, in European (Meuser et al., 2021; Tiemann et al., 2023) as well as Asian breeds (Yoshidome et al., 2021). Whether breeds are described as more or less fearful and more or less exploratory has fundamental implications for their husbandry environment. Just as there are precise management recommendations for high-performance breeds, the husbandry niche of a local breed should also be described. Breeds that show a high level of exploration may be better suited to free-range husbandry or require sufficient enrichment materials. In addition, local breeds often show less tolerance to large flock sizes, although one of the few studies does not show any impact of stocking density on production characteristics in a local breed (Tong et al., 2012). Breed differences are quantified, for example, in standardized behavioral tests. Studies on tonic immobility show that the response is breed-specific and that commercial breeds show less fear than local breeds (Campo & Muñoz, 2001; Rozempolska-Rucińska et al., 2020; Tiemann et al., 2023). The open field test also quantitatively measures exploratory ability, although the significance of the distance covered as a parameter is currently under discussion (Tiemann et al., 2022). The novel object test, avoidance distance, and handling tests assess similar characteristics, with the latter reflecting interactions with humans and showing that these characteristics are also breed-specific and that commercial lines may tolerate humans more than the local breeds that have been extensively kept and bred accordingly (Campo et al., 2005; Meuser et al., 2021; Moroi et al., 2019). These breed-specific profiles ultimately manifest themselves in the affective state (Oyeniran et al., 2025), and thus in positive animal welfare, and are therefore of current importance.

The use of zootechnical interventions in these less intensively farmed chicken breeds is not known. However, the issue of chick culling plays a particularly important role. Day-old chicks from laying lines are killed immediately after hatching if they do not meet quality standards or, in general, if they are male day-old chicks. These so-called brother roosters do not build muscle mass and are therefore not suitable for meat production. Some countries have begun to ban this practice of chick culling. As a result, sex determination in eggs plays a special role, but only in large hatcheries. Niche or local poultry production is therefore looking for alternatives, for example, in dual-purpose breeds. For this reason, profile descriptions and management manuals, which are rarely available for these alternatives, are becoming increasingly important and require a systematic investigation of diversity. However, this is difficult because comparisons of local origins usually remain local—in close proximity and adapted to the respective environment.

Dual-purpose breeds can contribute to the preservation of genetic diversity and the promotion of sustainable, regional value chains (Duijvesteijn & Perrault, 2019; Tiemann et al., 2020). Furthermore, in the wake of global epidemics and animal health issues, local breeds may take on a new role in the future. Recent studies show that these animals may have differentiated immune responses (Freier et al., 2025). Dual-purpose and local breeds may not perform as well as high-performance breeds in terms of feed conversion and resource efficiency, but they have other advantages in relation to sustainability strategies.

1. Farmers can reproduce pure breeds themselves if they use them.
2. Alternative breeds preserve and promote the biodiversity of domestic chickens.
3. Their lower performance may allow the use of lower-quality feed.
4. Their mortality may be lower and their longevity higher.
5. Alternative breeds ensure diversification, especially in regional value chains.

Breeding strategies and innovation potential

Selection strategies applied to conservation and valorization of local breeds strongly differ from those applied to the creation and production of commercial hybrid strains. Conservative selection follows a process of identification of the unique genetic make-up of local poultry breeds, focusing on the values of biodiversity and breed-specific genetic variability. One of the main goals is, following FAO protocols and indications, to create a consistent population characterized by reduced inbreeding coefficients. Local breeds play a fundamental role as a "unique gene reservoir." Local breeds' reproductive and productive performance are the results of the adaptive plasticity of the breed to a specific environment, nutrient availability, and management procedures. On the contrary, reproductive and productive performance of commercial hybrid strains, both for intensive production and slow-growing/

dual-purpose strains, is based on the improvement of the economic yield, maximizing profit, and reducing expenses. The results are obtained starting from pedigree birds to highly inbred lines that are mixed to obtain high hybrid vigor-related performance in commercial birds. So, the synthetic line shows the effects of its genetic make-up only when merged with a complementary one. Commercial hybrid strains are characterized by intense selection sweeps (Elferink et al., 2012; Wen et al., 2025). Whereas local breeds are specifically suitable for extensive "low input" production systems, high-performing strains, in contrast, need intensive facilities and a controlled environment to express their genetic potential (Restoux et al., 2022).

The production of local breeds is oriented to niche, high-quality, value-added markets; commercial hybrids are produced for mainstream markets (Barbut & Leishman, 2022; Hisasaga et al., 2020). Since the breeding of local and dual-purpose breeds is heterogeneous and not limited to a few breeding lines, and has local connections, it is often not economically viable for large breeding companies. Local communities taking over the breeding could therefore be a possible solution (Mueller et al., 2015). Community-based breeding programs offer an alternative, especially for low-input breeds, and are well established in other animal species, such as small ruminants. These breeding organizations do not have access to the full range of breeding methods and require alternatives in terms of financial stability. On the other hand, as a bottom-up process, they are close to the animals with knowledge of their advantages and disadvantages, and often have a broad network with corresponding socio-political support. Examples of intensive research into local breeds include the Italian Ermellinata di Rovigo, Robusta Maculata, Robusta Lionata, Pépoi, Padovana, and Polverara (Zanetti et al., 2010). Ethiopia is also actively supporting local origins and developing corresponding breeding programs (Ayalew et al., 2023; Dana et al., 2010). The study surveyed households that keep chickens and gathered information about the functional characteristics that are important to smallholders. These included not only different comb shapes and plumage colors depending on the region, but also the general desire for dual-purpose birds and the ability to successfully produce offspring based on the hens' brooding behavior. This reproductive assessment is particularly important as another characteristic that is common among local breeds, beneficial for self-sufficiency but detrimental to marketing, is the pronounced breeding propensity of certain breeds and animals (Jiang et al., 2005; Ohkubo et al., 1998).

A booster in selection strategies was supplied by the release of the chicken (*Gallus gallus*) genome assembly, where traits of interest like productivity and growth have been mapped (Yang et al., 2021). Genomic tools allow the prediction of breeders' genetic merit through the integration of dense genetic markers across the genome. Genome-wide association studies (GWAS) approaches allow the identification of trait-associated single nucleotide polymorphisms(SNPs), defining candidate genes for traits of interest. Accurate phenotyping is a fundamental step in genomic research; for this reason, trait measurement is very important and is nowadays based on cameras, integrated sensors, and artificial intelligence (AI), for high-resolution trait measurements. Precision farming plays a pivotal role in genomic selection evolution. In this scenario, the integration of GWAS and machine learning with selection algorithms like best linear unbiased prediction dramatically improves breeders' value prediction accuracy. Genomic approaches and AI are transforming trait selection procedures: how traits of interest are measured, selected, and optimized (Li et al., 2024; Yang et al., 2021). Genome- wide comparative analysis is effectively applied to heritage breed conservation and valorization studies. A clear example is given in a research paper about the study of genetic diversity and genetic structure of eight Italian chicken breeds: Ermellinata di Rovigo, Millefiori di Lonigo, Polverara Bianca, Polverara Nera, Padovana, Pepoi, Robusta Lionata, and Robusta Maculata. High-resolution genotyping data were analyzed using a genotyping array comprising more than 600,000 SNPs. This analysis revealed the unique genomic asset of the studied breeds as genetically separate populations. Six genomic regions were identified as hotspots of autozygosity, defined as regions of the genome where both alleles are identical by descent (inherited from the same ancestor), across the analyzed breeds, characterized by extended runs of homozygosity (recent inbreeding). These loci encompass putative candidate genes implicated in the regulation of morphometric traits, including pectoral muscle hypertrophy, intramuscular water content, and somatic growth (Cendron et al., 2021). These results are important for planning effective conservation and valorization strategies for local breeds in order to maintain a valuable reservoir of genetic diversity, primarily intended to preserve biodiversity and to support alternative production systems.

Conclusion

Specialty poultry breeds, including dual-purpose breeds, local and commercial alternatives, represent a significant genetic and cultural resource in poultry production. They have the potential to improve animal welfare, preserve biodiversity, and strengthen regional food systems. However, compared to highly specialized hybrids, they usually have lower performance and less favorable feed conversion rates. This results in a complex area of tension: on the one hand, they promote animal welfare and resilience, but on the other hand, they are less sustainable in terms of purely resource-economic criteria. They should therefore be seen not as a substitute, but rather as a complementary addition to high-performance breeds, enabling a more diverse and robust poultry sector.

The future role of local breeds depends on the balance between preserving genetic diversity, innovation in breeding, and marketability. Preserving local breeds ensures important characteristics such as climate resistance, disease resistance, and behavioral diversity. Innovation—for example, through genomic methods, precision breeding, or targeted crossbreeding—can improve their performance and adaptability. However, marketability remains crucial: without demand and economically viable niche markets, local breeds will remain limited to small initiatives. Only by combining conservation, innovation, and targeted marketing can they contribute to animal welfare-oriented and sustainable poultry systems.

Political guidelines have a decisive influence on the potential of local breeds. In the EU, animal welfare laws and the ban on chick culling have strengthened interest in dual-purpose breeds and alternative genetics. Growing consumer awareness of animal welfare and sustainability is also supporting this development. At the same time, value chains are based almost exclusively on high-performance production systems. Greater political support for the value of local breeds—especially for biodiversity, resilience, and social expectations—is needed to ensure their long-term establishment.

Research on local poultry breeds has been fragmented and always limited in time. Breeding programs are often too small to achieve genetic progress with the dynamics of commercial lines. There is a lack of long-term breeding populations, databases, and comprehensive networks. Research on genotype-environment interactions, animal-based welfare indicators, and consumer acceptance is particularly needed. The establishment of joint research infrastructures, including public-private partnerships, could strengthen the scientific basis and consolidate the position of local breeds in food production.

Digital technologies open up new perspectives for local breeds. Precision farming can record behavior, animal welfare, and performance even in smaller flocks, thus providing valuable data for breeding. Digital platforms also enable transparency along the value chain and strengthen consumer confidence. Regionalization also offers opportunities: specialty breeds fit in well with regional traditions, short supply chains, and the demand for authentic products. The combination of digitalization and regional marketing could significantly increase their competitiveness.

In future poultry systems, local breeds should not be evaluated solely on the basis of their productivity. Their contribution to animal welfare, biodiversity, and cultural values must also be taken into account. They serve as real-world laboratories for animal-friendly husbandry, breeding, and management approaches and embody an alternative vision of poultry production that combines efficiency with ethics, biodiversity, and resilience. Even though they will not and should not replace high-performance hybrids, they are indispensable for diversifying production systems and meeting society's expectations for high standards in animal welfare, product quality, and regionality.

Local breeds and dual-purpose breeds highlight the opportunities and challenges of balancing animal welfare and sustainability. Their lower efficiency shows the limitations of purely performance-oriented indicators, but at the same time, they ensure genetic diversity, cultural values, and high animal welfare potential. Political will, targeted research, and innovative market strategies are needed to strengthen their role. If conservation, innovation, and marketing can be successfully combined, local breeds can play a central role in the animal welfare-oriented and locally adapted poultry production of the future.

AI disclosure

During the preparation of this work, the author(s) used ConsensusApp & DeepL. in order to improve grammar and readability. The ConsensusApp adds to the basic reference search in Web of Science as it condenses key findings of the studies, especially in this very small research field. Deepl and ChatGPT were used for translation. After using this tool/service, the author(s) reviewed and edited the content as needed and take(s) full responsibility for the content of the publication.

References

Abdelqader, A., Wollny, C. B. A., & Gauly, M. (2007). Characterization of local chicken production systems and their potential under different levels of management practice in Jordan. *Tropical Animal Health and Production, 39*(3), 155–164. https://doi.org/10.1007/s11250-007-9000-x.

Alene, A. T., Mersha, G. B., & Woldegiorgis, W. E. (2024). Productive and reproductive performances, egg quality, and carcass traits among indigenous, exotic and crossbred chickens in Ethiopia. *World's Poultry Science Journal, 80*(2), 387–401. https://doi.org/10.1080/00439339.2023.2276735.

Ayalew, H., Chanie, D., Fentahun, T., Yinnesu, A., Dagnew, Y., & Moges, D. (2023). Agro ecological base differences of village based local chicken performance and household product consumption in Amhara Region, Ethiopia. *Cogent Food and Agriculture, 9*(1), 2164662. https://doi.org/10.1080/23311932.2022.2164662.

Baldinger, L., & Bussemas, R. (2021). Dual-purpose production of eggs and meat — Part 1: Cockerels of crosses between layer and meat breeds achieve moderate growth rates while showing unimpaired animal welfare. *Organic Agriculture, 11*(3), 489–498. https://doi.org/10.1007/s13165-021-00357-z.

Bamidele, O., Sonaiya, E. B., Adebambo, O. A., & Dessie, T. (2020). On-station performance evaluation of improved tropically adapted chicken breeds for smallholder poultry production systems in Nigeria. *Tropical Animal Health and Production, 52*(4), 1541–1548. https://doi.org/10.1007/s11250-019-02158-9.

Barbut, S., & Leishman, E. M. (2022). Quality and processability of modern poultry meat. *Animals, 12*(20), 2766. https://doi.org/10.3390/ani12202766.

Becker, S., Büscher, W., & Tiemann, I. (2023). The British Ixworth: Individual growth and egg production of a purebred dual-purpose chicken. *British Poultry Science, 64*(6), 659–669. https://doi.org/10.1080/00071668.2023.2246142.

Campo, J. L., & Muñoz, I. (2001). Relationships between dust bathing activity and stressor fear-related responses in white, tinted, and brown egg breeds of chickens. *European Poultry Science, 65*(2), 88–91. https://doi.org/10.1016/s0003-9098(25)00162-6.

Campo, J. L., Gil, M. G., & Dávila, S. G. (2005). Social aggressiveness, pecking at hands, and its relationships with tonic immobility duration and heterophil to lymphocyte ratio in chickens of different breeds. *European Poultry Science, 69*(1), 11–15. https://doi.org/10.1016/s0003-9098(25)01520-6.

Cedraz, H., Gromboni, J. G. G., Garcia, A. A. P. Junior, Farias Filho, R. V., Souza, T. M., Ribeiro de Oliveira, E., de Oliveira, E. B., Souza do Nascimento, C., Meneghetti, C., & Arias Wenceslau, A. (2017). , Heat stress induces expression of HSP genes in genetically divergent chickens. *PLoS One, 12*(10), e0186083. https://doi.org/10.1371/journal.pone.0186083.

Cendron, F., Mastrangelo, S., Tolone, M., Perini, F., Lasagna, E., & Cassandro, M. (2021). Genome-wide analysis reveals the patterns of genetic diversity and population structure of 8 Italian local chicken breeds. *Poultry Science, 100*(2), 441–451. https://doi.org/10.1016/j.psj.2020.10.023.

Cheng (2010). Breeding of tomorrow's chickens to improve well-being. *Poultry Science, 89*(4), 805–813. https://doi.org/10.3382/ps.2009-00361.

Choo, Y. K., Kwon, H. J., Oh, S. T., Um, J. S., Kim, B. G., Kang, C. W., Lee, S. K., & An, B. K. (2014). Comparison of growth performance, carcass characteristics and meat quality of korean local chickens and silky fowl. *Asian-Australasian Journal of Animal Sciences, 27*(3), 398–405. https://doi.org/10.5713/ajas.2013.13638.

Damme, K., Urselmans, & Schmidt, E. (2015). Economics of dual-purpose breeds–A comparison of meat and egg production using dual purpose breeds versus conventional broiler and layer strains. *Lohmann Information, 50*(2), 4–9. https://lohmann-breeders.com/lohmanninfo/economics-of-dual-purpose-breeds/.

Dana, N., van der Waaij, L. H., Dessie, T., & van Arendonk, J. A. M. (2010). Production objectives and trait preferences of village poultry producers of Ethiopia: Implications for designing breeding schemes utilizing indigenous chicken genetic resources. *Tropical Animal Health and Production, 42*(7), 1519–1529. https://doi.org/10.1007/s11250-010-9602-6.

Dana, N., vander Waaij, E. H., & van Arendonk, J. A. M. (2011). Genetic and phenotypic parameter estimates for body weights and egg production in Horro chicken of Ethiopia. *Tropical Animal Health and Production, 43*(1), 21–28. https://doi.org/10.1007/s11250-010-9649-4.

Desta, T. T. (2021). The genetic basis and robustness of naked neck mutation in chicken. *Tropical Animal Health and Production, 53*(1), 95. https://doi.org/10.1007/s11250-020-02505-1.

Domestic Animal Diversity Information System (DAD-IS). 2026. https://www.fao.org/dad-is/en/.

Duijvesteijn, N., & Perrault, L. (2019). 180 How dual-purpose chickens can help African smallholder farmers. *Journal of Animal Science, 97*(Supplement_3), 182. https://doi.org/10.1093/jas/skz258.375 182.

Dzungwe, J. T., Adjei-Mensah, B., Chrysostome, C. A. A. M., & Tozo, K. (2024). Effect of crossbreeding on growth performance, meat quality, and the economics of production of the pure and reciprocal crosses between the Sasso and Wassachie chickens. *Poultry Science, 103*(8), 103823. https://doi.org/10.1016/j.psj.2024.103823.

Eberhart, D. E., & Washburn, K. W. (1993). Assessing the effects of the naked neck gene on chronic heat stress resistance in two genetic populations. *Poultry Science, 72*(8), 1391–1399. https://doi.org/10.3382/ps.0721391.

Elferink, M. G., Megens, H.-J., Vereijken, A., Hu, X., Crooijmans, R. P. M. A., & Groenen, M. A. M. (2012). Signatures of selection in the genomes of commercial and non-commercial chicken breeds. *PLoS One, 7*(2), e32720. https://doi.org/10.1371/journal.pone.0032720.

Escobedo del Bosque, C. I., Grahl, S., Nolte, T., & Mörlein, D. (2022). Meat quality parameters, sensory properties and consumer acceptance of chicken meat from dual-purpose crossbreeds fed with regional faba beans. *Foods, 11*(8), 1074. https://doi.org/10.3390/foods11081074.

Escobedo del Bosque, C. I., Spiller, A., & Risius, A. (2021). Who wants chicken? Uncovering consumer preferences for produce of alternative chicken product methods. *Sustainability, 13*(5), 2440. https://doi.org/10.3390/su13052440.

FAO Gateway to poultry production and products. 2026. https://www.fao.org/poultry-production-products/production/poultry-species/chicken/en.

Fernandes, E., Raymundo, A., Martins, L. L., Lordelo, M., & de Almeida, A. M. (2023). The naked neck gene in the domestic chicken: A genetic strategy to mitigate the impact of heat stress in poultry production—A review. *Animals, 13*(6), 1007. https://doi.org/10.3390/ani13061007.

Freier, L., Stuff, J., Götzke, N., Preisinger, R., Grund, C., Tiemann, I., Weigend, S., & Blohm, U. (2025). Baseline immune profiles of local chicken breeds: Linking biodiversity, animal health, and vaccination response. *Poultry Science, 104*(10), 105565. https://doi.org/10.1016/j.psj.2025.105565.

Gangnat, I. D. M., Mueller, S., Messikommer, R. E., & Kreuzer, M. (2020). Performance, egg quality and resilience to nutritional challenges of Lohmann dual hens as opposed to layer and traditional dual-purpose types: A preliminary study. *European Poultry Science, 84*, 1–16. https://doi.org/10.1399/eps.2020.301.

Giersberg, M. F., & Kemper, N. (2018). Rearing male layer chickens: A German perspective. *Agriculture, 8*(11), 176. https://doi.org/10.3390/agriculture8110176.

Giersberg, M. F., Spindler, B., Rodenburg, B., & Kemper, N. (2020). The dual-purpose hen as a chance: Avoiding injurious pecking in modern laying hen husbandry. *Animals, 10*(1), 16. https://doi.org/10.3390/ani10010016.

Groeneveld, L. F., Lenstra, J. A., Eding, H., Toro, M. A., Scherf, B., Pilling, D., Negrini, R., Finlay, E. K., Jianlin, H., Groeneveld, E., & Weigend, S. (2010). Genetic diversity in farm animals – A review. *Animal Genetics, 41*(s1), 6–31. https://doi.org/10.1111/j.1365-2052.2010.02038.x.

Hammershøj, M., Kristiansen, G., & Steenfeldt, S. (2021). Dual-purpose poultry in organic egg production and effects on egg quality parameters. *Foods, 10*(4), 897. https://doi.org/10.3390/foods10040897.

Han, D., Wang, S., Hu, Y., Zhang, Y., Dong, X., Yang, Z., Wang, J., Li, J., & Deng, X. (2015). Hyperpigmentation results in aberrant immune development in silky fowl (*Gallus gallus domesticus* Brisson). *PLoS One, 10*(6), e0125686. https://doi.org/10.1371/journal.pone.0125686.

Haunshi, S., Niranjan, M., Shanmugam, M., Padhi, M. K., Reddy, M. R., Sunitha, R., Rajkumar, U., & Panda, A. K. (2011). Characterization of two Indian native chicken breeds for production, egg and semen quality, and welfare traits. *Poultry Science, 90*(2), 314–320. https://doi.org/10.3382/ps.2010-01013.

Hisasaga, C., Griffin, S. E., & Tarrant, K. J. (2020). Survey of egg quality in commercially available table eggs. *Poultry Science, 99*(12), 7202–7206. https://doi.org/10.1016/j.psj.2020.09.049.

Huang, R., Zhu, C., & Zhen, Y. (2024). Genetic diversity, demographic history, and selective signatures of Silkie chicken. *BMC Genomics, 25*(1), 754. https://doi.org/10.1186/s12864-024-10671-x.

Islam, M. A., & Nishibori, M. (2009). Indigenous naked neck chicken: A valuable genetic resource for Bangladesh. *World's Poultry Science Journal, 65*(1), 125–138. https://doi.org/10.1017/s0043933909000105.

Jaturasitha, S., Srikanchai, T., Kreuzer, M., & Wicke, M. (2008). Differences in carcass and meat characteristics between chicken indigenous to northern Thailand (Black-Boned and Thai

Native) and imported extensive breeds (Bresse and Rhode Island Red). *Poultry Science, 87*(1), 160–169. https://doi.org/10.3382/ps.2006-00398.

Jayasena, D. D., Ahn, D. U., Nam, K. C., & Jo, C. (2013). Flavour chemistry of chicken meat: A review. *Asian-Australasian Journal of Animal Sciences, 26*(5), 732–742. https://doi.org/10.5713/ajas.2012.12619.

Jiang, R.-S., Xu, G.-Y., Zhang, X.-Q., & Yang, N. (2005). Association of polymorphisms for prolactin and prolactin receptor genes with broody traits in chickens. *Poultry Science, 84*(6), 839–845. https://doi.org/10.1093/ps/84.6.839.

Jin, Y., Cui, H., Yuan, X., Liu, L., Liu, X., Wang, Y., Ding, J., Xiang, H., Zhang, X., Liu, J., Li, H., Zhao, G., & Wen, J. (2021). Identification of the main aroma compounds in Chinese local chicken high-quality meat. *Food Chemistry, 359*, 129930. https://doi.org/10.1016/j.foodchem.2021.129930.

Jung, L., Hillemacher, S., Tiemann, I., Lepke, M., & Hinrichs, D. (2024). Presence of keel bone damage in laying hens, pullets and roosters of local chicken breeds. *PLoS One, 19*(1), e0297586. https://doi.org/10.1371/journal.pone.0297586.

Kebede, F. G., Komen, H., Dessie, T., Hanotte, O., Kemp, S., Alemu, S. W., & Bastiaansen, J. W. M. (2024). Multi-environment performance analysis identifies more productive and widely adapted chicken breeds for smallholder farmers. *Frontiers in Sustainable Food Systems, 8*, 1441295. https://doi.org/10.3389/fsufs.2024.1441295.

Krautwald-Junghanns, M.-E., Cramer, K., Fischer, B., Förster, A., Galli, R., Kremer, F., Mapesa, E. U., Meissner, S., Preisinger, R., Preusse, G., Schnabel, C., Steiner, G., & Bartels, T. (2018). Current approaches to avoid the culling of day-old male chicks in the layer industry, with special reference to spectroscopic methods. *Poultry Science, 97*(3), 749–757. https://doi.org/10.3382/ps/pex389.

Kreuzer, M., Müller, S., Mazzolini, L., Messikommer, R. E., & Gangnat, I. D. M. (2020). Are dual-purpose and male layer chickens more resilient against a low-protein-low-soybean diet than slow-growing broilers? *British Poultry Science, 61*(1), 33–42. https://doi.org/10.1080/00071668.2019.1671957.

Lambertz, C., Wuthijaree, K., & Gauly, M. (2018). Performance, behavior, and health of male broilers and laying hens of 2 dual-purpose chicken genotypes. *Poultry Science, 97*(10), 3564–3576. https://doi.org/10.3382/ps/pey223.

Li, G., Xu, J., Chen, S., Tan, S., & Li, H. (2022). Pigment concentrations in eggshell and their related gene expressions in uterus of Changshun blue eggshell chickens. *British Poultry Science, 63*(3), 421–425. https://doi.org/10.1080/00071668.2021.1983919.

Li, X., Chen, X., Wang, Q., Yang, N., & Sun, C. (2024). Integrating bioinformatics and machine learning for genomic prediction in chickens. *Genes, 15*(6), 690. https://doi.org/10.3390/genes15060690.

Liao, X., Shi, X., Hu, H., Han, X., Jiang, K., Liu, Y., & Xiong, G. (2022). Comparative metabolomics analysis reveals the unique nutritional characteristics of breed and feed on muscles in chinese taihe black-bone silky fowl. *Metabolites, 12*(10), 914. https://doi.org/10.3390/metabo12100914.

Lin, H., Jiao, H. C., Buyse, J., & Decuypere, E. (2006). Strategies for preventing heat stress in poultry. *World's Poultry Science Journal, 62*(1), 71–86. https://doi.org/10.1079/wps200585.

Lordelo, M., Cid, J., Cordovil, C. M. D. S., Alves, S. P., Bessa, R. J. B., & Carolino, I. (2020). A comparison between the quality of eggs from indigenous chicken breeds and that from commercial layers. *Poultry Science, 99*(3), 1768–1776. https://doi.org/10.1016/j.psj.2019.11.023.

Manyelo, T. G., Selaledi, L., Hassan, Z. M., & Mabelebele, M. (2020). Local chicken breeds of Africa: Their description, uses and conservation methods. *Animals, 10*(12), 2257. https://doi.org/10.3390/ani10122257.

Mehlhorn, J., & Caspers, S. (2021). The effects of domestication on the brain and behavior of the chicken in the light of evolution. *Brain, Behavior and Evolution, 95*(6), 287–301. https://doi.org/10.1159/000516787.

Meira, M., Afonso, I. M., Casal, S., Lopes, J. C., Domingues, J., Ribeiro, V., Dantas, R., Leite, J. V., & Brito, N. V. (2022). Carcass and meat quality traits of males and females of the "Branca" portuguese autochthonous chicken breed. *Animals, 12*(19), 2640. https://doi.org/10.3390/ani12192640.

Menchetti, L., Birolo, M., Mugnai, C., Mancinelli, A. C., Xiccato, G., Trocino, A., & Castellini, C. (2024). Effect of genotype and nutritional and environmental challenges on growth curve dynamics of broiler chickens. *Poultry Science, 103*(10), 104095. https://doi.org/10.1016/j.psj.2024.104095.

Meuser, V., Weinhold, L., Hillemacher, S., & Tiemann, I. (2021). Welfare-related behaviors in chickens: Characterization of fear and exploration in local and commercial chicken strains. *Animals, 11*(3), 679. https://doi.org/10.3390/ani11030679.

Mokoena, K., Mbazima, V., & Tyasi, T. L. (2025). A literature review on effect of chicken crossbreeding on egg production traits. *World's Poultry Science Journal, 81*(1), 151–167. https://doi.org/10.1080/00439339.2024.2443184.

Moroi, S., Nishimura, K., Imai, N., Kunishige, K., Sato, S., & Goto, T. (2019). Rapid behavioral assay using handling test provides breed and sex differences in tameness of chickens. *Brain and Behavior, 9*(10), e01394. https://doi.org/10.1002/brb3.1394.

Mueller, J. P., Rischkowsky, B., Haile, A., Philipsson, J., Mwai, O., Besbes, B., Valle Zárate, A., Tibbo, M., Mirkena, T., Duguma, G., Sölkner, J., & Wurzinger, M. (2015). Community-based livestock breeding programmes: Essentials and examples. *Journal of Animal Breeding and Genetics, 132*(2), 155–168. https://doi.org/10.1111/jbg.12136.

Mueller, S., Kreuzer, M., Siegrist, M., Mannale, K., Messikommer, R. E., & Gangnat, I. D. M. (2018). Carcass and meat quality of dual-purpose chickens (Lohmann Dual, Belgian Malines, Schweizerhuhn) in comparison to broiler and layer chicken types. *Poultry Science, 97*(9), 3325–3336. https://doi.org/10.3382/ps/pey172.

Nolte, T., Jansen, S., Halle, I., Scholz, A. M., Simianer, H., Sharifi, A. R., & Weigend, S. (2020a). Egg production and bone stability of local chicken breeds and their crosses fed with faba beans. *Animals, 10*(9), 1480. https://doi.org/10.3390/ani10091480.

Nolte, T., Jansen, S., Weigend, S., Moerlein, D., Halle, I., Link, W., Hummel, J., Simianer, H., & Sharifi, A. R. (2020b). Growth performance of local chicken breeds, a high-performance genotype and their crosses fed with regional faba beans to replace soy. *Animals, 10*(4), 702. https://doi.org/10.3390/ani10040702.

Ohkubo, T., Tanaka, M., Nakashima, K., Talbot, R. T., & Sharp, P. J. (1998). Prolactin receptor gene expression in the brain and peripheral tissues in broody and nonbroody breeds of domestic hen. *General and Comparative Endocrinology, 109*(1), 60–68. https://doi.org/10.1006/gcen.1997.7008.

Oyeniran, V. J., Whittle, R. H., Orlowski, S. K., & Weimer, S. L. (2025). Judgment bias, fear, and stress responses of Red Junglefowl and Athens Canadian Random Bred chickens. *Frontiers in Animal Science, 6*, 1573847. https://doi.org/10.3389/fanim.2025.1573847.

Pellattiero, E., Tasoniero, G., Cullere, M., Gleeson, E., Baldan, G., Contiero, B., & Dalle Zotte, A. (2020). Are meat quality traits and sensory attributes in favor of slow-growing chickens? *Animals, 10*(6), 960. https://doi.org/10.3390/ani10060960.

Perini, F., Cendron, F., Lasagna, E., Cassandro, M., & Penasa, M. (2024). Genomic insights into shank and eggshell color in Italian local chickens. *Poultry Science, 103*(6), 103677. https://doi.org/10.1016/j.psj.2024.103677.

Prakash, A., Singh, Y., Chatli, M., Sharma, A., Acharya, P., & Singh, M. (2023). Review of the black meat chicken breeds: Kadaknath, Silkie, and Ayam Cemani. *World's Poultry Science Journal, 79*(4), 879–891. https://doi.org/10.1080/00439339.2023.2250322.

Rehkämper, G., Kart, E., Frahm, H. D., & Werner, C. W. (2003). Discontinuous variability of brain composition among domestic chicken breeds. *Brain, Behavior and Evolution, 61*(2), 59–69. https://doi.org/10.1159/000069352.

Restoux, G., Rognon, X., Vieaud, A., Guemene, D., Petitjean, F., Rouger, R., Brard-Fudulea, S., Lubac-Paye, S., Chiron, G., & Tixier-Boichard, M. (2022). Managing genetic diversity in breeding programs of small populations: The case of French local chicken breeds. *Genetics Selection Evolution, 54*(1), 56. https://doi.org/10.1186/s12711-022-00746-2.

Rieke, L., Spindler, B., Zylka, I., Kemper, N., & Giersberg, M. F. (2021). Pecking behavior in conventional layer hybrids and dual-purpose hens throughout the laying period. *Frontiers in Veterinary Science, 8*, 660400. https://doi.org/10.3389/fvets.2021.660400.

Rizzi, C. (2019). Growth and slaughtering performance, carcase fleshiness and meat quality according to the plumage colour in Padovana male chickens slaughtered at 18 weeks of age. *Italian Journal of Animal Science, 18*(1), 450–459. https://doi.org/10.1080/1828051x.2018.1532823.

Rizzi, C. (2023). A study on egg production and quality according to the age of four Italian chicken dual-purpose purebred hens reared outdoors. *Animals, 13*(19), 3064. https://doi.org/10.3390/ani13193064.

Rizzi, C., Cendron, F., Penasa, M., & Cassandro, M. (2023). Egg quality of Italian local chicken breeds: I. Yield performance and physical characteristics. *Animals, 13*(1), 148. https://doi.org/10.3390/ani13010148.

Rozempolska-Rucińska, I., Kasperek, K., Drabik, K., Zięba, G., & Ziemiańska, A. (2020). Behavioural variability in chicks vs. the pattern of behaviour in adult hens. *Animals, 10*(2), 269. https://doi.org/10.3390/ani10020269.

Shafiq, M., Khan, M. T., Rehman, M. S., Raziq, F., Bughio, E., Farooq, Z., Gondal, M. A., Rauf, M., Liaqat, S., Sarwar, F., Azad, A., Asad, T., Arslan, M., Azhar, M., Kamal, R. M. A., & Shakir, M. (2022). Assessing growth performance, morphometric traits, meat chemical composition and cholesterol content in four phenotypes of naked neck chicken. *Poultry Science, 101*(3), 101667. https://doi.org/10.1016/j.psj.2021.101667.

Sharma, R., Sehrawat, R., Ahlawat, S., Sharma, V., Thakur, M. S., Mishra, A. K., Arora, R., & Tantia, M. S. (2023). Functional quality characteristics of the meat from a dual-purpose poultry crossbreed suitable for backyard rearing in comparison to commercial broilers. *Foods, 12*(13), 2434. https://doi.org/10.3390/foods12132434.

Siebenmorgen, C., Mörlein, J., Strack, M., Tetens, J., & Mörlein, D. (2024). Enhancing agro-biodiversity in chicken: A sensory comparison of broths from German local chicken breeds and their crossbreeds. *Poultry Science, 103*(6), 103683. https://doi.org/10.1016/j.psj.2024.103683.

Siebenmorgen, C., Spreckelmeyer, A., Strack, M., & Mörlein, D. (2026). No parts left behind? Exploring pros and cons of cooking whole chicken dishes at homes in Germany. *Poultry Science, 105*(2), 106227. https://doi.org/10.1016/j.psj.2025.106227.

Siekmann, L., Meier-Dinkel, L., Janisch, S., Altmann, B., Kaltwasser, C., Sürie, C., & Krischek, C. (2018). Carcass quality, meat quality and sensory properties of the dual-purpose chicken Lohmann Dual. *Foods, 7*(10), 156. https://doi.org/10.3390/foods7100156.

Stefanetti, V., Mancinelli, A. C., Pascucci, L., Menchetti, L., Castellini, C., Mugnai, C., Fiorilla, E., Miniscalco, B., Chiattelli, D., Franciosini, M. P., & Proietti, P. C. (2023). Effect of rearing systems on immune status, stress parameters, intestinal morphology, and mortality in conventional and local chicken breeds. *Poultry Science, 102*(12), 103110. https://doi.org/10.1016/j.psj.2023.103110.

Tan, Y., Huang, Y., Xu, C., Huang, X., Li, S., & Yin, Z. (2024). Long noncoding RNAs and mRNAs profiling in ovary during laying and broodiness in Taihe Black-Bone Silky Fowls (*Gallus gallus domesticus Brisson*). *BMC Genomics, 25*(1), 357. https://doi.org/10.1186/s12864-024-10281-7.

Tiemann, I., Hillemacher, S., & Wittmann, M. (2020). Are dual-purpose chickens twice as good? Measuring performance and animal welfare throughout the fattening period. *Animals, 10*(11), 1980. https://doi.org/10.3390/ani10111980.

Tiemann, I., Becker, S., Büscher, W., & Meuser, V. (2022). Exploring animal genetic resources of the domestic chicken and their behavior in the open field. *Journal of Applied Poultry Research, 31*(2), 100237. https://doi.org/10.1016/j.japr.2022.100237.

Tiemann, I., Becker, S., Fournier, J., Damiran, D., Büscher, W., & Hillemacher, S. (2023). Differences among domestic chicken breeds in tonic immobility responses as a measure of fearfulness. *PeerJ, 11*, e14703. https://doi.org/10.7717/peerj.14703.

Tong, H. B., Lu, J., Zou, J. M., Wang, Q., & Shi, S. R. (2012). Effects of stocking density on growth performance, carcass yield, and immune status of a local chicken breed. *Poultry Science, 91*(3), 667–673. https://doi.org/10.3382/ps.2011-01597.

Torres, A., Muth, P. C., Capote, J., Rodríguez, C., Fresno, M., & Valle Zárate, A. (2019). Suitability of dual-purpose cockerels of 3 different genetic origins for fattening under free-range conditions. *Poultry Science, 98*(12), 6564–6571. https://doi.org/10.3382/ps/pez429.

Tudorache, M., Custura, I., Gheorghe, A., Habeanu, M., Lefter, N. A., Pogurschi, E. N., & Popa, D. C. (2022). Effects of genotype and diet on performance, carcass traits, and blood profiles of slow-growing chicks obtained by crosses of local breed with commercial genotype. *Agriculture, 12*(11), 1906. https://doi.org/10.3390/agriculture12111906.

van Niekerk T., Workamp J. 2022. Scenarios for addressing the dilemma of 'the culling of day-old male chicks of layer breeds'. Wageningen Livestock Research, Report 1381-UK. https://edepot.wur.nl/583326.

Vaarst, M., Steenfeldt, S., & Horsted, K. (2015). Sustainable development perspectives of poultry production. *World's Poultry Science Journal, 71*(4), 609–620. https://doi.org/10.1017/s0043933915002433.

Verdiglione, R., & Rizzi, C. (2018). A morphometrical study on the skull of Padovana chicken. *Italian Journal of Animal Science, 17*(3), 785–796. https://doi.org/10.1080/1828051x.2017.1412810.

Verrier, E., Tixier-Boichard, M., Bernigaud, R., & Naves, M. (2005). Conservation and value of local livestock breeds: Usefulness of niche products and/or adaptation to specific environments. *Animal Genetic Resources Information, 36*, 21–31. Available from https://doi.org/10.1017/S1014233900005538.

Wang, H., Cahaner, A., Lou, L., Zhang, L., Ge, Y., Li, Q., & Zhang, X. (2021). Genetics and breeding of a black-bone and blue eggshell chicken line. 1. Body weight, skin color, and their combined selection. *Poultry Science, 100*(5), 101035. https://doi.org/10.1016/j.psj.2021.101035.

Wang, H., Ge, Y., Zhang, L., Wei, Y., Li, Q., Zhang, X., & Pan, Y. (2023). The pigments in eggshell with different colour and the pigment regulatory gene expression in corresponding chicken's shell gland. *Animal, 17*(5), 100776. https://doi.org/10.1016/j.animal.2023.100776.

Wang, M.-S., Li, Y., Peng, M.-S., Zhong, L., Wang, Z.-J., Li, Q.-Y., Tu, X.-L., Dong, Y., Zhu, C.-L., Wang, L., Yang, M.-M., Wu, S.-F., Miao, Y.-W., Liu, J.-P., Irwin, D. M., Wang, W., Wu,

D.-D., & Zhang, Y.-P. (2015). Genomic analyses reveal potential independent adaptation to high altitude in tibetan chickens. *Molecular Biology and Evolution, 32*(7), 1880–1889. https://doi.org/10.1093/molbev/msv071.

Wattanachant, S., Benjakul, S., & Ledward, D. A. (2004). Composition, color, and texture of thai indigenous and broiler chicken muscles. *Poultry Science, 83*(1), 123–128. https://doi.org/10.1093/ps/83.1.123.

Wen, Z., Cai, X., Liu, Z., Tan, L., Kong, Y., Wang, Y., & Zhao, Y. (2025). Genomic analyses reveal a lack of widespread strong selection in indigenous chickens. *Poultry Science, 104*(5), 105081. https://doi.org/10.1016/j.psj.2025.105081.

Werner, D., Bussemas, R., & Baldinger, L. (2023). Crossing the old local breed Deutsches Lachshuhn with the layer breed white rock: Effects on laying performance of the females and fattening performance of the males. *Animals, 13*(19), 2999. https://doi.org/10.3390/ani13192999.

Xiao, L., Qi, L., Fu, R., Nie, Q., Zhang, X., & Luo, W. (2024). A large-scale comparison of the meat quality characteristics of different chicken breeds in South China. *Poultry Science, 103*(6), 103740. https://doi.org/10.1016/j.psj.2024.103740.

Xiong, G., Chen, W., Jiang, K., Liu, S., Li, J., & Liao, X. (2024). Integrated transcriptome and proteome analysis reveals the unique molecular features and nutritional components on the muscles in Chinese Taihe black-bone silky fowl chicken. *PLoS One, 19*(3), e0299385. https://doi.org/10.1371/journal.pone.0299385.

Xu, Y., Lai, X., Li, Z., Zhang, X., & Luo, Q. (2018). Effect of chronic heat stress on some physiological and immunological parameters in different breed of broilers. *Poultry Science, 97*(11), 4073–4082. https://doi.org/10.3382/ps/pey256.

Yang, N., & Jiang, R. S. (2005). Recent advances in breeding for quality chickens. *World's Poultry Science Journal, 61*(3), 373–524. https://doi.org/10.1079/WPS200463.

Yang, R., Xu, Z., Wang, Q., Zhu, D., Bian, C., Ren, J., Huang, Z., Zhu, X., Tian, Z., Wang, Y., Jiang, Z., Zhao, Y., Zhang, D., Li, N., & Hu, X. (2021). Genome-wide association study and genomic prediction for growth traits in yellow-plumage chicken using genotyping-by-sequencing. *Genetics Selection Evolution, 53*(1), 82. https://doi.org/10.1186/s12711-021-00672-9.

Yoshidome, K., Fukano, N., Ouchi, Y., Tomonaga, S., Cockrem, J. F., & Bungo, T. (2021). The use of behavioral tests of fearfulness in chicks to distinguish between the Japanese native chicken breeds, Tosa-Kukin and Yakido. *Animal Science Journal, 92*(1), e13507. https://doi.org/10.1111/asj.13507.

Zanetti, E., De Marchi, M., Dalvit, C., & Cassandro, M. (2010). Genetic characterization of local Italian breeds of chickens undergoing in situ conservation. *Poultry Science, 89*(3), 420–427. https://doi.org/10.3382/ps.2009-00324.

Zhang, H., Wu, C. X., Chamba, Y., & Ling, Y. (2007). Blood characteristics for high altitude adaptation in Tibetan chickens. *Poultry Science, 86*(7), 1384–1389. https://doi.org/10.1093/ps/86.7.1384.

Zhang, J., Liu, F., Cao, J., & Liu, X. (2015). Skin transcriptome profiles associated with skin color in chickens. *PLoS One, 10*(6), e0127301. https://doi.org/10.1371/journal.pone.0127301.

Zhao, R., Xu, G.-Y., Liu, Z.-Z., Li, J.-Y., & Yang, N. (2006). A study on eggshell pigmentation: Biliverdin in blue-shelled chickens. *Poultry Science, 85*(3), 546–549. https://doi.org/10.1093/ps/85.3.546.

Zhu, Y., Yakhkeshi, S., Yusuf, A., & Zhang, X. (2025). Frontiers and emerging topics in a century of Silkie chicken research: Insights, challenges, and opportunities. *Poultry Science, 104*(5), 105030. https://doi.org/10.1016/j.psj.2025.105030.

Chapter 9

Pasture-based systems: laying hens

Sabine G. Gebhardt-Henrich
Center for Proper Housing: Poultry and Rabbits (ZTHZ), Division of Animal Welfare, Vetsuisse Faculty, University of Bern, Zollikofen, Switzerland

In this chapter, housing systems for laying hens with an outdoor area next to the barn are discussed. These can be referred to as pasture-based or free-range systems, with terminology varying across global regions. The outdoor area is covered with vegetation, although high animal density might destroy plant cover, especially close to the barn (Lolli et al., 2019). In contrast to pastures, which are generally defined as grassland with some nongrass plants for grazing animals (https://en.wikipedia.org/wiki/Pasture), covered verandas, wintergardens, or patios are outdoor structures that are typically paved (https://en.wikipedia.org/wiki/Patio,5.10.2025). The latter can be part of pasture-based systems, but are not discussed in this chapter.

Description of pasture-based systems

Low-intensity pasture production

In countries with emerging economies, backyard flocks, also called village flocks, can contribute as a significant source of nutrition to rural communities (Alders & Pym, 2009; Mujyambere et al., 2022; Saleque, 1996, in Kitalyi, 1996). Most of the birds are indigenous chickens instead of modern breeds or hybrids. In the most extreme case, no barn and no formulated feed are provided, and mortality of the adult fowl and their offspring resembles that of wild birds (Mahoro et al., 2017). Often, women are involved in caring for the chickens, which can provide some income and may strengthen their role in the community, although production levels are low (Gueye, 2000). The natural conditions can be associated with predation, bad weather, and diseases (especially Newcastle), leading to high mortality rates.

In countries with advanced economies, backyard flocks are kept for eggs and meat for the family and friends, but their roles are more as companion animals and as a way to bring a part of nature to an urban environment than for production (Buddle et al., 2025; Elkhoraibi et al., 2014). Awareness and

Integrating Welfare and Sustainability in Poultry Systems. DOI: https://doi.org/10.1016/B978-0-443-40594-5.00014-6

measures of biosecurity are missing in about 50% of the households surveyed in the United States and the United Kingdom (Karabozhilova et al., 2012). Housing and management are largely unregulated, and if regulated, are rarely controlled. Similar to countries with emerging economies, mortality rates due to predation and diseases can be high in these backyard flocks (Madsen et al., 2013a, 2013b).

High-intensity pasture production

Low input and organic production

On a scale ranging from systems without a barn or formulated feed to intensive operations with tens of thousands of laying hens with outdoor access, commercial low-input production, including organic production, ranks in the middle (Fig. 9.1).

Commercial low-input production may involve special hybrids bred for pasture-based systems (Hammershøj et al., 2021) or high-producing hybrids selected for cage housing (Bestman & Wagenaar, 2014). A recent EU project is devoted to low-input and organic production of poultry and pigs (https://ppilow.eu/,27.9.2025), demonstrating that there is a place for this concept.

As an example of low-input systems, mobile housing for egg production is increasingly used in German-speaking countries in Europe. Mobile housing consists of a wagon-like barn providing litter, perches, nests, feeders, and drinkers, which is placed on the fenced pasture. Flock sizes may vary from about 100 to a few 1000 birds, and according to the size of the barn, it is more or less mobile (see examples in EFSA Panel on Animal Health and Animal Welfare [AHAW]) (Nielsen et al., 2023). Small barns may be moved by cars on roads; larger structures on skids are drawn a short distance by tractors. The idea is to provide fresh pasture for the hens throughout their life cycle (Bassler et al., 2000; Fürmetz et al., 2005a, Fürmetz et al., 2005b, in Knierim, 2006).

Organic production always involves access to pasture with additional regulations on feed, which must be organic, space allowance per bird on the pasture, and other specifications, which also differ depending on the country and the specific organic label within countries. For more details on organic production, see Chapter 10.

High input, large-scale commercial production

On the other extreme on scale of intensity, several tens of thousands of laying hens are kept in stationary aviary or floor-housed barns with access to a pasture or several rotating pastures. Due to the growing preference of consumers for eggs from hens with access to a pasture, most pasture-based laying hens are kept in these systems. Regulations of the system can be from the government

FIGURE 9.1 Pasture systems. Pasture systems for laying hens range from low to high intensity production.
From First row picture: Mass, A. (Photographer). Woman feeding chickens near a rustic coop [Photograph]. Pexels. https://www.pexels.com/photo/woman-feeding-chickens-near-rustic-coop-32112377/,

(e.g., EU, Switzerland, and Australia), NGOs through labeling schemes, or entirely left to the industry.

There is one EU regulation (2018/848) that directly applies to systems with access to a pasture. Regulations most often include a maximum stocking density on the pasture and in the barn, vegetation cover, a minimum duration of daily access to the pasture, the time of day when the pasture must be accessible, a minimum proportion of the lifetime of the hen when the pasture can be accessed, natural or artificial shelters, numbers and dimensions of popholes, and maximum flock sizes.

In conclusion, a multitude of pasture-based housing systems exist across and within countries, which poses a challenge to egg consumers when making purchasing decisions based on housing systems. The way audits are conducted also differs, which may lead to varying credibility of welfare-oriented certification labels.

Impact on the animal

Behavior

As housing systems, flock sizes, geographic regions, and genetic backgrounds of the hens vary greatly, the behavior of the laying hens in pasture-based systems also varies. Here, only the behavior relating to pasture use will be discussed.

In the most extensive systems involving feral chickens, sometimes without provision of shelter or feed, the hens' behavior probably resembles that of the wild ancestors most closely, meaning that foraging, including exploration, scratching, and pecking, takes up to 60%–80% of the day (Dawkins, 1989). Due to the presence of predators, these birds choose to spend the night perching on high branches of trees. Scientific studies on the behavior of domesticated laying hens living on an island without human intervention were conducted by Wood-Gush & Duncan (1976). In the first colonization attempt, almost all chickens were killed by predators (mainly mink), and predation remained the main source of mortality for the second colony, where some hens managed to reproduce. Even without feed provision by humans, two groups of chickens managed to survive, apprearing in good condition. The area used for foraging was 0.5 ha for one group and 2 ha for the other group.

As we move down to the end of the scale of Fig. 9.1, behavior may deviate more from wild and feral forms. Although provided with (part-time) access to a pasture, hens use the outdoor areas to varying extents and only during a relatively short portion of the day (Campbell et al., 2016; Chielo et al., 2016). In an epidemiological study of organic laying hens in Sweden, only 13% of the hens on average were seen outside simultaneously (Göransson et al., 2023), and this number is lower in larger flocks than in flock sizes that are common for organic production (Gebhardt-Henrich et al., 2014). In big flocks with large pastures, the density of hens declines with increasing distance from the barn (Chielo et al., 2016; Göransson et al., 2023). The behavior also differs according to the distance to the barn, with walking and foraging shown in the areas that are farthest away from the barn (Chielo et al., 2016).

Feeding predominantly takes place in the barn, and uptake of feed on the pasture is not desired out of fear of contamination with pollutants, parasites, and an unbalanced diet (Gautron et al., 2022; EFSA Panel on Contaminants in the Food Chain (CONTAM), et al., 2018). In support of this concern, eggs from flocks using the pasture might contain more dioxin than eggs from flocks accessing the pasture less (Kijlstra et al., 2007). Vegetation cover, the extent of ranging, and internal factors, such as age of the hens, influence the feed uptake on the pasture, which is estimated to be between 9% and 17% in old and young organic laying hens, respectively (Lorenz et al., 2013). However, studies estimating the uptake of nutrients from the pasture often refer to small organic or experimental flocks (Tosar et al., 2021) and might not be applicable to large commercial flocks with limited time on the pasture and little vegetation cover in the areas close to the barn where most hens are located. Despite the risk of an uptake of pollutants, positive effects of feeding on the pasture exist (Iqbal et al., 2018; Sibanda et al., 2020; Hammershøj & Johansen, 2016). In some systems with small flocks, for example, in organic systems, or in countries with fewer resources, feeding on the pasture is an essential nutritional input (Tufarelli et al., 2018).

Multiple internal (e.g., genetics, experience, age, and personality) and external (e.g., vegetation cover and weather) factors affect the number of hens accessing the outdoor spaces and the frequency and duration of pasture use (Pettersson et al., 2016). Vegetation cover and the type of vegetation (short or tall grass, bushes, or trees) vary widely among farms (Göransson et al., 2023) and may influence the extent to which hens use the pasture (Larsen et al., 2017; Pacher-Deutsch et al., 2025). Areas close to the barn are most heavily frequented by the birds with the least vegetation cover compared to areas further than 40 m from the barn (Lolli et al., 2019). It is well known that shelters made of vegetation or artificial structures increase the number of birds on the pasture (Deutsch et al., 2024; Hegelund et al., 2005; Zeltner & Hirt, 2003). When these structures are closer to the barn, their impact is greater. With an even distribution of shelters, the distribution of hens on the pasture can also be even (Zeltner & Hirt, 2008). Not surprisingly, the weather and the season also have an impact on the extent of ranging (Wurtz et al., 2022). Hot sunny weather might cause hens to seek shelter, so shaded pastures can still be used, but sunny areas might be avoided. Weather can also influence the distribution of birds on the pasture. In a study in Sweden, temperatures and humidities had more influence on the use of the outer areas of the pasture than on the areas closer to the barn (Göransson et al., 2023).

Flock sizes have an impact on the use of pasture (Gebhardt-Henrich et al., 2014). Although the number of laying hens that access the range does not differ among flock sizes between 2000 and 18,000 hens, the frequency and duration of visits to the pasture are higher in hens from smaller flocks (Gebhardt-Henrich et al., 2014). It is unclear whether flock size per se has such an effect. When the flock size is larger, the barns and covered verandas are larger and may cause hens in bigger flocks to spend more time inside than hens in small flocks with small barns. However, other aspects of flock size might contribute to this, and flock size and barn size are usually confounded.

The environment during rearing can influence the use of the pasture during the laying period (Campbell et al., 2020). The environment during rearing seldom includes access to outdoor areas and might be an obstacle to ranging behavior during the production phase. In support of this, exposure to outdoor environments during rearing increases the use of the pasture as well as lowers the fear level measured by tonic immobility in laying hens (Grigor et al., 1995). Thus there might be a mismatch between the rearing and production housing systems regarding access to pasture. When pullets have access to pasture, management techniques must be adapted (Du et al., 2022). This could be alleviated by providing access to covered verandas during rearing, as is commonly done in Switzerland, but experimental studies on the provision of covered verandas during rearing are lacking.

Besides external factors, genotype (Banerjee et al., 2014; Wurtz et al., 2022), age, experience (habituation), fearfulness, and injuries affect hens' likelihood of accessing the pasture. Brown hens (Lohmann Tradition [LT]) spend more time on the pasture, whereas white hens (Lohmann Selected Leghorn) move between barn and pasture more frequently than LT (Mahboub et al., 2004). Wurtz et al. (2022) also found differences between white and brown laying hens. Dekalb White initially used the pasture more than Bovans Brown, but after 2 months, the pattern switched. Interestingly, in an experimental study with different genetic lines and different protein content of the diet, both genotype and diet influenced the ranging behavior of laying hens (Elwinger et al., 2008).

Fearfulness and anxiety related to open spaces are thought to prevent hens from using pastures (Keeling et al., 1988). On the other hand, access to pasture might have an impact on fear levels. Controversial evidence exists on the fear levels of hens with and without access to a pasture, which might be due to different methodologies and different types of fear. Tonic immobility, which measures the fear towards humans, was shorter in hens with access to a pasture than in hens without access at three different ages (Gregory et al., 1991; Oke & Onagbesan, 2024). In support of this, at the level of the individual, hens with a higher fear level spent less time on the pasture than hens with lower fear levels (Campbell et al., 2016). Probably related to fear level, social factors may enhance ranging behavior as was shown for broilers (Ferreira et al., 2020), and the presence of a dominant or several unfamiliar birds can hinder ranging behavior (Grigor et al., 1995). A weak social network was also found by Gómez et al. (2022). As individual hens are consistent in the temporal pattern of ranging behavior, the same group may be on the pasture, and their relatively small number might enable them to recognize each other even in large flocks. Being together with familiar conspecifics might alleviate fear, which is called social buffering (Kikusui et al., 2006). Methylation patterns in the nidopallium explain different time patterns of using the pasture within groups of hens (Guerrero-Bosagna et al., 2020).

In most cases, ranging behavior increases with age, which is confounded with experience (e.g., Campbell et al. (2017) (2018) (2020), but see Hegelund et al. (2005)). The impact of injuries such as keel bone fractures or wounds on the comb on ranging is unclear. Whereas Richards et al. (2012) found birds with keel bone fractures to range less than intact hens, other studies did not detect an influence (Banerjee et al., 2014; Larsen et al., 2018; Wurtz et al., 2023) or even found more ranging behavior in hens with keel bone fractures (Bestman et al., 2019). The high prevalence of keel bone fractures found in flocks might prevent the detection of a relationship with the degree of ranging. However, as with most of the other cited studies, relationships are derived from observational studies and, therefore, causation remains unclear.

In conclusion, since pastures supposedly enable laying hens to forage and explore, dustbathe, and expose their skin to sunlight, the properties of the pastures determine whether the hens can use these areas for that purpose. So far, it is largely unclear how pasture use affects the movement patterns inside the barn. Correlational relationships between ranging and the use of zones in an aviary barn exist, but causal relationships are unclear (Kolakshyapati et al., 2020; Sibanda et al., 2020).

Health and animal welfare

Various definitions of animal welfare exist, which are summarized by Dawkins (2012) as a state where animals are healthy and they have what they want. Different definitions of health and welfare can lead to different outcomes for how noncage housing and access to pasture are evaluated in terms of animal welfare. The possibility of death through predation can be seen as a severe threat to animal welfare. Death by predation may not always be quick and painless, and can lead to fear and pain. When predation events are observed by conspecifics, it can lead to predation stress for many birds (EFSA Panel on AHAW et al., 2023).

Systems with access to pastures seem more natural and should enable more natural behavior, which the public associates with higher welfare (Bray & Ankeny, 2017). However, pastures may have bare areas and, in that case, do not resemble the natural habitat of the ancestors of chickens. When pastures lack features that are essential for chickens, such as shelters, shade, opportunities to forage, and soil that is useful for dustbathing, they might not enable natural behavior. In addition, not everything that is natural is automatically beneficial for animals, such as predation, starvation, weather extremes, etc. (Dawkins, 2023).

However, there might be a fundamental difference between natural stressors, such as predation, bad weather, etc., that laying hens are adapted to deal with through their evolutionary past, and chronic stressors due to unnatural housing conditions to which chickens have not been adapted in the past (Wingfield et al., 1997). There is some evidence that the distinction between natural stressors and stressors through housing and management may apply to laying hens. Natural stressors tend to be short-lived, and acute stress is regarded as less severe for health than chronic stress (López-Moraga et al., 2022) for rodents. Injurious abnormal behaviors, such as feather- and toe-pecking and cannibalism, lead to poor welfare in many laying hen flocks. Finally, piling of hens, including smothering, is a yet unexplained behavior which occurs inside barns, covered verandas, but also on pasture. Its prevalence varies among farms but can be a significant source of mortality (Chowdhury et al., 2024).

At best, access to a pasture both benefits the animals using the pasture and those that stay in the barn due to lower densities inside (Bestman et al., 2019). Studies have shown positive as well as negative effects on the health and wellbeing of laying hens in pasture-based systems (see review by Campbell et al. (2021)). This is not surprising considering the different barn and pasture systems, the different extent of the actual use of the pasture by the hens, and the different study designs, from epidemiological studies comparing cages with free-range (e.g., Coton et al., 2019) to epidemiological studies on pasture-based farms (e.g., Chowdhury et al., 2024) and comparisons of hens hardly using the pastures with flock mates using the pastures a lot (e.g., Oke & Onagbesan, 2024).

As mentioned before, predation by raptors, foxes, or other predators can cause instantaneous death (a problem for the farmer), but also injuries and pain, as well as predation stress for all birds on the pasture witnessing the incident. In a study in the Netherlands, 40% of the flocks experienced predation (Bestman & Wagenaar, 2014). Laying hens using the pasture more frequently had indications of raised stress levels compared to their flock mates using the pasture less (Armstrong et al., 2020). This result was based on neurogenesis in the hippocampus. This was a correlational study where all birds had access to the pasture, and variation in ranging reflected individual choices. In contrast, another study showed that hens with access to a pasture had lower stress levels than hens without access, as indicated by their H/L ratios (Fitra et al., 2021). However, behavioral tests revealed that hens with access to a pasture were less fearful than nonrangers (Campbell et al., 2019), which was confirmed by Bari et al. (2021), Campo et al. (2008) in two Spanish breeds, and in small experimental groups when fear level was assessed by tonic immobility (Hartcher et al., 2016). Although the farmer can influence predation through the position and structure of the pasture to some extent

(Dal Bosco et al., 2014), predation is a possible threat to health and welfare in all flocks. Recently, a system has been developed to detect hawks, which could be a possible tool for farmers to reduce bird-of-prey risks (Smith et al., 2024).

In addition to death or injury by predation, the presence of wild animals on the pasture (e.g., wild birds) also poses a threat to the health of the hens due to viral and bacterial infections, as well as exo and endoparasites (see review by Bonnefous et al., 2022). Rotation of pastures can reduce the rate of worm infections (Bethonico Terra et al., 2021). However, Bestman & Wagenaar (2014) did not find differences in mortality levels between flocks with and without access to a pasture. Biosecurity is an issue in systems with and without access to a pasture, but might be used as an argument against pastures (Galvão et al., 2018). In areas with a risk of avian influenza, wild infected birds may infect free-range hens (Bestman et al., 2018). This may cause states or regions to restrict the use of pastures for laying hens in general or during times of increased risk. Removing pasture access during disease outbreaks could have negative welfare impacts on birds that are accustomed to this outdoor resource, but to date, there is limited evidence quantifying this. Unpublished data (Gebhardt-Henrich et al.) showed that the litter area in the barn serves as a substitute for the covered veranda when the latter is closed and may mitigate welfare impacts under necessary lockdown practices.

Despite these risks to animal health and welfare (Elson, 2015), pastures can have a positive effect as perceived by the public and farmers (Göransson et al., 2023). In support of this, high pasture use correlates with less feather pecking at a flock level (El-Lethey et al., 2000; Green et al., 2000), which is plausible under the assumption that feather pecking is a redirected form of foraging behavior (Rodenburg et al., 2013). At an individual level, comparing hens using the outdoor space and those that do not within small flocks, ranging birds have less feather damage and less footpad dermatitis (Bari et al., 2020; Rodriguez-Aurrekoetxea & Estevez, 2016). Likewise, in an experimental flock in Australia, laying hens using the pasture had fewer injuries on the comb and better plumage (Bari et al., 2020; Michel et al., 2012) than hens not using the pasture, which was also shown in Dutch flocks (Bestman & Wagenaar, 2014; Bestman et al., 2019) but not in small experimental groups (Hartcher et al., 2016). Within French flocks, hens using the pasture more had better feather coverage than hens using the pasture less (Coton et al., 2019). Within a flock, better plumage was seen in the birds that used the pasture farther away from the barn than in birds that were on the pasture close to the barn (Chielo et al., 2016).

Access to a pasture gives hens a level of agency because they can decide to leave the barn and go outside with different light levels and different temperatures. Moving to the pasture and back into the barn implies activity that may benefit the hen. However, studies showing health benefits of hens using the pasture are correlational studies, and hens in good condition may use the pasture more than hens in poor condition. Alternatively, using the pasture may boost the immune system and lead to better body condition. As footpad dermatitis is related to moist litter, spending time on the pasture may indeed make the hen less susceptible to footpad dermatitis, depending on the quality of the pasture. Nothing is known about the effect of fresh air and exposure to the sun. Pasture access may have implications for respiratory health as well: ammonia levels can be high inside the barn, which poses a health risk to chickens and workers in the barn. However, even hens spending some time outside will still be exposed to the noxious gases inside the barn most of the time. Birds can synthesize vitamin D when exposed to sunlight, but vitamin D and its precursors are normally sufficiently provided in the feed.

Surprisingly, when flocks of laying hens are given access to a pasture during the production period, the composition of the microbiota in the cloaca does not change (Schreuder et al., 2021). This result could be due to the small sample size of two flocks with and two flocks without access in the study design, the stability of the microbiome after the intervention, and the limited feed uptake on the pasture. Furthermore, the actual time on the pasture of the tested chickens was unknown and might have been too limited to make a change. When first given access, laying hens are known to be hesitant to use a pasture (Campbell et al., 2020). Another study demonstrated an influence of ranging on cecal microbiota (Bari et al., 2022), but overall, research on microbiota and range use is limited (Pires et al., 2024).

Despite the documented advantages of access to pastures for the health and welfare of laying hens, the risks of pastures for health and welfare led Holt et al. (2011); Holt (2021) to conclude that the risks outweighed the advantages and that access to pasture was bad for the hens as well as for the quality of the eggs. Finally, the weighing of the positive and negative factors and their severity for the specific production system and flock determines the outcome of the trade-off between positive and negative impacts on animal welfare.

Impact on production of a pastured system

Pasture-based systems can have negative impacts on overall system output for several reasons. First, the land devoted to the laying hens cannot be used for other purposes easily. Orchards seem ideal habitats for laying hens, but the hens might eat the fruits, which would be otherwise sold for a high price. On an egg-producing farm in Switzerland with cherry trees (espalier fruit), the hens were observed to consume the cherries up to a height of 80 cm. The hens' feed consumption dropped when the cherries were ripe, but the value of the cherries was higher than that of the chicken feed (personal observation). Double usage of pastures must take biosecurity issues into account. When plots are rotated, even more land must be reserved for egg production, with other limited options. There is not much hard evidence for the transmission of (zoonotic) diseases among different species of livestock, although this is still a theoretically potential threat. However, in a study with a common pasture for chickens and pigs, no increased prevalence of *E. coli* and *Campylobacter spp.* wes detected (Kaeder et al., 2022). For more discussion on silvopasture systems, see Chapter 11.

Pasture management and the management of the outdoor flock take time, which could be used for other activities. Predation decreases the number of hens and thus the number of eggs. Access to a pasture may lead to higher activity levels in some of the flock, which will increase feed consumption and feed conversion rate, leading to higher feed costs and reduced economic returns. Accordingly, hens using the pasture had lower body weights than hens using the pasture less (Bari et al., 2020; Karsten et al., 2010; Michel et al., 2012), but ranging birds were heavier in an experimental study (Iqbal et al., 2018). Hens might have to be treated for ecto and endoparasites that they pick up outside, which also leads to expenses in money and time. Epidemics such as avian flu can lead to the culling of an entire flock. On an individual farm level, a multitude of factors influence production. Thus the number of eggs produced does not differ in any general way between flocks with and without access to a free range (Häne et al., 2000).

Regardless of discussions of quality, pastured/free-range eggs can be chemically (Barbosa et al., 2014) and physically (Huang et al., 2024; Yang et al., 2020) distinguished from cage eggs. The chemical elements Cd, Mn, Mg, Se, and K are distinctly different between the shells of cage eggs and free-range eggs with a reliability of above 90% (Dekker et al., 2011; Xie et al., 2023). However, it is unclear whether barn eggs and free-range eggs can be distinguished in general because not all hens in a free-range system use the pasture. More related to quality differences between eggs from pastured hens and hens without access to a pasture, eggs from hens that consume grasses on the pasture have more vitamin E and omega-3 fatty acids compared to eggs from hens fed commercial hen mash that is not specifically fortified with omega-3 fatty acids (Karsten et al., 2010). The eggs from hens using the pasture also contain more carotenoids than eggs from hens without access to a pasture (Fitra et al., 2021). However, the uptake of grasses cannot be controlled in hens with access to a pasture, and the addition of omega-3 fatty acids to the feed will achieve the same concentrations in the eggs. A thorough review of egg quality in different housing systems revealed that studies controlling for confounding factors such as genetics and age were lacking (Pires et al., 2021).

In summary, the production cost of a pastured egg is higher than that of a barn or cage egg. This can be covered by higher egg prices or by state subsidies, as in Switzerland, where the government and thus the taxpayers provide a fixed subsidy per hen per year when a covered veranda, and even more when a pasture, is available. The rising popularity of free-range eggs shows that consumers and (Swiss) taxpayers are willing to pay these extra costs. In Australia, the majority of supermarket eggs are from free-range systems—an increasing trend driven by consumer preferences.

Emissions

Pasture-based egg production means that barns are not sealed off from the outside climate, but there are popholes where cold air can enter the barn. This can lead to moist litter, which favors the production of ammonia in the litter. The scratching hens release the ammonia into the air, which can then leave the barn. As a result, ventilation schemes must be adapted for barns with popholes (Chen et al., 2021). Similarly, a build-up of feces on the pasture can lead to increased levels of phosphorus and nitrogen in the soil, exceeding fertilization standards (Dekker et al., 2012; Elbe et al., 2005; McGregor et al., 2002; Zoli et al., 2023). Feces and litter in the barn are regularly removed, but the soil next to the barn with the highest density of ranging hens (e.g., Deutsch et al., 2024) is not. Thus the nutrient load on the pasture near the barn often exceeds legal limits (Aarnink et al., 2006). Likewise, ammonia levels in the barns of cage-free housing systems can exceed legal limits and harm the animals (David et al., 2015) and human workers (Snyder et al., 2003) alike. For barns with access to a pasture, these problems are more severe due to the presence of popholes.

Sustainability of pastured hens

The most well-respected definition of sustainability is the ability to meet the needs of the present without compromising the ability of future generations to meet their own needs (World Commission on Environment and Development, 1987), which is not applicable in that form for livestock or animal products that are consumed (see discussion by Rigby and Cáceres, 2001). More practically, sustainability covers the principles of One Health, One Welfare, and the Protection of the Environment. All three principles are directly related to egg production involving hens in pasture-based systems.

One Health means that healthy food for humans can only be produced by healthy animals, which includes positive welfare (Rault et al., 2025). The nutritional quality of the egg is only marginally influenced by the feed for the chickens and not at all by the housing system. As stated before, actual time spent on the pasture is variable among flocks, seasons, and individual hens, and the uptake of vegetation is also low. So, the effects of access to pasture on egg quality and taste are minimal at best. This could be increased by using certain genetic strains bred for ranging.

Food safety, that is, the production of healthy eggs that are free from zoonoses, such as Salmonella and (chemical) pollutants, can be affected using pasture-based systems. Eggs from free-range hens can be contaminated with noxious chemicals when the soil is contaminated (EFSA, 2018). More bacteria are found in the cloaca and on the eggshells of hens with access to pasture compared with caged hens in the United States (Sharma et al., 2022).

However, free-range eggs are not more likely to be infected by Salmonella than cage eggs (Jones et al., 2012, 2015; Wilson et al., 2021).

Most importantly, free-range eggs score high regarding their ethical value. Consumers demand eggs that are produced in animal welfare-friendly systems (Gautron et al., 2021), and pasture-based systems are considered one of them (Campbell et al., 2021). Therefore pasture-based systems are increasing in number (Gautron et al., 2022) in Western societies and in China (Raj & Jayanthi, 2018; Yan et al., 2022).

The impact of egg production on the environment is equally two-sided, such as the issues around food safety and ethics. Concerning the comparatively low CO_2 footprint, eggs are the best protein source for humans (Molnár & Szőllősi, 2020) and therefore score high regarding sustainability in general. However, pasture-based and organic systems require more resources and thus likely have a larger CO_2 footprint than eggs from systems without access to a pasture. Hens with access to pasture have a higher feed conversion rate (Table 1 in Gautron et al., 2022) and thus use up more phosphorus and nitrogen. Since part of these chemicals is deposited on the pasture, this part is lost for production (Dekker et al., 2012). The impact of laying hens on the level of biodiversity in the soil and on the biodiversity of arthropods depends on the farming systems. Laying hens on pastures can contribute to pest and weed control and fertilize the soil. In horticultural fields, they increase the abundance of earthworms, but in orchards, they decrease the number of spiders (Soares et al., 2023).

Conclusion

This chapter aimed to provide an overview of the multiple aspects of pasture-based systems that affect the level of animal welfare for laying hens. It was not intended to be a systematic review, and more studies can be found by reading the cited publications. Ultimately, two main points influence the level of animal welfare in pasture-based systems. First, farm and flock-specific attributes largely determine animal welfare and access to a pasture, access to a pasture does not automatically lead to good welfare but provides the opportunity for positive welfare. Secondly, individual weighing of agency, health, and predation risk might cause different views on the animal welfare level of pasture-based egg production. Finally, avoiding the trade-offs between agency, health, and predation risk is only possible by optimizing management, genetics, and the attributes of the barns and the pastures to achieve optimal welfare for laying hens in pasture-based systems.

References

Aarnink, A. J. A., Hol, J. M. G., & Beurskens, A. G. C. (2006). Ammonia emission and nutrient load in outdoor runs of laying hens. *NJAS: Wageningen Journal of Life Sciences, 54*(2), 223–234. https://doi.org/10.1016/s1573-5214(06)80024-2.

Alders, R. G., & Pym, R. A. E. (2009). Village poultry: Still important to millions, eight thousand years after domestication. *World's Poultry Science Journal, 65*(2), 181–190. https://doi.org/10.1017/s0043933909000117.

Armstrong, E. A., Voelkl, B., Voegeli, S., Gebhardt-Henrich, S. G., Guy, J. H., Sandilands, V., Boswell, T., Toscano, M. J., & Smulders, T. V. (2020). Cell proliferation in the adult chicken hippocampus correlates with individual differences in time spent in outdoor areas and tonic immobility. *Frontiers in Veterinary Science, 7*, 587. https://doi.org/10.3389/fvets.2020.00587.

Banerjee, D., Daigle, C. L., Dong, B., Wurtz, K., Newberry, R. C., Siegford, J. M., & Biswas, S. (2014). Detection of jumping and landing force in laying hens using wireless wearable sensors. *Poultry Science, 93*(11), 2724–2733. https://doi.org/10.3382/ps.2014-04006.

Barbosa, R. M., Nacano, L. R., Freitas, R., Batista, B. L., & Barbosa, F. (2014). The use of decision trees and naïve bayes algorithms and trace element patterns for controlling the authenticity of free-range-pastured hens' eggs. *Journal of Food Science, 79*(9), C1672–C1677. https://doi.org/10.1111/1750-3841.12577.

Bari, M. S., Allen, S. S., Mesken, J., Cohen-Barnhouse, A. M., & Campbell, D. L. M. (2021). Relationship between range use and fearfulness in free-range hens from different rearing enrichments. *Animals, 11*(2), 300. https://doi.org/10.3390/ani11020300.

Bari, M. S., Kheravii, S. K., Bajagai, Y. S., Wu, S.-B., Keerqin, C., & Campbell, D. L. M. (2022). Cecal microbiota of free-range hens varied with different rearing enrichments and ranging patterns. *Frontiers in Microbiology, 12*, 797396. https://doi.org/10.3389/fmicb.2021.797396.

Bari, M. S., Laurenson, Y. C. S. M., Cohen-Barnhouse, A. M., Walkden-Brown, S. W., & Campbell, D. L. M. (2020). Effects of outdoor ranging on external and internal health parameters for hens from different rearing enrichments. *PeerJ, 8*(3), e8720. https://doi.org/10.7717/peerj.8720.

Bassler, A., Ciszuk, P., & Sjelin, K. (2000). Management of laying hens in mobile houses: A review of experiences. In J. E. Hermansen, V. Lund, & E. Thuen (Eds.). *Ecological animal husbandry in the Nordic countries: Proceedings of Seminar No. 303 of the Nordic Association of Agricultural Scientists (NJF)* (pp. 45–50). Horsens, Denmark: Danish Research Centre for Organic Farming, September 16–17, 1999.

Bestman, M., de Jong, W., Wagenaar, J.-P., & Weerts, T. (2018). Presence of avian influenza risk birds in and around poultry free-range areas in relation to range vegetation and openness of surrounding landscape. *Agroforestry Systems, 92*(4), 1001–1008. https://doi.org/10.1007/s10457-017-0117-2.

Bestman, M., Verwer, C., van Niekerk, T., Leenstra, F., Reuvekamp, B., Amsler-Kepalaite, Z., & Maurer, V. (2019). Factors related to free-range use in commercial laying hens. *Applied Animal Behaviour Science, 214*, 57–63. https://doi.org/10.1016/j.applanim.2019.02.015.

Bestman, M., & Wagenaar, J.-P. (2014). Health and welfare in Dutch organic laying hens. *Animals, 4*(2), 374–390. https://doi.org/10.3390/ani4020374.

Bethonico Terra, M. T., Pacheco, W. J., Harrison, M., McCrea, B. A., & Hauck, R. (2021). A survey of coccidia and nematodes in pastured poultry in the state of Georgia. *Avian Diseases, 65*(2), 250–256. https://doi.org/10.1637/aviandiseases-d-20-00120.

Bonnefous, C., Collin, A., Guilloteau, L. A., Guesdon, V., Filliat, C., Réhault-Godbert, S., Rodenburg, T. B., Tuyttens, F. A. M., Warin, L., Steenfeldt, S., Baldinger, L., Re, M., Ponzio, R., Zuliani, A., Venezia, P., Väre, M., Parrott, P., Walley, K., Niemi, J. K., & Leterrier, C. (2022). Welfare issues and potential solutions for laying hens in free range and organic production systems: A review based on literature and interviews. *Frontiers in Veterinary Science, 9*, 952922. https://doi.org/10.3389/fvets.2022.952922.

Bray, H. J., & Ankeny, R. A. (2017). Happy chickens lay tastier eggs: Motivations for buying free-range eggs in Australia. *Anthrozoos, 30*(2), 213–226. https://doi.org/10.1080/08927936.2017.1310986.

Buddle, E. A., Stevens, K., Bray, H. J., & Ankeny, R. A. (2025). The chicken for the egg: Australian motivations for raising backyard chickens. *Anthrozoös, 38*(2), 299–310. https://doi.org/10.1080/08927936.2024.2418698.

Campbell, D. L. M., Dickson, E. J., & Lee, C. (2019). Application of open field, tonic immobility, and attention bias tests to hens with different ranging patterns. *PeerJ, 7*(11), e8122. https://doi.org/10.7717/peerj.8122.

Campbell, D. L. M., Dyall, T. R., Downing, J. A., Cohen-Barnhouse, A. M., & Lee, C. (2020). Rearing enrichments affected ranging behavior in free-range laying hens. *Frontiers in Veterinary Science, 7*, 446. https://doi.org/10.3389/fvets.2020.00446.

Campbell, D. L. M., Hinch, G. N., Downing, J. A., & Lee, C. (2016). Fear and coping styles of outdoor-preferring, moderate-outdoor and indoor-preferring free-range laying hens. *Applied Animal Behaviour Science, 185*, 73–77. https://doi.org/10.1016/j.applanim.2016.09.004.

Campbell, D. L. M., Bari, M. S., & Rault, J.-L. (2021). Free-range egg production: Its implications for hen welfare. *Animal Production Science, 61*(10), 848–855. https://doi.org/10.1071/an19576.

Campbell, D. L. M., Hinch, G. N., Downing, J. A., & Lee, C. (2018). Early enrichment in free-range laying hens: Effects on ranging behaviour, welfare and response to stressors. *Animal, 12*(3), 575–584. https://doi.org/10.1017/s1751731117001859.

Campbell, D. L. M., Hinch, G. N., Dyall, T. R., Warin, L., Little, B. A., & Lee, C. (2017). Outdoor stocking density in free-range laying hens: Radio-frequency identification of impacts on range use. *Animal, 11*(1), 121–130. https://doi.org/10.1017/s1751731116001154.

Campo, J. L., Prieto, M. T., & Dávila, S. G. (2008). Effects of housing system and cold stress on heterophil-to-lymphocyte ratio, fluctuating asymmetry, and tonic immobility duration of chickens. *Poultry Science, 87*(4), 621–626. https://doi.org/10.3382/ps.2007-00466.

Chen, L., Fabian-Wheeler, E. E., Cimbala, J. M., Hofstetter, D., & Patterson, P. (2021). Computational fluid dynamics analysis of alternative ventilation schemes in cage-free poultry housing. *Animals, 11*(8), 2352. https://doi.org/10.3390/ani11082352.

Chielo, L., Pike, T., & Cooper, J. (2016). Ranging behaviour of commercial free-range laying hens. *Animals, 6*(5), 28. https://doi.org/10.3390/ani6050028.

Chowdhury, P., Hemsworth, P. H., Fisher, A. D., Rice, M., Galea, R. Y., Taylor, P. S., & Stevenson, M. (2024). Descriptive epidemiology of smothering in Australian commercial free-range layer hen farms. *Preventive Veterinary Medicine, 223*, 106098. https://doi.org/10.1016/j.prevetmed.2023.106098.

Coton, J., Guinebretière, M., Guesdon, V., Chiron, G., Mindus, C., Laravoire, A., Pauthier, G., Balaine, L., Descamps, M., Bignon, L., Huneau-Salaün, A., & Michel, V. (2019). Feather pecking in laying hens housed in free-range or furnished-cage systems on French farms. *British Poultry Science, 60*(6), 617–627. https://doi.org/10.1080/00071668.2019.1639137.

Dal Bosco, A., Mugnai, C., Rosati, A., Paoletti, A., Caporali, S., & Castellini, C. (2014). Effect of range enrichment on performance, behavior, and forage intake of free-range chickens. *Journal of Applied Poultry Research, 23*(2), 137–145. https://doi.org/10.3382/japr.2013-00814.

David, B., Mejdell, C., Michel, V., Lund, V., & Moe, R. O. (2015). Air quality in alternative housing systems may have an impact on laying hen welfare. Part II-ammonia. *Animals, 5*(3), 886–896. https://doi.org/10.3390/ani5030389.

Dawkins, M. S. (2012). *Why animals matter. Animal consciousness, animal welfare, and human wellbeing*. Oxford: Oxford University Press.

Dawkins, M. S. (1989). Time budgets in Red Junglefowl as a baseline for the assessment of welfare in domestic fowl. *Applied Animal Behaviour Science, 24*(1), 77–80. https://doi.org/10.1016/0168-1591(89)90126-3.

Dawkins, M. S. (2023). Natural behaviour is not enough: Farm animal welfare needs modern answers to tinbergen's four questions. *Animals, 13*(6), 988. https://doi.org/10.3390/ani13060988.

Dekker, S. E. M., Aarnink, A. J. A., de Boer, I. J. M., & Koerkamp, P. W. G. G. (2011). Emissions of ammonia, nitrous oxide, and methane from aviaries with organic laying hen husbandry. *Biosystems Engineering, 110*(2), 123–133. https://doi.org/10.1016/j.biosystemseng.2011.07.006.

Dekker, S. E. M., Aarnink, A. J. A., De Boer, I. J. M., & Koerkamp, P. W. G. G. (2012). Total loss and distribution of nitrogen and phosphorus in the outdoor run of organic laying hens. *British Poultry Science, 53*(6), 731–740. https://doi.org/10.1080/00071668.2012.749342.

Deutsch, M., Dalseg, I. K., Kuchling, S., Sefc, K. M., Erregger, B., & Kalcher-Sommersguter, E. (2024). Space use of free-range laying hens on two outdoor ranges with different amounts of vegetation cover. *Applied Animal Behaviour Science, 277*, 106353. https://doi.org/10.1016/j.applanim.2024.106353.

Du, X., Qin, P., Liu, Y., Amevor, F. K., Shu, G., Li, D., & Zhao, X. (2022). Effects of key farm management practices on pullets welfare—A review. *Animals, 12*(6), 729. https://doi.org/10.3390/ani12060729.

EFSA Panel on Contaminants in the Food Chain (CONTAM), Knutsen, H. K., Alexander, J., Barregård, L., Bignami, M., Brüschweiler, B., Ceccatelli, S., Cottrill, B., Dinovi, M., Edler, L., Grasl-Kraupp, B., Hogstrand, C., Nebbia, C. S., Oswald, I. P., Petersen, A., Rose, M., Roudot, A.-C., Schwerdtle, T., Vleminckx, C., Vollmer, G., ... Hoogenboom, L. R. (2018). Risk for animal and human health related to the presence of dioxins and dioxin-like PCBs in feed and food. *EFSA Journal, 16*(11), e05333. https://doi.org/10.2903/j.efsa.2018.5333.

EFSA Panel on Animal Health and Animal Welfare (AHAW), Nielsen, S. S., Alvarez, J., Bicout, D. J., Calistri, P., Canali, E., Drewe, J. A., Garin-Bastuji, B., Gonzales Rojas, J. L., Gortázar Schmidt, C., Herskin, M., Miranda Chueca, M.Á., Padalino, B., Pasquali, P., Roberts, H. C., Spoolder, H., Stahl, K., Velarde, A., Viltrop, A., Winckler, C., ... Michel, V. (2023). Welfare of laying hens on farm. *EFSA Journal, 21*(2), e07789. https://doi.org/10.2903/j.efsa.2023.7789.

Elbe, U., Ross, A., Steffens, G., Van den Weghe, H., & Winckler, C. (2005). Ökologische Legehennenhaltung in grossen Herden: Spezifische Auslaufnutzung und Nährstoffeintrag [Organic layers in large flocks: Use of the outdoor run and accumulation of nutrients in the soil]. In J. Heß & G. Rahmann (Eds.), *8. Wissenschaftstagung Ökologischer Landbau* (pp. 307–310). Kassel University Press.

Elkhoraibi, C., Blatchford, R. A., Pitesky, M. E., & Mench, J. A. (2014). Backyard chickens in the United States: A survey of flock owners. *Poultry Science, 93*(11), 2920–2931. https://doi.org/10.3382/ps.2014-04154.

El-Lethey, H., Aerni, V., Jungi, T. W., & Wechsler, B. (2000). Stress and feather pecking in laying hens in relation to housing conditions. *British Poultry Science, 41*(1), 22–28. https://doi.org/10.1080/00071660086358.

Elson, H. A. (2015). Poultry welfare in intensive and extensive production systems. *World's Poultry Science Journal, 71*(3), 449–460. https://doi.org/10.1017/s0043933915002172.

Elwinger, K., Tufvesson, M., Lagerkvist, G., & Tauson, R. (2008). Feeding layers of different genotypes in organic feed environments. *British Poultry Science, 49*(6), 654–665. https://doi.org/10.1080/00071660802491519.

Ferreira, V. H. B., Barbarat, M., Lormant, F., Germain, K., Brachet, M., Løvlie, H., Calandreau, L., & Guesdon, V. (2020). Social motivation and the use of distal, but not local, featural cues are related to ranging behavior in free-range chickens (*Gallus gallus* domesticus). *Animal Cognition, 23*(4), 769–780. https://doi.org/10.1007/s10071-020-01389-w.

Fitra, D., Ulupi, N., Arief, I. I., Mutia, R., Abdullah, L., & Erwan, E. (2021). Development of chicken production on free-range system. *Indonesian Bulletin of Animal and Veterinary Sciences, 31*(4), 175. https://doi.org/10.14334/wartazoa.v31i4.2683.

Fürmetz, A., Keppler, C., Knierim, U., Deerberg, F., & Heß, J. (2005a). Legehennen in einem mobilen Stallsystem—Auslaufnutzung und Flächenzustand. In J. Heß, & G. Rahmann (Eds.). *Ende der Nische. Beiträge zur 8. Wissenschaftstagung Ökologischer Landbau* (pp. 302–313). Kassel: Kassel University Press, 1–4 March 2005.

Fürmetz, A., Keppler, C., Knierim, U., Deerberg, F., & Heß, J. (2005b). Legehennen in einem mobilen Stallsystem—Flächenmanagement und resultierende Stickstoffgehalte im Auslauf. In J. Heß, & G. Rahmann (Eds.). *Ende der Nische. Beiträge zur 8. Wissenschaftstagung Ökologischer Landbau* (pp. 299–302). Kassel: Kassel University Press, 1–4 March 2005.

Galvão, J. A., Biondo, A. W., Possebon, F. S., Spina, T. L. B., Correia, L. B. N., Zuim, C. V., Guerra Filho, J. B. P., Pantoja, J. C. F., & Pinto, J. P. D. A. N. (2018). Microbiological vulnerability of eggs and environmental conditions in conventional and free-range housing systems. *Semina: Ciências Agrárias, 39*(1), 133. https://doi.org/10.5433/1679-0359.2018v39n1p133.

Gautron, J., Dombre, C., Nau, F., Feidt, C., & Guillier, L. (2022). Review: Production factors affecting the quality of chicken table eggs and egg products in Europe. *Animal, 16*, 100425. https://doi.org/10.1016/j.animal.2021.100425.

Gautron, J., Réhault-Godbert, S., Van de Braak, T. G. H., & Dunn, I. C. (2021). Review: What are the challenges facing the table egg industry in the next decades and what can be done to address them? *Animal, 15*, 100282. https://doi.org/10.1016/j.animal.2021.100282.

Gebhardt-Henrich, S. G., Toscano, M. J., & Fröhlich, E. K. F. (2014). Use of outdoor ranges by laying hens in different sized flocks. *Applied Animal Behaviour Science, 155*, 74–81. https://doi.org/10.1016/j.applanim.2014.03.010.

Gómez, Y., Berezowski, J., Jorge, Y. A., Gebhardt-Henrich, S. G., Vögeli, S., Stratmann, A., Toscano, M. J., & Voelkl, B. (2022). Similarity in temporal movement patterns in laying hens increases with time and social association. *Animals, 12*(5), 555. https://doi.org/10.3390/ani12050555.

Göransson, L., Abeyesinghe, S., Gunnarsson, S., & Yngvesson, J. (2023). Easier said than done! Organic farmers consider free-ranging important for laying hen welfare but outdoor areas need more shelter – important gaps between research and practice. *British Poultry Science, 64*(5), 544–551. https://doi.org/10.1080/00071668.2023.2220650.

Green, L. E., Lewis, K., Kimpton, A., & Nicol, C. J. (2000). Cross-sectional study of the prevalence of feather pecking in laying hens in alternative systems and its associations with management and disease. *Veterinary Record, 147*(9), 233–238. https://doi.org/10.1136/vr.147.9.233.

Gregory, N. G., Wilkins, L. J., Kestin, S. C., Belyavin, C. G., & Alvey, D. M. (1991). Effect of husbandry system on broken bones and bone strength in hens. *The Veterinary Record, 128*(17), 397–399. https://doi.org/10.1136/vr.128.17.397.

Grigor, P. N., Hughes, B. O., & Appleby, M. C. (1995). Social inhibition of movement in domestic hens. *Animal Behaviour, 49*(5), 1381–1388. https://doi.org/10.1006/anbe.1995.0168.

Guerrero-Bosagna, C., Pértille, F., Gomez, Y., Rezaei, S., Gebhardt-Henrich, S. G., Vögeli, S., Stratmann, A., Voelkl, B., & Toscano, M. J. (2020). DNA methylation variation in the brain of laying hens in relation to differential behavioral patterns. *Comparative Biochemistry and*

Physiology Part D: Genomics and Proteomics, 35, 100700. https://doi.org/10.1016/j.cbd.2020.100700.

Gueye, E. H. F. (2000). Women and family poultry production in rural Africa. *Development in Practice, 10*(1), 98–102. https://doi.org/10.1080/09614520052565.

Hammershøj, M., & Johansen, N. F. (2016). Review: The effect of grass and herbs in organic egg production on egg fatty acid composition, egg yolk colour and sensory properties. *Livestock Science, 194*, 37–43. https://doi.org/10.1016/j.livsci.2016.11.001.

Hammershøj, M., Kristiansen, G., & Steenfeldt, S. (2021). Dual-purpose poultry in organic egg production and effects on egg quality parameters. *Foods, 10*(4), 897. https://doi.org/10.3390/foods10040897.

Häne, M., Huber-Eicher, B., & Fröhlich, E. (2000). Survey of laying hen husbandry in Switzerland. *World's Poultry Science Journal, 56*(1), 21–31. https://doi.org/10.1079/wps20000003.

Hartcher, K. M., Hickey, K. A., Hemsworth, P. H., Cronin, G. M., Wilkinson, S. J., & Singh, M. (2016). Relationships between range access as monitored by radio frequency identification technology, fearfulness, and plumage damage in free-range laying hens. *Animal, 10*(5), 847–853. https://doi.org/10.1017/s1751731115002463.

Hegelund, L., Sørensen, J. T., Kjær, J. B., & Kristensen, I. S. (2005). Use of the range area in organic egg production systems: Effect of climatic factors, flock size, age and artificial cover. *British Poultry Science, 46*(1), 1–8. https://doi.org/10.1080/00071660400023813.

Holt, P. S. (2021). Centennial review: A revisiting of hen welfare and egg safety consequences of mandatory outdoor access for organic egg production. *Poultry Science, 100*(12), 101436. https://doi.org/10.1016/j.psj.2021.101436.

Holt, P. S., Davies, R. H., Dewulf, J., Gast, R. K., Huwe, J. K., Jones, D. R., Waltman, D., & Willian, K. R. (2011). The impact of different housing systems on egg safety and quality. *Poultry Science, 90*(1), 251–262. https://doi.org/10.3382/ps.2010-00794.

Huang, M. C., Lin, Q., Cai, H., & Ni, H. (2024). Fast recognition of table eggs from different farming systems using physical traits and multi-layer perceptron. *Revista Brasileira de Ciencia Avicola / Brazilian Journal of Poultry Science, 26*(3). https://doi.org/10.1590/1806-9061-2023-1895.

Iqbal, Z., Roberts, J., Perez-Maldonado, R. A., Goodarzi Boroojeni, F., Swick, R. A., & Ruhnke, I. (2018). Pasture, multi-enzymes, benzoic acid and essential oils positively influence performance, intestinal organ weight and egg quality in free-range laying hens. *British Poultry Science, 59*(2), 180–189. https://doi.org/10.1080/00071668.2017.1403566.

Jones, D. R., Anderson, K. E., & Guard, J. Y. (2012). Prevalence of coliforms, Salmonella, Listeria, and Campylobacter associated with eggs and the environment of conventional cage and free-range egg production. *Poultry Science, 91*(5), 1195–1202. https://doi.org/10.3382/ps.2011-01795.

Jones, D. R., Cox, N. A., Guard, J., Fedorka-Cray, P. J., Buhr, R. J., Gast, R. K., Abdo, Z., Rigsby, L. L., Plumblee, J. R., Karcher, D. M., Robison, C. I., Blatchford, R. A., & Makagon, M. M. (2015). Microbiological impact of three commercial laying hen housing systems. *Poultry Science, 94*(3), 544–551. https://doi.org/10.3382/ps/peu010.

Kaeder, E., Dorn-In, S., Gareis, M., & Schwaiger, K. (2022). Symbiotic husbandry of chickens and pigs does not increase pathogen transmission risk. *Foods, 11*(19), 3126. https://doi.org/10.3390/foods11193126.

Karabozhilova, I., Wieland, B., Alonso, S., Salonen, L., & Häsler, B. (2012). Backyard chicken keeping in the greater London urban area: Welfare status, biosecurity and disease control issues. *British Poultry Science, 53*(4), 421–430. https://doi.org/10.1080/00071668.2012.707309.

Karsten, H. D., Patterson, P. H., Stout, R., & Crews, G. (2010). Vitamins A, E and fatty acid composition of the eggs of caged hens and pastured hens. *Renewable Agriculture and Food Systems, 25*(1), 45–54. https://doi.org/10.1017/s1742170509990214.

Keeling, L. J., Hughes, B. O., & Dun, P. (1988). Performance of free-range laying hens in a polythene house and their behavior on range. *Farm Building Progress, 94*, 21–28 S.

Kijlstra, A., Traag, W. A., & Hoogenboom, L. A. P. (2007). Effect of flock size on dioxin levels in eggs from chickens kept outside. *Poultry Science, 86*(9), 2042–2048. https://doi.org/10.1093/ps/86.9.2042.

Kikusui, T., Winslow, J. T., & Mori, Y. (2006). Social buffering: Relief from stress and anxiety. *Philosophical Transactions of the Royal Society B: Biological Sciences, 361*(1476), 2215–2228. https://doi.org/10.1098/rstb.2006.1941.

Kitalyi, A. J. (1996). *Village chicken production systems in developing countries: What does the future hold?*. https://www.fao.org/4/w6437t/w6437t07.htm.

Knierim, U. (2006). Animal welfare aspects of outdoor runs for laying hens: a review. *NJAS: Wageningen Journal of Life Sciences, 54*(2), 133–145. https://doi.org/10.1016/s1573-5214(06)80017-5.

Kolakshyapati, M., Taylor, P. S., Hamlin, A., Sibanda, T. Z., Vilela, J. D. S., & Ruhnke, I. (2020). Frequent visits to an outdoor range and lower areas of an aviary system is related to curiosity in commercial free-range laying hens. *Animals, 10*(9), 1706. https://doi.org/10.3390/ani10091706.

Larsen, H., Cronin, G., Smith, C. L., Hemsworth, P., & Rault, J. L. (2017). Behaviour of free-range laying hens in distinct outdoor environments. *Animal Welfare, 26*(3), 255–264. https://doi.org/10.7120/09627286.26.3.255.

Larsen, H., Hemsworth, P. H., Cronin, G. M., Gebhardt-Henrich, S. G., Smith, C. L., & Rault, J.-L. (2018). Relationship between welfare and individual ranging behaviour in commercial free-range laying hens. *Animal, 12*(11), 2356–2364. https://doi.org/10.1017/s1751731118000022.

Lolli, S., Grilli, G., Ferrari, L., Ferrari, P., & Ferrante, V. (2019). Effect of range use on endo- and ectoparasite infestation in italian organic egg production. *Italian Journal of Animal Science, 18*(1), 690–695. https://doi.org/10.1080/1828051x.2018.1564377.

López-Moraga, A., Beckers, T., & Luyten, L. (2022). The effects of stress on avoidance in rodents: An unresolved matter. *Frontiers in Behavioral Neuroscience, 16*, 983026. https://doi.org/10.3389/fnbeh.2022.983026.

Lorenz, C., Kany, T., & Grashorn, M. A. (2013). Method to estimate feed intake from pasture in broilers and laying hens. *European Poultry Science, 77*(3), 160–165. https://doi.org/10.1016/s0003-9098(25)01524-3.

Madsen, J. M., Zimmermann, N. G., Timmons, J., & Tablante, N. L. (2013a). Evaluation of Maryland backyard flocks and biosecurity practices. *Avian Diseases, 57*(2), 233–237. https://doi.org/10.1637/10428-101912-Reg.1.

Madsen, J. M., Zimmermann, N. G., Timmons, J., & Tablante, N. L. (2013b). Prevalence and differentiation of diseases in Maryland backyard flocks. *Avian Diseases, 57*(3), 587–594. https://doi.org/10.1637/10423-101612-Reg.1.

Mahboub, H. D. H., Müller, J., & Von Borell, E. (2004). Outdoor use, tonic immobility, heterophil/lymphocyte ratio and feather condition in free-range laying hens of different genotype. *British Poultry Science, 45*(6), 738–744. https://doi.org/10.1080/00071660400014267.

Mahoro, J., Muasya, T. K., Mbuza, F., Habimana, R., & Kahi, A. K. (2017). Characterization of indigenous chicken production systems in Rwanda. *Poultry Science, 96*(12), 4245–4252. https://doi.org/10.3382/ps/pex240.

McGregor, I. S., Schrama, L., Ambermoon, P., & Dielenberg, R. A. (2002). Not all 'predator odours' are equal: Cat odour but not 2,4,5 trimethylthiazoline (TMT; fox odour) elicits specific defensive behaviours in rats. *Behavioural Brain Research, 129*(1-2), 1–16. https://doi.org/10.1016/s0166-4328(01)00324-2.

Michel, V., Prampart, E., Mirabito, L., Allain, V., Arnould, C., Huonnic, D., Le Bouquin, S., & Albaric, O. (2012). Histologically-validated footpad dermatitis scoring system for use in chicken processing plants. *British Poultry Science, 53*(3), 275–281. https://doi.org/10.1080/00071668.2012.695336.

Molnár, S., & Szőllősi, L. (2020). Sustainability and quality aspects of different table egg production systems: A literature review. *Sustainability, 12*(19), 7884. https://doi.org/10.3390/su12197884.

Mujyambere, V., Adomako, K., Olympio, S. O., Ntawubizi, M., Nyinawamwiza, L., Mahoro, J., & Conroy, A. (2022). Local chickens in East African region: Their production and potential. *Poultry Science, 101*(1), 101547. https://doi.org/10.1016/j.psj.2021.101547.

Oke, O. E., & Onagbesan, O. M. (2024). Impacts of access to legume- or grass-based pasture on behaviour, physiological responses and bacterial load of laying hens. *Heliyon, 10*(14), e34780. https://doi.org/10.1016/j.heliyon.2024.e34780.

Pacher-Deutsch, M., Meyer, P., Meimberg, H., & Gierus, M. (2025). Enhancing range use in free-range laying hen systems: The impact of vegetation cover over time. *Animals, 15*(9), 1204. https://doi.org/10.3390/ani15091204.

Pettersson, I. C., Weeks, C. A., Wilson, L. R. M., & Nicol, C. J. (2016). Consumer perceptions of free-range laying hen welfare. *British Food Journal, 118*(8), 1999–2013. https://doi.org/10.1108/bfj-02-2016-0065.

Pires, P. G.da S., Bavaresco, C., Prato, B. S., Wirth, M. L., & Moraes, P. D. O. (2021). The relationship between egg quality and hen housing systems - A systematic review. *Livestock Science, 250*, 104597. https://doi.org/10.1016/j.livsci.2021.104597.

Pires, Oliveira, McManus, Santos, & Moraes (2024). Impact of housing system on intestinal microbiota of laying hens – A systematic review. *Research in Veterinary Science, 170*, 105184. https://doi.org/10.1016/j.rvsc.2024.105184.

Raj, A. A. G., & Jayanthi, J. G. (2018). IoT-based real-time poultry monitoring and health status identification. In *Proceedings of the 11th international symposium on mechatronics and its applications (ISMA 2018)* (pp. 1–7). https://doi.org/10.1109/ISMA.2018.8330139.

Rault, J.-L., Bateson, M., Boissy, A., Forkman, B., Grinde, B., Gygax, L., Harfeld, J. L., Hintze, S., Keeling, L. J., Kostal, L., Lawrence, A. B., Mendl, M. T., Miele, M., Newberry, R. C., Sandøe, P., Špinka, M., Taylor, A. H., Webb, L. E., Whalin, L., & Jensen, M. B. (2025). A consensus on the definition of positive animal welfare. *Biology Letters, 21*(1), 20240382. https://doi.org/10.1098/rsbl.2024.0382.

Richards, G. J., Wilkins, L. J., Knowles, T. G., Booth, F., Toscano, M. J., Nicol, C. J., & Brown, S. N. (2012). Pop hole use by hens with different keel fracture status monitored throughout the laying period. *Veterinary Record, 170*(19), 494. https://doi.org/10.1136/vr.100489.

Rigby, D., & Cáceres, D. (2001). Organic farming and the sustainability of agricultural systems. *Agricultural Systems, 68*(1), 21–40. https://doi.org/10.1016/s0308-521x(00)00060-3.

Rodenburg, T. B., Van Krimpen, M. M., De Jong, I. C., De Haas, E. N., Kops, M. S., Riedstra, B. J., Nordquist, R. E., Wagenaar, J. P., Bestman, M., & Nicol, C. J. (2013). The prevention and control of feather pecking in laying hens: Identifying the underlying principles. *World's Poultry Science Journal, 69*(2), 361–374. https://doi.org/10.1017/s0043933913000354.

Rodriguez-Aurrekoetxea, A., & Estevez, I. (2016). Use of space and its impact on the welfare of laying hens in a commercial free-range system. *Poultry Science, 95*(11), 2503–2513. https://doi.org/10.3382/ps/pew238.

Saleque, M.A. (1996). *Introduction to a poultry development model applied to landless women in Bangladesh* [Conference paper]. Integrated Farming in Human Development, Development Workers' Course, Tune, Denmark.

Schreuder, J., Velkers, F. C., Bossers, A., Bouwstra, R. J., de Boer, W. F., van Hooft, P., Stegeman, J. A., & Jurburg, S. D. (2021). Temporal dynamics of cloacal microbiota in adult laying chickens with and without access to an outdoor range. *Frontiers in Microbiology, 11*, 626713. https://doi.org/10.3389/fmicb.2020.626713.

Sharma, M. K., McDaniel, C. D., Kiess, A. S., Loar, R. E., & Adhikari, P. (2022). Effect of housing environment and hen strain on egg production and egg quality as well as cloacal and eggshell microbiology in laying hens. *Poultry Science, 101*(2), 101595. https://doi.org/10.1016/j.psj.2021.101595.

Sibanda, T. Z., Walkden-Brown, S. W., Kolakshyapati, M., Dawson, B., Schneider, D., Welch, M., Iqbal, Z., Cohen-Barnhouse, A., Morgan, N. K., Boshoff, J., & Ruhnke, I. (2020). Flock use of the range is associated with the use of different components of a multi-tier aviary system in commercial free-range laying hens. *British Poultry Science, 61*(2), 97–106. https://doi.org/10.1080/00071668.2019.1686123.

Smith, A., Small, J., & Lee, B. K. (2024). Implications for designing hawks detection with data augmentation and network optimizations. In: Deligiannidis, L., Ghareh Mohammadi, F., Shenavarmasouleh, F., Amirian, S., Arabnia, H.R. (eds) Image Processing, Computer Vision, and Pattern Recognition and Information and Knowledge Engineering. CSCE 2024. Communications in Computer and Information Science, vol 2262. Springer, Cham. https://doi.org/10.1007/978-3-031-85933-5_17.

Snyder, M. C., Leopold, D. A., Chiu, B. C., G Von Essen, S., & Liebentritt, N. (2003). The relationship between agricultural environments and olfactory dysfunction. *Journal of Agricultural Safety and Health, 9*(3), 211–219. https://doi.org/10.13031/2013.13686.

Soares, P. R., Guilherme, R., Conceição, A., & Galhano, C. (2023). Soil macrofauna under laying hens' grazed fields in two different agroecosystems in Portugal. *Open Agriculture, 8*(1), 20220172. https://doi.org/10.1515/opag-2022-0172.

Tosar, V., Rousseau, G., Decruyenaere, V., Beckers, Y., Larondelle, Y., & Froidmont, E. (2021). Red clover-rich grassland increases equol concentration in eggs from free-range laying hens. *British Poultry Science, 62*(6), 840–845. https://doi.org/10.1080/00071668.2021.1929069.

Tufarelli, V., Ragni, M., & Laudadio, V. (2018). Feeding forage in poultry: A promising alternative for the future of production systems. *Agriculture, 8*(6), 81. https://doi.org/10.3390/agriculture8060081.

Wilson, A., Chandry, P. S., Turner, M. S., Courtice, J. M., & Fegan, N. (2021). Comparison between cage and free-range egg production on microbial composition, diversity and the presence of Salmonella enterica. *Food Microbiology, 97*, 103754. https://doi.org/10.1016/j.fm.2021.103754.

Wingfield, J. C., Hunt, K., Breuner, C., Dunlap, K., Fowler, G. S., Freed, L., & Lepson, J. (1997). Environmental stress, field endocrinology, and conservation biology. In J. R. Clemmons & R. Buchholz (Eds.), *Behavioral approaches to conservation in the wild* (pp. 95–131). Cambridge University Press.

Wood-Gush, D. G. M., & Duncan, I. J. H. (1976). Some behavioural observations on domestic fowl in the wild. *Applied Animal Ethology, 2*, 255–260. https://doi.org/10.1016/0304-3762(76)90057-2.

World Commission on Environment and Development (1987). *Our Common Future*. United Nations: Oxford University Press.

Wurtz, K. E., Tahamtani, F. M., Foldager, L., Thodberg, K., & Riber, A. B. (2023). Associations of range use with individual behaviour, clinical welfare indicators, fear response and gastrointestinal characteristics of two laying hen hybrids. *Applied Animal Behaviour Science, 265*, 105978. https://doi.org/10.1016/j.applanim.2023.105978.

Wurtz, K. E., Thodberg, K., Berenjian, A., Foldager, L., Tahamtani, F. M., & Riber, A. B. (2022). Commercial layer hybrids kept under organic conditions: A comparison of range use, welfare, and egg production in two layer strains. *Poultry Science, 101*(9), 102005. https://doi.org/10.1016/j.psj.2022.102005.

Xie, S., Hai, C., He, S., Lu, H., Xu, L., Fu, H., & Buccolieri, A. (2023). Discrimination of free-range and caged eggs by chemometrics analysis of the elemental profiles of eggshell. *Journal of Analytical Methods in Chemistry, 2023, 1271409*, 1–8. https://doi.org/10.1155/2023/1271409.

Yan, S., Yang, C., Zhu, L., & Xue, Y. (2022). The potential of understory production systems to improve laying hen welfare. *Animals, 12*(17), 2305. https://doi.org/10.3390/ani12172305.

Yang, Mo, Adetula, Elokil, Akbar Bhuiyan, Huang, & Li (2020). Bilateral apex pubis distance: A novel index for follicular development and egg laying status in domestic hens (*Gallus gallus* domesticus). *British Poultry Science, 61*(2), 195–199. https://doi.org/10.1080/00071668.2019.1697429.

Zeltner, E., & Hirt, H. (2003). Effect of artificial structuring on the use of laying hen runs in a free-range system. *British Poultry Science, 44*(4), 533–537. https://doi.org/10.1080/00071660310001616264.

Zeltner, E., & Hirt, H. (2008). Factors involved in the improvement of the use of hen runs. *Applied Animal Behaviour Science, 114*(3-4), 395–408. https://doi.org/10.1016/j.applanim.2008.04.007.

Zoli, M., Mantovi, P., Ferrari, P., Ferrari, L., & Ferrante, V. (2023). Soil organic matter and nutrient levels in outdoor runs in organic laying farms. *Animals, 13*(3), 401. https://doi.org/10.3390/ani13030401.

Chapter 10

Organic systems: laying hens and broiler chickens

Lina Elmersson
Department of Applied Animal Science and Welfare, Swedish University of Agricultural Sciences (SLU), Skara, Sweden

Introduction

In response to the preceding intensification of agriculture, the organic movement emerged in the early 20th century. This was in pursuit of more sustainable farming practices and a more natural way of living, to promote human health as well as safeguard rural communities from the pressures of urbanization and industrialization (Padel et al., 2004; Vaarst et al., 2004). Public concern over environmental issues further accelerated the development and adoption of alternative agricultural methods during the following decades (Padel et al., 2004).

The organic vision entails an integrated farming system that is adapted and tailored according to the local ecological context, and which relies on internal farm resources rather than external (Box 10.1). It hinges upon the four principles of organic agriculture—health, ecology, fairness and care (Box 10.2)—promoting the health and well-being of soils, plants, animals, humans and the planet as one and indivisible (IFOAM, n.d.).

Although animals were initially regarded primarily as functional components within integrated farming systems, concerns for their welfare gradually emerged as a fundamental aspect of organic livestock production (Padel et al., 2004). Today, high welfare standards are central to organic agriculture. There is particular emphasis on allowing animals to live naturally and express species-specific behaviors as well as a strong focus on disease prevention and the promotion of animal health through the use of appropriate breeds and good living conditions (Vaarst & Alrøe, 2012).

Organic farming may be seen as offering potential for higher animal welfare compared to conventional systems (Murphy & Legrand, 2023), an aspect that resonates strongly with consumers' preferences (Ammann et al., 2024). However, the overall environmental sustainability and capacity of

Integrating Welfare and Sustainability in Poultry Systems. DOI: https://doi.org/10.1016/B978-0-443-40594-5.00009-2

BOX 10.1 Definition of organic agriculture (IFOAM, n.d.)

"Organic Agriculture is a production system that sustains the health of soils, ecosystems, and people. It relies on ecological processes, biodiversity, and cycles adapted to local conditions, rather than the use of inputs with adverse effects. Organic Agriculture combines tradition, innovation, and science to benefit the shared environment and promote fair relationships and good quality of life for all involved."

BOX 10.2 The four principles of organic agriculture (IFOAM, n.d.)

Principle of health: Organic agriculture should sustain and enhance the health of soil, plant, animal, human, and planet as one and indivisible. This principle points out that the health of individuals and communities cannot be separated from the health of ecosystems—healthy soils produce healthy crops that foster the health of animals and people. Health is the wholeness and integrity of living systems. It is not simply the absence of illness, but the maintenance of physical, mental, social, and ecological well-being.

Principle of ecology: Organic agriculture should be based on living ecological systems and cycles, work with them, emulate them, and help sustain them. This principle states that production is to be based on ecological processes. Inputs should be reduced by reuse, recycling, and efficient management of materials and energy. Organic agriculture should attain ecological balance through the design of farming systems, establishment of habitats, and maintenance of genetic and agricultural diversity.

Principle of fairness: Organic agriculture should build on relationships that ensure fairness with regard to the common environment and life opportunities. Fairness is characterized by equity, respect, justice, and stewardship of the shared world, both among people and in their relations to other living beings. Organic agriculture should provide everyone involved with a good quality of life and contribute to food sovereignty and the reduction of poverty.

Principle of care: Organic agriculture should be managed in a precautionary and responsible manner to protect the health and well-being of current and future generations and the environment. Practitioners of organic agriculture can enhance efficiency and increase productivity, but this should not be at the risk of jeopardizing health and well-being. This principle states that precaution and responsibility are the key concerns in management, development, and technology choices in organic agriculture.

organic systems to feed a growing global population have been questioned (Connor, 2021), due to their generally lower yields relative to more intensive nonorganic systems (Clark & Tilman, 2017; Muller et al., 2017; Röös et al., 2018). As such, the relatively high animal welfare standards in organic production may both support and, at times, conflict with other aspects of sustainability.

The focus of this chapter is on some of the key principles of organic meat chicken- and egg production as they relate to animal welfare—including both explicit welfare requirements and those that may have indirect impacts on poultry well-being—and how these intersect with the economic, environmental, and social dimensions of sustainable agriculture.

Some of these key principles are highlighted in Tables 10.1, 10.2, exemplified by the minimum requirements for organic meat chicken and egg production according to the EU regulations on organic production, the USDA Organic Foods Production Act Provisions, and the Australian Certified Organic Standard, alongside the corresponding minimum requirements for conventional production as stipulated in relevant legislation and industry standards for reference.

Organic poultry farming practices: animal welfare and sustainability

Indoor housing and space requirements

Lower stocking density

The stocking density in organic production is, in general, lower compared to conventional production (Tables 10.1, 10.2), and should allow the birds to move freely and engage in species-specific behaviors such as foraging and dust bathing (IFOAM, 2019). Although there is no specific threshold stocking density above which overall bird welfare is clearly compromised (Dawkins et al., 2004), there is a substantial amount of research, especially in broilers, concluding that relatively high stocking densities are, in general, associated with direct and indirect welfare impairments (EFSA, 2023a). For example, in studies of fast-growing broilers comparing 24–25 kg/m^2 (or 8–10 birds/m^2) with higher stocking densities, lower stocking densities have been associated with several welfare benefits such as better gait and skeletal bone health (Sun et al., 2013), improved litter quality (van der Eijk et al., 2023), better foot pad health and lower prevalence of hock burns (Sun et al., 2013; Ventura et al., 2010), higher activity levels (Ventura et al., 2012) and enhanced immune system status (Gomes et al., 2014). Similar welfare improvements have also been observed in slower-growing hybrids commonly used in organic production (van der Eijk et al., 2022, 2023).

While corresponding research in laying hens in noncage systems has not been conducted to the same extent, a few studies comparing stocking densities within the range of 4–12 birds/m^2 indicate that relatively lower stocking densities may improve housing conditions in terms of litter moisture and ammonia emissions, increase free-ranging in hens with outdoor access (Gilani et al., 2014; Steenfeldt & Nielsen, 2015), as well as reduce stress levels (Kang

TABLE 10.1 Minimum standards for laying hens according to Council Directive 1999/74/EC, Regulation (EU) 2018/848, and Commission Implementing Regulation (EU) 2020/464; the US United Egg Producers (UEP) industry guidelines for cage-free housing (2024) and the USDA Organic Foods Production Act Provisions (7 CFR Part 205); the Australian Animal Welfare Standards and Guidelines for Poultry (Dept. of Agriculture, Fisheries and Forestry, 2022) and the Australian Certified Organic Standard (ACOS) 2023. (Göransson & Lundmark Hedman, 2024).

	EU		US		Australia	
	Conventional	**Organic**	**Conventional**	**Organic**	**Conventional**	**Organic**
Cages	Enriched cages permitted[a]	Prohibited	Permitted[b]	Prohibited	Permitted	Prohibited
Stocking density	Max. 9 hens/m^2 usable indoor area	Max. 6 hens/m^2 usable indoor area	At placement, min. 1 ft^2/hen (≅11 hens/m^2)	Max. 4.5 lb/ft^2 or min. 1.5 ft^2/hen (≅7 hens/m^2)	Max. 30 kg/m^2	Max. 16 kg/m^2 incl. roosting areas
Perch space (per hen)	15 cm	18 cm	6 in (=15.24 cm)	6 in (=15.24 cm)	15 cm	18 cm
Lighting environment	Light levels sufficient to see one another, investigate the surroundings and display normal activity levels	Buildings shall permit plentiful natural light to enter	Light intensity min. 0.5 fc (=5.38 lux) at feeders, drinkers, perches, and litter throughout the barn during the daytime lighting period	Artificial light spectrum may not be manipulated to increase feed intake and growth rate	Light intensity must be adequate to allow poultry and equipment to be inspected and any problems to be identified. During light periods this must be min. 10 lux at bird level	Living conditions must consider the natural needs of the animals, for example, direct unfiltered sunlight. Artificial lighting may supplement natural daylight

TABLE 10.1 (Continued)

	EU		US		Australia	
	Conventional	**Organic**	**Conventional**	**Organic**	**Conventional**	**Organic**
Nocturnal rest (hrs per day without artifical light)	About one third of day	Min. 8 h continuous	Lighting period must be no greater than 18 continuous hrs in closed barns or natural day length in open barns. The dark period may be interrupted for one nightly feeding	Min. 8 h continuous	From 16 weeks of age, lighting schedules must provide min. 4 h of continuous darkness and min. 6 h in total. Some exceptions allowed, for example, during first week of life	Min. 8 h continuous
Outdoor access	No requirement	One third of life and 4 m^2/hen[c]	No requirement	Year-round access and min. 1 ft^2 per every 2.25 lb of bird, or min. 3 ft^2/bird (≅0.3 m^2/hen)[c]	No requirement	Fully feathered birds shall have access to pastured areas during the majority of daylight hours. Max. 2500 birds/ha where stock or forage rotations are practiced, or max. 1500 birds/ha for set stocking systems[c]

(Continued)

TABLE 10.1 (Continued)

	EU		US		Australia	
	Conventional	**Organic**	**Conventional**	**Organic**	**Conventional**	**Organic**
Roughage	No requirement	Permanent access to sufficient quantities when kept indoors	No requirement	No requirement besides year-round outdoor access. Plastic pellets as roughage not permitted	Poultry (excl. cage systems) must be provided reasonable access to appropriate substrate for pecking, foraging, and scratching	Roughage, fresh or dried fodder, or silage shall be added to the daily ration for poultry
Prophylactic antibiotics	Prohibited as feed additives, except coccidiostats or histomonostats[d]	Substances to promote growth or production (incl. antibiotics and coccidiostats) shall not be used for preventive treatment	Judicious and responsible use of antibiotics for prevention of a specific disease is permitted, under veterinary oversight, in accordance with policies established by the AVMA and FDA[e]	Prohibited, including ionophorous coccidiostats	Responsible use of antimicrobials as prescribed by a veterinarian in accordance with the national APVA code of practice for the use of antimicrobials in the poultry industry, as aligned with relevant federal and state legislations[f]	Antibiotics, coccidiostats, hormonal growth promoters, or any other substances intended to stimulate growth or production are prohibited
Dietary synthetic amino acids	Permitted	Prohibited	Permitted	Prohibited, with the exception of DL-methionine for poultry (max. 2 lb/ton of feed)	Permitted	Prohibited, with the exception of methionine for poultry

Mutilations (beak trimming)	Permitted to prevent feather pecking and cannibalism (chicks <10 days old)	Prohibited (permitted only as an exception in chicks ≤3 days old)	Permitted	De-beaking prohibited. Beak trimming in chicks >10 days of age is prohibited	Permitted in a hatchery if using an infrared beam within 24 h of take-off (i.e., the removal of hatched chicks from the incubator into a container for transport). Hot-blade beak trimming not permitted, except during outbreaks of injurious feather pecking (only under veterinary advice)	Animal modification involving surgical removal of animal organs or parts shall be kept to a minimum, and carried out with minimization of suffering with appropriate analgesia or anesthetics where required. Systematic debeaking is prohibited, but can be allowed under certain circumstances

[a] *But see the European Citizens' Initiative "End the Cage Age."*
[b] *UEP Animal Welfare Guidelines for U.S. Egg Laying Flocks, available at: https://uepcertified.com/wp-content/uploads/2024/09/2025-UEP-Cage-Guidelines_Final.pdf.*
[c] *Temporary confinement allowed in case of, for example, inclement weather, for nest box training, or when temporary restrictions have been imposed.*
[d] *Stipulated in Regulation (EC) No 1831/2003 of the European Parliament and of the Council of 22 September 2003 on additives for use in animal nutrition.*
[e] *In accordance with FDA policy, available at: https://www.fda.gov/regulatory-information/search-fda-guidance-documents/cvm-gfi-213-new-animal-drugs-and-new-animal-drug-combination-products-administered-or-medicated-feed.*
[f] *Australasian Veterinary Poultry Association (AVPA) code of practice for the use of antimicrobials in the poultry industry (2021 ed.), available at: https://www.avpa.asn.au/wp-content/uploads/files/AVPA-code-of-practice-for-the-use-of-antimicrobials-in-the-poultry-industry-2021.pdf.*
Source: Modified from Göransson, L., & Lundmark Hedman, F. (2024). The perks of being an organic chicken – Animal welfare science on the key features of organic poultry production. Frontiers in Animal Science, 5, 1400384. https://doi.org/10.3389/fanim.2024.1400384.

TABLE 10.2 Minimum standards for broilers according to Council Directive 2007/43/EC, Regulation (EU) 2018/848, and Commission Implementing Regulation (EU) 2020/464; US National Chicken Council Broiler Welfare Guidelines (2022) and the USDA Organic Foods Production Act Provisions (7 CFR Part 205); the Australian Animal Welfare Standards and Guidelines for Poultry (Dept. of Agriculture, Fisheries and Forestry, 2022) and the ACOS 2023.

	EU		US		Australia	
	Conventional	**Organic**	**Conventional**	**Organic**	**Conventional**	**Organic**
Growth rate	Fast-growing breeds permitted	Slow-growing breeds, or reared to a minimum age of 81 days[a]	Fast-growing breeds permitted	Fast-growing breeds permitted	Fast-growing breeds permitted	Preference should be given to slower-growing hybrids. As a guide, meat chicken species should be grown to a minimum age of 70 days
Stocking density (maximum)	33 kg/m^2[b]	21 kg/m^2	Range from 6.5 lb/ft^2 if max. bird weight < 4.5 lb (≅31.7 kg/m^2 if max. bird weight <2 kg) to 9 lb/ft^2 if max. bird weight >7.5 lb (≅ 43.9 kg/m^2 if max. bird weight >3.4 kg)	5 lb of bird per ft^2 (≅24.4 kg/m^2) or min. 2 ft^2/bird	38 kg/m^2 in tunnel ventilated or extractive systems; 38 kg/m^2 (winter) or 36 kg/m^2 (summer) in other mechanically ventilated systems; 28 kg/m^2 in nonmechanically ventilated systems	25 kg/m^2 over the usable area of the shed

Perches and/or raised sitting areas	No requirement	5 cm perch and/or 25 cm^2 raised sitting level per chicken	No requirement	No requirement	No requirement	Chickens, pullets, and laying hens shall have access to sufficient perches to enable normal roosting for roosting birds (no specific minimum requirements for broilers)
Lighting environment	Min. 20 lux during the lighting periods, measured at bird eye level and illuminating min. 80% of the useable area[c]	Buildings shall permit plentiful natural light to enter	There must be a written plan in place for lighting programs appropriate to bird age, size, and activity level[d]	For all fully feathered birds: artificial light spectrum may not be manipulated to increase feed intake and growth rate	Light intensity must be adequate to allow poultry and equipment to be inspected and any problems to be identified. During light periods this must be min. 10 lux at bird level	Living conditions must consider the natural needs of the animals, for example, direct unfiltered sunlight. Artificial lighting may supplement natural daylight
Nocturnal rest (hrs per day without artificial light)	Min. 6 h of which 4 h continuous	Min. 8 h continuous	Min 4 h within a 24-h period, except for the first and last week of growout. While ideal that these four hours are consecutive, they may be divided	Min. 8 h continuous for all fully feathered birds	Min. 6 h of which 4 h continuous, from 7 days of age (except during very hot weather)	Min. 8 h continuous

(Continued)

TABLE 10.2 (Continued)

	EU		US		Australia	
	Conventional	**Organic**	**Conventional**	**Organic**	**Conventional**	**Organic**
Outdoor access	No requirement	One third of life and 4 m^2/chicken[e]	No requirement	Year-round access and min 1 ft^2 per every 5 lb of bird, or min. 2 ft^2/bird (≅0.19 m^2/chicken)[e]	No requirement	Fully feathered birds shall have access to pastured areas during the majority of daylight hours. Max. 4800 birds/ha where stock or forage rotations are practiced, or max. 2500 birds/ha for set stocking systems[e]
Roughage	No requirement	Permanent access to sufficient quantities when kept indoors	No requirement	No requirement besides year-round outdoor access. Plastic pellets as roughage not permitted	Poultry must be provided reasonable access to appropriate substrate for pecking, foraging, and scratching	Roughage, fresh or dried fodder, or silage shall be added to the daily ration for poultry

Prophylactic antibiotics	Prohibited as feed additives, except coccidiostats or histomonostats[f]	Substances to promote growth or production (incl. antibiotics and coccidiostats) shall not be used for preventive treatment	Judicious and responsible use of antibiotics for prevention of a specific disease permitted, under veterinary oversight, in accordance with policies established by the AVMA and FDA[g]	Prohibited, including ionophorous coccidiostats	Responsible use of antimicrobials as prescribed by a veterinarian in accordance with the national APVA code of practice for the use of antimicrobials in the poultry industry as aligned with relevant federal and state legislations[h]	Antibiotics, coccidiostats, hormonal growth promoters, or any other substances intended to stimulate growth or production are prohibited
Mutilations (beak trimming)	Permitted to prevent feather pecking and cannibalism (chicks <10 days old)	Prohibited (permitted only as an exception in chicks ≤3 days old)	Permitted	De-beaking prohibited. Beak trimming in chicks >10 days of age prohibited	Prohibited, unless under veterinary advice	Animal modification involving surgical removal of animal organs or parts shall be kept to a minimum, and carried out with minimization of suffering with appropriate analgesia or anesthetics where required.

(*Continued*)

TABLE 10.2 (Continued)

	EU		US		Australia	
	Conventional	**Organic**	**Conventional**	**Organic**	**Conventional**	**Organic**
						Systematic debeaking is prohibited, but can be allowed under certain circumstances

[a] *The competent authority in each member state shall define the criteria for slow-growing strains.*
[b] *Up to 42 kg/m^2 may be allowed, provided that certain requirements are complied with.*
[c] *A temporary reduction in the lighting level may be allowed when necessary following veterinary advice.*
[d] *Recognizing that there is no optimum light intensity that may be uniformly applied to all broiler chicken strains, house types, lighting sources, or growout programs. Consultation with a veterinarian or poultry welfare specialist is required for the design of a flock lighting program.*
[e] *Temporary confinement allowed in case of, for example, inclement weather or when temporary restrictions have been imposed.*
[f] *Stipulated in Regulation (EC) No 1831/2003 Of The European Parliament and of the Council of 22 September 2003 on additives for use in animal nutrition.*
[g] *In accordance with FDA policy, available at: https://www.fda.gov/regulatory-information/search-fda-guidance-documents/cvm-gfi-213-new-animal-drugs-and-new-animal-drug-combination-products-administered-or-medicated-feed.*
[h] *AVPA code of practice for the use of antimicrobials in the poultry industry (2021 ed.), available at: https://www.avpa.asn.au/wp-content/uploads/files/AVPA-code-of-practice-for-the-use-of-antimicrobials-in-the-poultry-industry-2021.pdf.*
Source: Modified from Göransson, L., & Lundmark Hedman, F. (2024). The perks of being an organic chicken – Animal welfare science on the key features of organic poultry production. Frontiers in Animal Science, 5, 1400384. https://doi.org/10.3389/fanim.2024.1400384.

et al., 2016) and severe feather pecking (Schwarzer et al., 2021). Moreover, EFSA (2023b) recommended a maximum stocking density of four adult laying hens per m^2 to improve welfare. Although the abovementioned health and welfare improvements may, in turn, lead to financial benefits, it has been argued that these are unlikely to offset the revenue losses caused by reduced throughput in broiler production (Jones et al., 2020), as a reduced stocking density means increased costs per m^2 (van Horne, 2020). Some studies demonstrate adverse effects of higher stocking densities on, for example, daily body weight gain, final body weight, and feed intake (Gomes et al., 2014; Sun et al., 2013; van der Eijk et al., 2023), as well as egg production (Kang et al., 2016). Consequently, improved production performance under lower stocking densities may, to some extent, compensate for the associated revenue losses. However, contrasting results have been reported (Bergeron et al., 2020; Ventura et al., 2010), likely reflecting the influence of other housing and management factors on production outcomes (Bergeron et al., 2020; Dawkins et al., 2004).

Regarding the environmental impact of broiler production, Karlsson et al. (2025) found that although lower stocking densities increase per-bird greenhouse gas emissions from sources such as heating and ventilation, the overall carbon footprint was only marginally affected. They therefore concluded that reducing the stocking density from 36 to 24 kg/m^2 would lead to relatively small environmental trade-offs for moderate welfare improvements (Karlsson et al., 2025).

Perch space for laying hens and raised sitting areas for broilers

Several studies show that both laying hens and broilers are highly motivated to perch, that is, to sit in an elevated position, especially during the nighttime, for example (Malchow et al., 2019; Olsson & Keeling, 2000). Some organic standards require additional perch space for laying hens, as well as raised sitting areas for broilers (Tables 10.1, 10.2). For example, the EU regulations on organic production and the Australian Certified Organic Standards require that laying hens are provided a minimum of 18 cm of perch per bird, that is, an additional 3 cm in comparison with corresponding requirements in conventional egg production (Table 10.1). Studies have shown that the horizontal space requirement for perching may be between 15 and 22 cm depending on the size of the laying hen hybrid (Giersberg et al., 2019; Riddle et al., 2018), and it has been argued that at least 18–20 cm perch space per hen is necessary to ensure synchronized perching within a flock (Cook, 2011; Newberry et al., 2001; Savory et al., 2002). For laying hens, especially larger hybrids, some additional perch space may thus enhance welfare, but there is limited research on whether this increase confers measurable health or production benefits (EFSA, 2015).

The EU regulations on organic production also stipulate that broilers must have access to perches or other raised sitting areas, such as platforms

(Table 10.2), although current space allowances permit merely a small proportion of the flock to perch simultaneously (Baxter et al., 2020; Göransson & Lundmark Hedman, 2024). Raised sitting areas have been associated with higher activity levels as well as improved, undisturbed resting in broilers (Bergmann et al., 2017; Elsayed et al., 2024; Lourenço da Silva et al., 2021). Some studies also indicate additional health benefits, such as improved leg health (Bailie et al., 2013; Kaukonen et al., 2017) and reduced prevalence of hock burns and foot pad dermatitis (Lourenço da Silva et al., 2021; Mocz et al., 2022). However, others have failed to demonstrate clear health improvements (de Jong & Gunnink, 2019; de Jong et al., 2021; Elsayed et al., 2024; Jones et al., 2020), which may be due to factors such as quantity and design of the elevated structures, as well as other variables such as bird strain and stocking density.

It has been suggested that the provision of environmental enrichment will, in general, have a negative impact on farm economics due to costs of materials, installation, and management (Jones et al., 2020; van Horne, 2020), and there are also biosecurity risks to consider (Riber et al., 2018). Moreover, high quantities of raised sitting areas would lead to loss of floor space and thus reduced throughput (Jones et al., 2020). While Elsayed et al. (2024) found no impact of straw bales and platforms on production performance, Mocz et al. (2022) reported positive effects on broiler body weight. However, overall knowledge remains limited regarding the impact of raised sitting areas on production performance and whether the additional costs can be balanced by potential health benefits, such as improved leg health (Riber et al., 2018).

Natural light and eight-hours nocturnal rest period

Ultraviolet light

Animals in organic production should also be provided with natural daylight (IFOAM, 2019) through outdoor access and natural light inlets (Tables 10.1, 10.2). Because poultry can perceive UVA wavelengths (315–400 nm) (Prescott & Wathes, 1999b), the ultraviolet light in natural light has important welfare implications, for example, foraging and normal social behaviors (Maddocks et al., 2001; Prescott & Wathes, 1999a). UV light has also been shown to lower stress and fear levels in broilers and laying hens (House et al., 2020; James et al., 2020; Sobotik et al., 2020). James et al. (2020) suggested that this could have in part accounted for the lower mortality and increased growth rate found in broilers reared under UVA light. However, more research is needed on the effects of UV light and production performance (Linhoss et al., 2025).

Similarly, there is limited evidence regarding the effects of UV light supplementation on egg production in laying hens (see Rana & Campbell (2021) for a review). It has, however, been suggested that such effects might be more pronounced in free-range hens exposed to direct sunlight and thus higher UV intensities (England & Ruhnke, 2020). It has also been suggested, although more research is needed, that exposure to UVB light (280–315 nm) could support skeletal health in laying hens (Rana & Campbell, 2021), particularly towards the end of the production cycle (Rana & Campbell, 2021; Wei et al., 2020). Hence, access to natural daylight may be an important factor in supporting extended laying cycles (England & Ruhnke, 2020), which is considered important for promoting sustainable egg production (Bain et al., 2016).

Light intensity

Another important aspect of natural light is light intensity, with levels shown to be over 600 times higher in poultry houses with natural light inlets compared to those using only artificial lighting (Linhoss et al., 2022), which may have important implications for poultry welfare (de Jong & Gunnink, 2019; Prescott & Wathes, 1999b). While empirical evidence on welfare as well as production effects of high light intensities comparable to natural light remains limited (Linhoss et al., 2022), the strategic use of natural light inlets, as seen, for example, on organic broiler farms in Sweden (Göransson et al., 2020), has been proposed as a way to achieve energy savings by reducing reliance on artificial lighting (Linhoss et al., 2025).

Length of scotoperiod

Broilers and laying hens in organic poultry production should be provided a minimum of eight uninterrupted hours of darkness for nighttime rest (Tables 10.1, 10.2). Most research on the effects of photoperiod and scotoperiod length on poultry welfare and production has focused on broilers, and findings demonstrate that a minimum of four consecutive hours of darkness per 24 hours is important for their welfare as it supports the development of a clear circadian rhythm (Bayram & Özkan, 2010; Schwean-Lardner et al., 2014). Dark periods of at least four hours have also been associated with improved leg and foot pad health (Schwean-Lardner et al., 2013; Shynkaruk et al., 2022), enhanced immunocompetence (Hofmann et al., 2020), and reduced fearfulness (Bayram & Özkan, 2010). However, longer dark periods of 8 to 12 hours have been associated with more pronounced welfare benefits (Jiang et al., 2023; Schwean-Lardner et al., 2014). In line with this, EFSA (2023a) recommends providing a minimum of seven to eight hours of darkness for broilers.

Regarding the influence of scotoperiod length on production performance, Jiang et al. (2023) reported higher final body weight and average daily weight gain in broilers provided at least eight hours of darkness, compared to those provided with six hours or fewer. In contrast, Shynkaruk et al. (2022) found that body weight and feed intake were highest in broilers provided a dark period of four hours, compared to both shorter and longer durations. However, feed efficiency improved when broilers were given 7 or 10 hours of darkness.

Thus overall, there is some evidence indicating that the health and welfare benefits of adequate night rest may also contribute positively to production outcomes.

Outdoor access

Outdoor access is a fundamental aspect of organic agriculture, serving not only to promote animal welfare and naturalness but also to support the holistic philosophy of organic farming through reliance on internal farm resources and integration of livestock into the agricultural system (Padel et al., 2004). Free-ranging can offer greater opportunities for important species-specific behaviors, such as foraging and dust bathing (Ferreira et al., 2022; Thuy Diep et al., 2018). It has also been associated with a reduced risk of feather pecking in laying hens (Bestman et al., 2017; Lambton et al., 2010) and improved gait in broilers (Taylor et al., 2018). However, free-range systems may also introduce welfare risks, such as infectious disease transmission and predators (Bonnefous et al., 2022).

Organic standards typically require a minimum outdoor area per bird (Tables 10.1, 10.2), yet studies indicate that actual use of the free-range is often limited, with most hens and chickens staying close to the house—especially when free-range areas lack trees, bushes, or artificial shelters that provide protection and encourage free-ranging behavior (Dawkins et al., 2003; Gilani et al., 2014). Limited free-ranging leads to issues such as inefficient land use and increased nitrogen leaching due to excessive nutrient loading in the soil (Zoli et al., 2023), raising concerns regarding both environmental sustainability and the fulfillment of organic farming values. Additionally, this may also have implications for consumer trust, given their expectations of outdoor access in organic production (Bayer et al., 2023).

For a more comprehensive discussion on the welfare implications of outdoor access for poultry, and how this relates to other aspects of sustainability, please see Chapter 9.

Poultry health and disease management

Prohibition of prophylactic veterinary medicinal treatment (coccidiostats as an example)

Another fundamental principle of organic farming is disease prevention that does not rely on prophylactic use of chemically synthesized allopathic veterinary medicinal products. Instead, good animal health should be promoted through, for example, the selection of suitable breeds and the implementation of appropriate housing and husbandry practices (IFOAM, 2019). For example, organic standards prohibit the routine administration of coccidiostats (Tables 10.1, 10.2). Coccidiosis is a disease of major animal welfare and economic concern in modern large-scale poultry production, causing diarrhea, reduced weight gain, poor feed conversion rate, and increased mortality (Mesa-Pineda et al., 2021). Consequently, some argue that the prophylactic use of coccidiostats is essential to ensure sustainability in modern poultry production systems, safeguarding poultry health and welfare as well as economic viability (Martins et al., 2022). Others contend that a system dependent on routine medicinal treatments cannot, however, be considered genuinely sustainable to begin with (Vaarst et al., 2015).

Anticoccidial vaccines may offer an alternative strategy for controlling coccidiosis in organic production (Jansson et al., 2022). However, the widespread use of these is currently limited by, for example, high costs and the lack of a suitable commercially available recombinant vaccine (see Lee et al. (2022); Martins et al. (2022) for reviews). Other alternative strategies are currently being explored, including different probiotics, prebiotics, phytogenics, and organic acids (Lee et al., 2022). Granstad et al. (2020) found that, overall, such alternative treatments had a positive impact on intestinal health and/or production performance—although the beneficial effects were less pronounced than compared to the commonly used ionophorous coccidiostat narasin.

The animal welfare as well as the economic implications of not using coccidiostats must, however, be considered within an organic poultry production context, where, for example, lower stocking densities and the use of slower-growing broilers are important aspects of health management and disease prevention. For instance, experimental research findings indicate that rapid body weight gain is a predisposing factor for necrotic enteritis in broiler chickens, and that slower-growing hybrids exhibit a more robust immune response to *Eimeria* spp. infection (Dierick et al., 2019; Giles et al., 2019). Furthermore, the routine use of certain coccidiostats in poultry production also raises important public health considerations, particularly concerning the potential contribution of ionophores to the development of antimicrobial resistance to important antibiotics used in human medicine (Simm et al., 2019). However, some argue that this risk is low and unlikely to pose a significant threat to human health (Simjee & Tice, 2023).

Regarding the overall prevalence of antimicrobial-resistant bacteria in poultry production, there are a number of scientific reviews comparing organic and conventional systems (e.g., Ager et al. 2023; Mak et al., 2022; van Wagenberg et al., 2017). Although in general, the levels of antimicrobial resistance seem to be lower in the former. Ager et al. (2023) strongly emphasize that broad-scale conclusions about general patterns of antimicrobial resistance are seriously hampered by a lack of data as well as substantial variations regarding resistance patterns across geographic contexts, pathogens, and hosts. What is evident, however, is that antimicrobial resistance constitutes a growing and widespread problem in organic production despite the prohibition of prophylactic use and restricted therapeutic administration of antibiotics (Ager et al., 2023; Mak et al., 2022).

Poultry breeds appropriate for organic production

Another important aspect of health management and disease prevention in organic production is the choice of animal breeds. These should conform to the principles of organic production, including exhibiting a high degree of genetic diversity, disease resistance, longevity, and the ability to adapt to local, site-specific conditions (IFOAM, 2019). Some organic standards specify that slower-growing hybrids should be used in chicken meat production (Table 10.2). Since their welfare, and any trade-offs and synergies in relation to economic, environmental, and social sustainability, have been covered in Chapter 3, this section will focus on laying hen hybrids in organic production.

In contradiction to the organic ideology and standards intended to safeguard animal health and welfare, organic egg production currently relies on the same high-yielding laying hen hybrids used in commercial nonorganic systems (Bayer et al., 2023). Although these hybrids may exhibit strain-specific differences in traits such as feather pecking propensity (Brinker et al., 2014), susceptibility to keel bone damage (Stratmann et al., 2016), immunocompetence (Hofmann et al., 2021; Schmucker et al., 2021), fearfulness (Brown et al., 2022; Nelson et al., 2020), and use of outdoor areas (Wurtz et al., 2022), no strain clearly meets the dual criteria of high production performance and enhanced animal welfare. Consequently, there is, at present, no commercially available "higher welfare breed" (Fernyhough et al., 2020) that might be more suited for organic egg production.

Over the last decades, genetic selection in laying hens has significantly improved production traits such as egg-laying rate, egg size and quality, and feed conversion rate, while at the same time contributing to a reduced environmental footprint in commercial egg production (Castro et al., 2023). However, breeding for high productivity and early onset of lay has been associated with welfare issues, including keel bone fractures (Thøfner et al., 2021), impaired immunocompetence, and lower disease resistance (Schmucker et al., 2021; Stehr et al., 2019). Hence, the use of alternative laying hen hybrids may offer animal welfare advantages, although their lower production performance would entail trade-offs in yield, resource efficiency, and production costs (Röös et al., 2018). Nevertheless, since the use of hybrids not subject to intense performance-driven physiological demands may lead to improved health outcomes, this could, in turn, enhance economic viability (Röös et al., 2018; Stehr et al., 2019). The financial gains due to improvements in traits such as disease resistance remain to be quantified.

It has been argued that enhancing animal robustness, in accordance with organic standards, requires a fundamental shift in breeding strategies—that emphasizes the preservation of genetic diversity and the promotion of diverse genotypes, to improve livestock adaptability to varying environmental conditions and increase resilience to emerging diseases (Phocas et al., 2016). This, in turn, could support better health and welfare of laying hens throughout extended production cycles, thereby contributing to the long-term environmental and economic sustainability of commercial egg production (England & Ruhnke, 2020).

No mutilations (beak trimming)

Mutilations or physical alterations such as beak trimming are, in general, not allowed in organic production, although beak trimming may sometimes be performed as an exception (Tables 10.1, 10.2). The beak is highly innervated and very sensitive (Gentle, 1989) and thus beak trimming, that is, removing the tip of the beak, which is performed on a routine basis in many poultry production systems, constitutes a severe welfare issue (Glatz & Underwood, 2021). The traditional hot-blade technique has been associated with both acute and chronic pain, and has therefore been largely replaced with infrared beak trimming, which is considered to have less pronounced negative welfare consequences (EFSA, 2023b; Glatz & Underwood, 2021). It has also been shown that an intact beak is important for reducing ectoparasite infestations, and that beak-trimmed hens have reduced ability to remove, for example, lice or mites, which can pose a risk to hen health and welfare (Vezzoli et al., 2015). Beak trimming is implemented as a preventative strategy against severe feather pecking and cannibalism, and plumage damage is generally not as severe or as prevalent in flocks with beak-trimmed hens compared to hens with intact beaks (e.g., Lambton et al. (2010; Sepeur et al., 2015; van Staaveren et al., 2021)). Relatively lower mortality rates have also been observed in the former (Riber & Hinrichsen, 2017; Sepeur et al., 2015). Therefore although the prohibition of beak trimming in organic poultry production protects the birds from the direct pain of the procedure, it may also be considered problematic from an animal welfare perspective (Riber & Hinrichsen, 2017). Moreover, poor plumage cover impairs thermal insulation in hens, which can lead to poorer feed conversion rates and reduced resource efficiency. As such, severe feather pecking has both negative economic and environmental implications (Nicol et al., 2013).

Although beak trimming may mitigate the resulting damage of severe feather pecking, the procedure does not eliminate the behavior (Lambton et al., 2010, 2013). Nor does it address its underlying cause(s), which are more appropriately managed through other preventative strategies, including various housing and management practices such as providing high-quality foraging substrates (EFSA, 2023b). In line with this, a study on public perceptions of mutilations in farm animals found that the acceptance of beak trimming was generally low, as this procedure was perceived as painful and at the same time considered to be unnecessary (Connor & Cowan, 2020).

Organic feed and provision of roughage

Prohibition of dietary synthetic amino acids

In addition to the requirement that animals in organic livestock production systems must be fed organically produced feedstuffs, certain feed additives and supplements are prohibited—including the use of synthetic amino acids (IFOAM, 2019), which is particularly relevant in the context of organic poultry production (van Krimpen et al., 2016). Methionine, and also lysine and tryptophan, for example, are essential amino acids that need to be provided in the poultry diet as these are vital for normal feather synthesis, immune system function, intestinal development and gut health, and protection against oxidative stress (see Alagawany et al. (2020) for a review). Furthermore, amino acid imbalances and inadequate protein levels have been shown to increase the risk of severe feather pecking in laying hens (Mens, van Krimpen & Kwakkel, 2020; Van Krimpen et al., 2005). While some organic standards currently permit the use of synthetic methionine as a feed additive for poultry, others impose a complete ban (Table 10.1). Considering the aforementioned welfare implications, a complete ban on synthetic amino acids poses a significant challenge in terms of safeguarding poultry welfare. However, it can also be argued that permitting dietary synthetic amino acids may have adverse animal welfare implications in a more long-term perspective—as it would enable higher production performance and thereby contribute to the intensification of poultry systems, which would conflict with core organic principles (NAHWOA, 2002).

Environmental sustainability concerns have also been raised in relation to the prohibition of synthetic amino acids in organic poultry feed. This restriction complicates the formulation of nutritionally well-balanced diets that meet the birds' nutritional requirements without the use of excessively high crude protein levels (van Krimpen et al., 2016). High dietary crude protein intake leads to increased nitrogen excretion in poultry manure, which in turn contributes to increased nitrate leaching and thus aquatic and terrestrial eutrophication and acidification (Detzel et al., 2018). The inclusion of synthetic amino acids in poultry diets allows for a reduction in total crude protein content, and can thereby diminish nitrogen emissions and mitigate associated environmental impacts (Detzel et al., 2018; Mosnier et al., 2011). Moreover, although synthetic amino acids have relatively high greenhouse gas emissions per kilogram of product, largely due to energy-intensive manufacturing processes, studies have shown that their overall contribution to the environmental footprint of poultry feed is low, given their relatively low inclusion rates (Benavides et al., 2020; Mosnier et al., 2011).

Alternative protein sources that fulfill the essential amino acid requirements of poultry have gained increased attention in organic production systems. Fishmeal, for example, is a high-quality protein source that contains relatively high methionine levels (Fanatico et al., 2018). However, the availability of organically certified fishmeal is often limited (Chalova et al., 2016), and the production of fishmeal is not always considered to be sustainable (Fanatico et al., 2018). Moreover, the use of fishmeal in laying hen diets has been linked to alarmingly high levels of per- and polyfluoroalkyl substances in organic eggs, raising public health concerns (Granby et al., 2024).

Roughage provision

Some organic standards require that poultry be provided with roughage, fresh or dried fodder, or silage (Tables 10.1, 10.2). Such provisions enhance opportunities for foraging behavior in poultry (Bach et al., 2019; Baxter et al., 2018; Holt et al., 2023). High-fiber diets have also been shown to reduce stress (El-Lethey et al., 2000) as well as performance of severe feather pecking and cannibalism in laying hens (El-Lethey et al. (2000; Patt et al., 2022; Wen et al., 2025), but see Schreiter et al. (2019) for a review). Moreover, several studies have demonstrated that dietary fiber can also stimulate intestinal development and support beneficial gut microbiota, thereby promoting gut health in both laying hens and broilers (see Desbruslais et al. (2021); Jha & Mishra (2021) for reviews). These prebiotic effects may improve nutrient digestibility and absorption, enhance growth performance, and offer protection against enteric pathogens such as *Eimeria* spp. (Jha & Mishra, 2021). However, some contrasting findings suggest that dietary fiber can also act as an antinutritional factor, functioning primarily as a dietary diluent. If so, fiber inclusion may reduce feed intake and impair nutrient retention, leading to decreased production performance in addition to increased feed costs (Jones et al., 2020, but see Jha & Mishra (2021) for a review). Ultimately, the potential health benefits, production outcomes, and economic consequences of dietary fiber inclusion in poultry diets under commercial conditions are highly dependent on several factors, including fiber type, quantity provided, and feed form (Jha & Mishra, 2021).

Animal welfare and sustainability in organic farming—concluding remarks

In organic production systems, the strong emphasis on natural living and providing greater opportunities for animals to engage in highly motivated, species-specific behaviors not only helps mitigate certain negative welfare outcomes but also fosters positive and pleasurable experiences for poultry (Göransson & Lundmark Hedman, 2024). This is consistent with the concept of "positive animal welfare" (Rault et al., 2025). Although there remains room for improvement in actual welfare outcomes, including within organic poultry production, several of the aforementioned housing and management practices are supported by animal welfare science as contributing to a higher welfare potential in these systems (Göransson & Lundmark Hedman, 2024). However, as discussed throughout this chapter, the relatively higher welfare standards in organic poultry systems involve complex trade-offs, as well as potential synergies, across the economic, environmental, and social dimensions of sustainability.

While certain production practices in organic poultry farming—such as lower stocking densities, the use of slower-growing hybrids, and the provision of raised sitting areas for broilers—are associated with improved animal welfare, they can also lead to increased per-bird production costs due to, for example, higher material costs, labor requirements, as well as reduced yields and throughput (van Horne, 2020; van Wagenberg et al., 2017). Moreover, these welfare-oriented practices, along with requirements for outdoor access and organic feed compositions, for instance, may also introduce trade-offs in terms of environmental impact (Bokkers & de Boer, 2009; Castellini et al., 2012; van Wagenberg et al., 2017). Nevertheless, as previously discussed, these economic and environmental trade-offs may be at least partially offset by potential health and welfare benefits, as some housing conditions and management practices can, for example, contribute to lower stress levels and enhanced immunocompetence in birds (El-Lethey et al., 2000; Gomes et al., 2014; Hofmann et al., 2020; James et al., 2020; Sobotik et al., 2020). This may, in turn, positively influence production performance and overall resource efficiency. However, further research is needed to quantify both the financial and environmental benefits associated with such health and welfare improvements.

The challenges of incorporating welfare-related benefits into environmental sustainability assessments of animal farming systems have been emphasized, particularly because the effects of improved animal welfare may be difficult to quantify (Lanzoni et al., 2023; Tallentire et al., 2019). Moreover, life cycle assessments tend to place significant emphasis on productivity and production performance, which often leads to a favorable evaluation of more intensive farming systems in terms of environmental sustainability (Lanzoni et al., 2023; Rocchi et al., 2021). It has therefore been suggested that alternative approaches may be more appropriate for capturing a more holistic understanding of the sustainability of systems such as organic production—by incorporating broader considerations, such as impacts on biodiversity and animal welfare (Lanzoni et al., 2023; Rocchi et al., 2021).

From an environmental sustainability perspective, it may also be argued that certain animal welfare improvements represent a more urgent priority than further marginal gains in specific environmental indicators—particularly given the relatively low overall environmental impact of broiler production compared to other meat production systems (Karlsson et al., 2025). This is especially important considering that feed production has been identified as the main contributor to the environmental impact of both organic and conventional poultry systems, as well as a major component of input costs, where the greatest potential for meaningful improvements therefore lies (Castellini et al., 2012, 2020; Kyriazakis et al., 2025). For a more comprehensive discussion on poultry feed and sustainability, please see Chapter 14.

In terms of economic sustainability, any additional costs associated with meeting some of the aforementioned higher animal welfare standards in organic production can also be partially offset by premium pricing and higher farmgate prices (van Wagenberg et al., 2017). Research shows that many consumers are willing to pay more for animal products from higher-welfare systems (Janssen et al., 2016), and that aspects such as outdoor access and lower stocking densities are particularly valued by consumers and citizens (Ammann et al., 2024; Janssen et al., 2016). Indeed, organic poultry systems often perform well in terms of social sustainability in multicriteria analyses comparing different production systems, not least due to the welfare benefits associated with the use of slower-growing broilers (Castellini et al., 2012; Rocchi et al., 2021). Nevertheless, there are also studies showing that consumers' stated willingness to pay for higher animal welfare does not always translate into actual purchasing behavior (Lai & Yue, 2020) and that there is a discrepancy between consumers' willingness to pay and the actual price premiums of organic food products (Smoluk-Sikorska et al., 2024).

Moreover, consumer expectations regarding organic farming practices and animal welfare outcomes may not always align with practical on-farm realities—for example, the continued use of high-yielding laying hen hybrids—and such mismatched perceptions pose a potential risk to consumer trust in organic labels (Bayer et al., 2023). In addition, the consumer perspective on animal welfare in organic farming could potentially have implications for farmer well-being, as it has been suggested that a "feeling of social responsibility" can contribute to improved mental health among organic farmers (Brigance et al., 2018). Furthermore, given that organic farmers in particular tend to place strong emphasis on natural living as a key component of animal welfare (Bergstra et al., 2017; Bock & van Huik, 2007; Molnár & Fraser, 2021), adherence to organic animal welfare principles may not only benefit the animals but also support farmers' occupational well-being, as a growing body of scientific literature demonstrates a link between the two (Johanssen et al., 2023; Vallance et al., 2023). However, more research is needed to explore this link specifically within the context of organic farming.

Finally, as previously discussed, some organic animal welfare requirements, such as the prohibition of prophylactic administration of coccidiostats and the use of fishmeal in poultry diets, may have implications for public health (Granby et al., 2024; Simm et al., 2019). It has also been suggested that access to pasture and the provision of roughage can lead to higher levels of omega-3 fatty acids in poultry meat and eggs (Mie et al., 2017), and that UV light exposure in laying hens may enhance vitamin D content of eggs (Wei et al., 2020). Although many consumers perceive organic foods as a healthier option (Siegrist & Hartmann, 2019), more research is needed to substantiate any such nutritional advantages of organic poultry production (Mie et al., 2017).

Challenges and future directions

Organic farming has undergone considerable transformations since its origins in the early 20th century. While this evolution has been necessary for the sector's expansion and integration into modern agriculture, it has also prompted concerns about whether current organic practices remain true to the founding principles of the movement (Padel, 2007). Balancing the foundational values of organic agriculture with the practical realities of contemporary farming and market dynamics continues to be a central challenge for the future development and growth of the organic sector (Vaarst et al., 2004). For example, efforts to increase yields and enhance production efficiency may be seen as necessary for organic agriculture to be able to sustainably meet growing global food demands and remain economically competitive with nonorganic systems (Muller et al., 2017; Röös et al., 2018). However, such intensification would undermine and conflict with fundamental organic values. Some have concluded that achieving global food security through large-scale organic farming would ultimately require a substantial reduction in the overall consumption of animal-sourced foods, if the organic system is to remain consistent with its core principles—including high animal welfare standards (Muller et al., 2017; Röös et al., 2018).

References

Ager, E. O., Carvalho, T., Silva, E. M., Ricke, S. C., & Hite, J. L. (2023). Global trends in antimicrobial resistance on organic and conventional farms. *Scientific Reports, 13*(1), 22608. https://doi.org/10.1038/s41598-023-47862-7.

Alagawany, M., Elnesr, S. S., Farag, M. R., Tiwari, R., Yatoo, M. I., Karthik, K., Michalak, I., & Dhama, K. (2020). Nutritional significance of amino acids, vitamins and minerals as nutraceuticals in poultry production and health–A comprehensive review. *Veterinary Quarterly, 41*(1), 1–29. https://doi.org/10.1080/01652176.2020.1857887.

Ammann, J., Mack, G., El Benni, N., Jin, S., Newell-Price, P., Tindale, S., Hunter, E., Vicario-Modroño, V., Gallardo-Cobos, R., Sánchez-Zamora, P., Miškolci, S., & Frewer, L. J. (2024). Consumers across five European countries prioritise animal welfare above environmental sustainability when buying meat and dairy products. *Food Quality and Preference, 117*, 105179. https://doi.org/10.1016/j.foodqual.2024.105179.

Bach, M. H., Tahamtani, F. M., Pedersen, I. J., & Riber, A. B. (2019). Effects of environmental complexity on behaviour in fast-growing broiler chickens. *Applied Animal Behaviour Science, 219*, 104840. https://doi.org/10.1016/j.applanim.2019.104840.

Bailie, C. L., Ball, M. E. E., & O'Connell, N. E. (2013). Influence of the provision of natural light and straw bales on activity levels and leg health in commercial broiler chickens. *Animal, 7*(4), 618–626. https://doi.org/10.1017/s1751731112002108.

Bain, M. M., Nys, Y., & Dunn, I. C. (2016). Increasing persistency in lay and stabilising egg quality in longer laying cycles. What are the challenges? *British Poultry Science, 57*(3), 330–338. https://doi.org/10.1080/00071668.2016.1161727.

Baxter, M., Richmond, A., Lavery, U., & O'Connell, N. E. (2020). Investigating optimal levels of platform perch provision for windowed broiler housing. *Applied Animal Behaviour Science, 225*, 104967. https://doi.org/10.1016/j.applanim.2020.104967.

Baxter, M., Bailie, C. L., & O'Connell, N. E. (2018). Evaluation of a dustbathing substrate and straw bales as environmental enrichments in commercial broiler housing. *Applied Animal Behaviour Science, 200*, 78–85. https://doi.org/10.1016/j.applanim.2017.11.010.

Bayer, E., von Meyer-Höfer, M., & Kühl, S. (2023). Hotspot analysis for organic laying hen husbandry—Identification of sustainability problems as potential risk points to lose consumers' trust. *Organic Agriculture, 13*(2), 261–292. https://doi.org/10.1007/s13165-023-00426-5.

Bayram, A., & Özkan, S. (2010). Effects of a 16-hour light, 8-hour dark lighting schedule on behavioral traits and performance in male broiler chickens. *Journal of Applied Poultry Research, 19*(3), 263–273. https://doi.org/10.3382/japr.2009-00026.

Benavides, P. T., Cai, H., Wang, M., & Bajjalieh, N. (2020). Life-cycle analysis of soybean meal, distiller-dried grains with solubles, and synthetic amino acid-based animal feeds for swine and poultry production. *Animal Feed Science and Technology, 268*, 114607. https://doi.org/10.1016/j.anifeedsci.2020.114607.

Bergeron, S., Pouliot, E., & Doyon, M. (2020). Commercial poultry production stocking density influence on bird health and performance indicators. *Animals, 10*(8), 1253. https://doi.org/10.3390/ani10081253.

Bergmann, S., Schwarzer, A., Wilutzky, K., Louton, H., Bachmeier, J., Schmidt, P., Erhard, M., & Rauch, E. (2017). Behavior as welfare indicator for the rearing of broilers in an enriched husbandry environment—A field study. *Journal of Veterinary Behavior, 19*, 90–101. https://doi.org/10.1016/j.jveb.2017.03.003.

Bergstra, T. J., Hogeveen, H., & Stassen, E. N. (2017). Attitudes of different stakeholders toward pig husbandry: A study to determine conflicting and matching attitudes toward animals, humans and the environment. *Agriculture and Human Values, 34*(2), 393–405. https://doi.org/10.1007/s10460-016-9721-4.

Bestman, M., Verwer, C., Brenninkmeyer, C., Willett, A., Hinrichsen, L. K., Smajlhodzic, F., Heerkens, J. L. T., Gunnarsson, S., & Ferrante, V. (2017). Feather-pecking and injurious pecking in organic laying hens in 107 flocks from eight european countries. *Animal Welfare, 26*(3), 355–363. https://doi.org/10.7120/09627286.26.3.355.

Bock, B. B., & van Huik, M. M. (2007). Animal welfare: The attitudes and behaviour of European pig farmers. *British Food Journal, 109*(11), 931–944. https://doi.org/10.1108/00070700710835732.

Bokkers, E. A. M., & de Boer, I. J. M. (2009). Economic, ecological, and social performance of conventional and organic broiler production in the Netherlands. *British Poultry Science, 50*(5), 546–557. https://doi.org/10.1080/00071660903140999.

Bonnefous, C., Collin, A., Guilloteau, L. A., Guesdon, V., Filliat, C., Réhault-Godbert, S., Rodenburg, T. B., Tuyttens, F. A. M., Warin, L., Steenfeldt, S., Baldinger, L., Re, M., Ponzio, R., Zuliani, A., Venezia, P., Väre, M., Parrott, P., Walley, K., Niemi, J. K., & Leterrier, C. (2022). Welfare issues and potential solutions for laying hens in free range and organic production systems: A review based on literature and interviews. *Frontiers in Veterinary Science, 9*, 952922. https://doi.org/10.3389/fvets.2022.952922.

Brigance, C., Soto Mas, F., Sanchez, V., & Handal, A. J. (2018). The mental health of the organic farmer: Psychosocial and contextual actors. *Workplace Health & Safety, 66*(12), 606–616. https://doi.org/10.1177/2165079918783211.

Brinker, T., Bijma, P., Visscher, J., Rodenburg, T. B., & Ellen, E. D. (2014). Plumage condition in laying hens: Genetic parameters for direct and indirect effects in two purebred layer lines. *Genetics Selection Evolution, 46*(33). https://doi.org/10.1186/1297-9686-46-33.

Brown, A. A., Sobotik, E. B., House, G. M., Nelson, J. R., & Archer, G. S. (2022). Differences in fear response strategy and stress susceptibility amongst four different commercial layer strains reared cage free. *Frontiers in Physiology, 13*, 943471. https://doi.org/10.3389/fphys.2022.943471.

Castellini, C., Boggia, A., Cortina, C., Dal Bosco, A., Paolotti, L., Novelli, E., & Mugnai, C. (2012). A multicriteria approach for measuring the sustainability of different poultry production systems. *Journal of Cleaner Production, 37*, 192–201. https://doi.org/10.1016/j.jclepro.2012.07.006.

Castro, F. L. S., Chai, L., Arango, J., Owens, C. M., Smith, P. A., Reichelt, S., DuBois, C., & Menconi, A. (2023). Poultry industry paradigms: Connecting the dots. *Journal of Applied Poultry Research, 32*(1), 100310. https://doi.org/10.1016/j.japr.2022.100310.

Chalova, V. I., Kim, J., Patterson, P. H., Ricke, S. C., & Kim, W. K. (2016). Reduction of nitrogen excretion and emission in poultry: A review for organic poultry. *Journal of Environmental Science and Health, Part B, 51*(4), 230–235. https://doi.org/10.1080/03601234.2015.1120616.

Clark, M., & Tilman, D. (2017). Comparative analysis of environmental impacts of agricultural production systems, agricultural input efficiency, and food choice. *Environmental Research Letters, 12*(6), 064016. https://doi.org/10.1088/1748-9326/aa6cd5.

Connor, D. J. (2021). What is the real productivity of organic farming systems? *Outlook on Agriculture, 50*(2), 125–129. https://doi.org/10.1177/00307270211017151.

Connor, M., & Cowan, S. L. (2020). Consumer evaluation of farm animal mutilations. *Research in Veterinary Science, 128*, 35–42. https://doi.org/10.1016/j.rvsc.2019.10.006.

Cook, N. J. (2011). Minimally-invasive assessments of the behavioural and physiological effects of enriched colony cages on laying hens. *The Open Agriculture Journal, 5*(1), 10–18. https://doi.org/10.2174/1874331501105010010.

Costantini, M., Lovarelli, D., Orsi, L., Ganzaroli, A., Ferrante, V., Febo, P., Guarino, M., & Bacenetti, J. (2020). Investigating on the environmental sustainability of animal products: The case of organic eggs. *Journal of Cleaner Production, 274*, 123046. https://doi.org/10.1016/j.jclepro.2020.123046.

Dawkins, M. S., Cook, P. A., Whittingham, M. J., Mansell, K. A., & Harper, A. E. (2003). What makes free-range broiler chickens range? In situ measurement of habitat preference. *Animal Behaviour, 66*(1), 151–160. https://doi.org/10.1006/anbe.2003.2172.

Dawkins, M. S., Donnelly, C. A., & Jones, T. A. (2004). Chicken welfare is influenced more by housing conditions than by stocking density. *Nature, 427*(6972), 342–344. https://doi.org/10.1038/nature02226.

de Jong, I. C., Blaauw, X. E., van der Eijk, J. A. J., Souza da Silva, C., van Krimpen, M. M., Molenaar, R., & van den Brand, H. (2021). Providing environmental enrichments affects activity and performance, but not leg health in fast- and slower-growing broiler chickens. *Applied Animal Behaviour Science, 241*, 105375. https://doi.org/10.1016/j.applanim.2021.105375.

de Jong, I. C., & Gunnink, H. (2019). Effects of a commercial broiler enrichment programme with or without natural light on behaviour and other welfare indicators. *Animal, 13*(2), 384–391. https://doi.org/10.1017/s1751731118001805.

Desbruslais, A., Wealleans, A., Gonzalez-Sanchez, D., & di Benedetto, M. (2021). Dietary fibre in laying hens: A review of effects on performance, gut health and feather pecking. *World's Poultry Science Journal, 77*(4), 797–823. https://doi.org/10.1080/00439339.2021.1960236.

Detzel A., Busch M., & Krüger M. (2018). Life cycle assessment of organic feed for laying hens and broilers, taking into account an increasing supplementation with DL-methionine. https://www.ifeu.de/fileadmin/uploads/LCA_Oekolog_Gefluegelhaltung_ifeu_Endbericht-Extended-Summary_Englisch.pdf.

Dierick, E., Hirvonen, O. P., Haesebrouck, F., Ducatelle, R., Van Immerseel, F., & Goossens, E. (2019). Rapid growth predisposes broilers to necrotic enteritis. *Avian Pathology, 48*(5), 416–422. https://doi.org/10.1080/03079457.2019.1614147.

EFSA (2015). Scientific opinion on welfare aspects of the use of perches for laying hens. *EFSA Journal, 13*(6). https://doi.org/10.2903/j.efsa.2015.4131.

EFSA (2023a). Welfare of broilers on farm. *EFSA Journal, 21*(2). https://doi.org/10.2903/j.efsa.2023.7788.

EFSA (2023b). Welfare of laying hens on farm. *EFSA Journal, 21*(2). https://doi.org/10.2903/j.efsa.2023.7789.

El-Lethey, H., Aerni, V., Jungi, T. W., & Wechsler, B. (2000). Stress and feather pecking in laying hens in relation to housing conditions. *British Poultry Science, 41*(1), 22–28. https://doi.org/10.1080/00071660086358.

Elsayed, R. A., Mohammed, A. S., Abou-Elnaga, A. F., Abou-Ismail, U. A., Darwish, R. A., & Mohammed Fouda, M. (2024). Effects of different types of environmental enrichment on behaviour, performance, welfare, carcass characteristics and lipid profile of broiler chickens. *Mansoura Veterinary Medical Journal, 25*(1), 2. https://doi.org/10.35943/2682-2512.1225.

England, A., & Ruhnke, I. (2020). The influence of light of different wavelengths on laying hen production and egg quality. *World's Poultry Science Journal, 76*(3), 443–458. https://doi.org/10.1080/00439339.2020.1789023.

Fanatico, A. C., Arsi, K., Upadhyaya, I., Morales Ramos, J., Donoghue, D., & Donoghue, A. M. (2018). Sustainable fish and invertebrate meals for methionine and protein feeds in organic poultry production. *Journal of Applied Poultry Research, 27*(4), 437–448. https://doi.org/10.3382/japr/pfy037.

Fernyhough, M., Nicol, C. J., van de Braak, T., Toscano, M. J., & Tønnessen, M. (2020). The ethics of laying hen genetics. *Journal of Agricultural and Environmental Ethics, 33*(1), 15–36. https://doi.org/10.1007/s10806-019-09810-2.

Ferreira, V. H. B., Simoni, A., Germain, K., Leterrier, C., Lansade, L., Collin, A., Mignon-Grasteau, S., Le Bihan-Duval, E., Guettier, E., Leruste, H., Løvlie, H., Calandreau, L., & Guesdon, V. (2022). Foraging behavior shows individual-consistency over time, and predicts range use in slow-growing free-range male broiler chickens. *Frontiers in Veterinary Science, 9*, 814054. https://doi.org/10.3389/fvets.2022.814054.

Gentle, M. J. (1989). Cutaneous sensory afferents recorded from the nervus intramandibularis of *Gallus gallus* var domesticus. *Journal of Comparative Physiology A, 164*(6), 763–774. https://doi.org/10.1007/BF00616748.

Giersberg, M. F., Spindler, B., & Kemper, N. (2019). Linear space requirements and perch use of conventional layer hybrids and dual-purpose hens in an aviary system. *Frontiers in Veterinary Science, 6*, 231. https://doi.org/10.3389/fvets.2019.00231.

Gilani, A.-M., Knowles, T. G., & Nicol, C. J. (2014). Factors affecting ranging behaviour in young and adult laying hens. *British Poultry Science, 55*(2), 127–135. https://doi.org/10.1080/00071668.2014.889279.

Giles, T., Sakkas, P., Belkhiri, A., Barrow, P., Kyriazakis, I., & Foster, N. (2019). Differential immune response to *Eimeria maxima* infection in fast- and slow-growing broiler genotypes. *Parasite Immunology, 41*(9), e12660. https://doi.org/10.1111/pim.12660.

Glatz, P. C., & Underwood, G. (2021). Current methods and techniques of beak trimming laying hens, welfare issues and alternative approaches. *Animal Production Science, 61*(10), 968–989. https://doi.org/10.1071/AN19673.

Gomes, A. V. S., Quinteiro-Filho, W. M., Ribeiro, A., Ferraz-de-Paula, V., Pinheiro, M. L., Baskeville, E., Akamine, A. T., Astolfi-Ferreira, C. S., Ferreira, A. J. P., & Palermo-Neto, J. (2014). Overcrowding stress decreases macrophage activity and increases *Salmonella* Enteritidis invasion in broiler chickens. *Avian Pathology, 43*(1), 82–90. https://doi.org/10.1080/03079457.2013.874006.

Göransson, L., & Lundmark Hedman, F. (2024). The perks of being an organic chicken – Animal welfare science on the key features of organic poultry production. *Frontiers in Animal Science, 5*, 1400384. https://doi.org/10.3389/fanim.2024.1400384.

Göransson, L., Yngvesson, J., & Gunnarsson, S. (2020). Bird health, housing and management routines on Swedish organic broiler chicken farms. *Animals, 10*(11), 2098. https://doi.org/10.3390/ani10112098.

Granby, K., Ersbøll, B. K., Olesen, P. T., Christensen, T., & Sørensen, S. (2024). Per- and polyfluoroalkyl substances in commercial organic eggs via fishmeal in feed. *Chemosphere, 346*, 140553. https://doi.org/10.1016/j.chemosphere.2023.140553.

Granstad, S., Kristoffersen, A. B., Benestad, S. L., Sjurseth, S. K., David, B., Sørensen, L., Fjermedal, A., Edvardsen, D. H., Sanson, G., Løvland, A., & Kaldhusdal, M. (2020). Effect of feed additives as alternatives to in-feed antimicrobials on production performance and intestinal clostridium perfringens counts in broiler chickens. *Animals, 10*(2). https://doi.org/10.3390/ani10020240.

Hofmann, T., Schmucker, S., Sommerfeld, V., Huber, K., Rodehutscord, M., & Stefanski, V. (2021). Immunomodulatory effects of dietary phosphorus and calcium in two strains of laying hens. *Animals, 11*(1), 129. https://doi.org/10.3390/ani11010129.

Hofmann, T., Schmucker, S. S., Bessei, W., Grashorn, M., & Stefanski, V. (2020). Impact of housing environment on the immune system in chickens: A review. *Animals, 10*(7), 1–26. https://doi.org/10.3390/ani10071138.

Holt, R. V., Vas, J., Vasdal, G., & Newberry, R. C. (2023). A buffet of litters – Broiler chickens behave differently according to litter type and freshness. *Applied Animal Behaviour Science, 260*, 105856. https://doi.org/10.1016/j.applanim.2023.105856.

House, G. M., Sobotik, E. B., Nelson, J. R., & Archer, G. S. (2020). Effect of the addition of ultraviolet light on broiler growth, fear, and stress response. *Journal of Applied Poultry Research, 29*(2), 402–408. https://doi.org/10.1016/j.japr.2020.01.003.

IFOAM. (2019). IFOAM-Organics International Germany The IFOAM Standard for Organic Production and Processing. In: The IFOAM NORMS for Organic Production and Processing, version 2014. Germany. Accessed June 24, 2025.

IFOAM. (n.d.). 2025 6 24 IFOAM. https://www.ifoam.bio/why-organic/shaping-agriculture/four-principles-organic.

James, C., Wiseman, J., & Asher, L. (2020). The effect of supplementary ultraviolet wavelengths on the performance of broiler chickens. *Poultry Science, 99*(11), 5517–5525. https://doi.org/10.1016/j.psj.2020.07.018.

Janssen, M., Rödiger, M., & Hamm, U. (2016). Labels for animal husbandry systems meet consumer preferences: Results from a meta-analysis of consumer studies. *Journal of*

Agricultural and Environmental Ethics, 29(6), 1071–1100. https://doi.org/10.1007/s10806-016-9647-2.

Jansson, D. S., Höglund, J., Bagge, E., Jinnerot, T., & Kaldhusdal, M. (2022). Anticoccidial vaccination is associated with improved intestinal health in organic chickens. *Veterinary Sciences, 9*(7), 347. https://doi.org/10.3390/vetsci9070347.

Jha, R., & Mishra, P. (2021). Dietary fiber in poultry nutrition and their effects on nutrient utilization, performance, gut health, and on the environment: A review. *Journal of Animal Science and Biotechnology, 12*(1). https://doi.org/10.1186/s40104-021-00576-0.

Jiang, S., Fu, Y., & Cheng, H. W. (2023). Daylight exposure and circadian clocks in broilers: Part I—Photoperiod effect on broiler behavior, skeletal health, and fear response. *Poultry Science, 102*(12), 103162. https://doi.org/10.1016/j.psj.2023.103162.

Johanssen, J. R. E., Kvam, G.-T., Logstein, B., & Vaarst, M. (2023). Interrelationships between cows, calves, and humans in cow-calf contact systems—An interview study among Norwegian dairy farmers. *Journal of Dairy Science, 106*(9), 6325–6341. https://doi.org/10.3168/jds.2022-22999.

Jones, P. J., Tahamtani, F. M., Pedersen, I. J., Niemi, J. K., & Riber, A. B. (2020). The productivity and financial impacts of eight types of environmental enrichment for broiler chickens. *Animals, 10*(3), 378. https://doi.org/10.3390/ani10030378.

Kang, H. K., Park, S. B., Kim, S. H., & Kim, C. H. (2016). Effects of stock density on the laying performance, blood parameter, corticosterone, litter quality, gas emission and bone mineral density of laying hens in floor pens. *Poultry Science, 95*(12), 2764–2770. https://doi.org/10.3382/ps/pew264.

Karlsson, L., Keeling, L., & Röös, E. (2025). What is a better chicken? Exploring trade-offs between animal welfare and greenhouse gas emissions in higher-welfare broiler systems. *Sustainable Production and Consumption, 55*, 203–216. https://doi.org/10.1016/j.spc.2025.02.015.

Kaukonen, E., Norring, M., & Valros, A. (2017). Perches and elevated platforms in commercial broiler farms: Use and effect on walking ability, incidence of tibial dyschondroplasia and bone mineral content. *Animal, 11*(5), 864–871. https://doi.org/10.1017/s1751731116002160.

Kyriazakis, I., Dokou, S., Taylor, J., Giannenas, I., & Murphy, E. (2025). A meta-analysis of the sources of variation in the environmental impacts of different broiler production systems. *British Poultry Science, 66*(2), 193–205. https://doi.org/10.1080/00071668.2024.2409192.

Lai, Y., & Yue, C. (2020). Consumer willingness to pay for organic and animal welfare product attributes: Do experimental results align with market data? *Journal of Agricultural and Resource Economics, 45*(3), 462–483. Available from https://www.jstor.org/stable/27154079.

Lambton, S. L., Knowles, T. G., Yorke, C., & Nicol, C. J. (2010). The risk factors affecting the development of gentle and severe feather pecking in loose housed laying hens. *Applied Animal Behaviour Science, 123*(1-2), 32–42. https://doi.org/10.1016/j.applanim.2009.12.010.

Lambton, S. L., Nicol, C. J., Friel, M., Main, D. C. J., McKinstry, J. L., Sherwin, C. M., Walton, J., & Weeks, C. A. (2013). A bespoke management package can reduce levels of injurious pecking in loose-housed laying hen flocks. *Veterinary Record, 172*(16), 423. https://doi.org/10.1136/vr.101067.

Lanzoni, L., Whatford, L., Atzori, A. S., Chincarini, M., Giammarco, M., Fusaro, I., & Vignola, G. (2023). Review: The challenge to integrate animal welfare indicators into the Life Cycle Assessment. *Animal, 17*(5), 100794. https://doi.org/10.1016/j.animal.2023.100794.

Lee, Y., Lu, M., & Lillehoj, H. S. (2022). Coccidiosis: Recent progress in host immunity and alternatives to antibiotic strategies. *Vaccines, 10*(2), 215. https://doi.org/10.3390/vaccines10020215.

Linhoss, J. E., Davis, J. D., Campbell, J. C., Purswell, J. L., Griggs, K. G., & Edge, C. M. (2022). Comparison of commercial broiler house lighting programs using LED and natural light: Part 1—Spatial and temporal analysis of light intensity. *Journal of Applied Poultry Research, 31*(3), 100272. https://doi.org/10.1016/j.japr.2022.100272.

Linhoss, J. E., Falana, O. B., Davis, J. D., Purswell, J. L., Edge, C. M., Olanrewaju, H. A., Baker-Cook, B. I., & Hanlon, C. (2025). An updated review on the effect of lighting on broilers and the environment of commercial houses. *World's Poultry Science Journal, 81*(2), 373–401. https://doi.org/10.1080/00439339.2024.2446298.

Lourenço da Silva, M. I., de Lima Almeida Paz, I. C., Chaves, G. H. C., de Lima Almeida, I. C., dos Ouros, C. C., de Souza, S. R. L., Milbradt, E. L., Caldara, F. R., Satin, A. J. G., da Costa, G. A., & Glavina, A. S. G. (2021). Behaviour and animal welfare indicators of broiler chickens housed in an enriched environment. *PLoS One, 16*(9), e0256963. https://doi.org/10.1371/journal.pone.0256963.

Maddocks, S. A., Cuthill, I. C., Goldsmith, A. R., & Sherwin, C. M. (2001). Behavioural and physiological effects of absence of ultraviolet wavelengths for domestic chicks. *Animal Behaviour, 62*(5), 1013–1019. https://doi.org/10.1006/anbe.2001.1842.

Mak, P. H. W., Rehman, M. A., Kiarie, E. G., Topp, E., & Diarra, M. S. (2022). Production systems and important antimicrobial resistant-pathogenic bacteria in poultry: A review. *Journal of Animal Science and Biotechnology, 13*(1), 148. https://doi.org/10.1186/s40104-022-00786-0.

Malchow, J., Berk, J., Puppe, B., & Schrader, L. (2019). Perches or grids? What do rearing chickens differing in growth performance prefer for roosting? *Poultry Science, 98*(1), 29–38. https://doi.org/10.3382/ps/pey320.

Martins, R. R., Silva, L. J. G., Pereira, A. M. P. T., Esteves, A., Duarte, S. C., & Pena, A. (2022). Coccidiostats and poultry: A comprehensive review and current legislation. *Foods, 11*(18), 2738. https://doi.org/10.3390/foods11182738.

Mens, A. J. W., van Krimpen, M. M., & Kwakkel, R. P. (2020). Nutritional approaches to reduce or prevent feather pecking in laying hens: Any potential to intervene during rearing? *World's Poultry Science Journal, 76*(3), 591–610. https://doi.org/10.1080/00439339.2020.1772024.

Mesa-Pineda, C., Navarro-Ruíz, J. L., López-Osorio, S., Chaparro-Gutiérrez, J. J., & Gómez-Osorio, L. M. (2021). Chicken coccidiosis: From the parasite lifecycle to control of the disease. *Frontiers in Veterinary Science, 8*, 787653. https://doi.org/10.3389/fvets.2021.787653.

Mie, A., Andersen, H. R., Gunnarsson, S., Kahl, J., Kesse-Guyot, E., Rembiałkowska, E., Quaglio, G., & Grandjean, P. (2017). Human health implications of organic food and organic agriculture: A comprehensive review. *Environmental Health: A Global Access Science Source, 16*, 111. https://doi.org/10.1186/s12940-017-0315-4.

Mocz, F., Michel, V., Janvrot, M., Moysan, J. P., Keita, A., Riber, A. B., & Guinebretière, M. (2022). Positive effects of elevated platforms and straw bales on the welfare of fast-growing broiler chickens reared at two different stocking densities. *Animals, 12*(5), 542. https://doi.org/10.3390/ani12050542.

Molnár, M., & Fraser, D. (2021). Animal welfare during a period of intensification: The views of confinement and alternative pig producers. *Animal Welfare, 30*(2), 121–129. https://doi.org/10.7120/09627286.30.2.121.

Mosnier, E., Van Der Werf, H. M. G., Boissy, J., & Dourmad, J. Y. (2011). Evaluation of the environmental implications of the incorporation of feed-use amino acids in the manufacturing of pig and broiler feeds using Life Cycle Assessment. *Animal, 5*(12), 1972–1983. https://doi.org/10.1017/S1751731111001078.

Muller, A., Schader, C., El-Hage Scialabba, N., Brüggemann, J., Isensee, A., Erb, K.-H., Smith, P., Klocke, P., Leiber, F., Stolze, M., & Niggli, U. (2017). Strategies for feeding the world more sustainably with organic agriculture. *Nature Communications, 8*, 1290. https://doi.org/10.1038/s41467-017-01410-w.

Murphy, E., & Legrand, A. (2023). Introduction to the concept of "welfare potential" of production systems and its practical relevance to welfare labelling. *Frontiers in Animal Science, 4*, 1225839. https://doi.org/10.3389/fanim.2023.1225839.

NAHWOA, Network for Animal Health and Welfare in Organic Agriculture. (2002). Statement concerning the ban on inclusion of synthetic amino acids in the diets of organically reared monogastric (poultry and pigs) livestock in the EU Regulation 1804/99. http://www.veeru.reading.ac.uk/organic/AAStatementAppend.pdf.

Nelson, J. R., Settar, P., Berger, E., Wolc, A., O'Sullivan, N., & Archer, G. S. (2020). Brown and white egg-layer strain differences in fearfulness and stress measures. *Applied Animal Behaviour Science, 231*, 105087. https://doi.org/10.1016/j.applanim.2020.105087.

Newberry, R. C., Estevez, I., & Keeling, L. J. (2001). Group size and perching behaviour in young domestic fowl. *Applied Animal Behaviour Science, 73*(2), 117–129. https://doi.org/10.1016/S0168-1591(01)00135-6.

Nicol, C. J., Bestman, M., Gilani, A.-M., De Haas, E. N., De Jong, I. C., Lambton, S., Wagenaar, J. P., Weeks, C. A., & Rodenburg, T. B. (2013). The prevention and control of feather pecking: Application to commercial systems. *World's Poultry Science Journal, 69*(4), 775–788. https://doi.org/10.1017/s0043933913000809.

Olsson, I. A. S., & Keeling, L. J. (2000). Night-time roosting in laying hens and the effect of thwarting access to perches. *Applied Animal Behaviour Science, 68*(3), 243–256. https://doi.org/10.1016/S0168-1591(00)00097-6.

Padel, S. (2007, May 13–15). *How do the ethical values of organic agriculture relate to standards and to current practice?* In *Proceedings of the 7th EurSafe Congress*. Vienna, Austria.

Padel, S., Schmid, O., & Vonne, L. (2004). *Animal health and welfare in organic agriculture: Organic livestock standards*. In: M. Vaarst, S. Roderick, V. Lund, & W. Lockeretz (Eds.). CAB International.

Patt, A., Halle, I., Dudde, A., Olbrich, A., Sieburg-Rockel, J., & Krause, E. T. (2022). Influence of different dietary fibre contents in the diet on feather pecking, locomotor activity and performance of laying hens. *British Poultry Science, 63*(5), 571–580. https://doi.org/10.1080/00071668.2022.2076212.

Phocas, F., Belloc, C., Bidanel, J., Delaby, L., Dourmad, J. Y., Dumont, B., Ezanno, P., Fortun-Lamothe, L., Foucras, G., Frappat, B., González-García, E., Hazard, D., Larzul, C., Lubac, S., Mignon-Grasteau, S., Moreno, C. R., Tixier-Boichard, M., & Brochard, M. (2016). Review: Towards the agroecological management of ruminants, pigs and poultry through the development of sustainable breeding programmes: I-selection goals and criteria. *Animal, 10*(11), 1749–1759. https://doi.org/10.1017/s1751731116000926.

Prescott, N. B., & Wathes, C. M. (1999). Reflective properties of domestic fowl (*Gallus g. domesticus*), the fabric of their housing and the characteristics of the light environment in environmentally controlled poultry houses. *British Poultry Science, 40*(2), 185–193. https://doi.org/10.1080/00071669987584.

Prescott, N. B., & Wathes, C. M. (1999). Spectral sensitivity of the domestic fowl (*Gallus g. domesticus*). *British Poultry Science, 40*(3), 332–339. https://doi.org/10.1080/00071669987412.

Rana, M. S., & Campbell, D. L. M. (2021). Application of ultraviolet light for poultry production: A review of impacts on behavior, physiology, and production. *Frontiers in Animal Science, 2*, 699262. https://doi.org/10.3389/fanim.2021.699262.

Rault, J.-L., Bateson, M., Boissy, A., Forkman, B., Grinde, B., Gygax, L., Harfeld, J. L., Hintze, S., Keeling, L. J., Kostal, L., Lawrence, A. B., Mendl, M. T., Miele, M., Newberry, R. C., Sandøe, P., Špinka, M., Taylor, A. H., Webb, L. E., Whalin, L., & Jensen, M. B. (2025). A consensus on the definition of positive animal welfare. *Biology Letters, 21*(1). https://doi.org/10.1098/rsbl.2024.0382.

Riber, A. B., & Hinrichsen, L. K. (2017). Welfare consequences of omitting beak trimming in barn layers. *Frontiers in Veterinary Science, 4*, 222. https://doi.org/10.3389/fvets.2017.00222.

Riber, A. B., van de Weerd, H. A., de Jong, I. C., & Steenfeldt, S. (2018). Review of environmental enrichment for broiler chickens. *Poultry Science, 97*(2), 378–396. https://doi.org/10.3382/ps/pex344.

Riddle, E. R., Ali, A. B. A., Campbell, D. L. M., & Siegford, J. M. (2018). Space use by 4 strains of laying hens to perch, wing flap, dust bathe, stand and lie down. *PLoS One, 13*(1), e0190532. https://doi.org/10.1371/journal.pone.0190532.

Rocchi, L., Cartoni Mancinelli, A., Paolotti, L., Mattioli, S., Boggia, A., Papi, F., & Castellini, C. (2021). Sustainability of rearing system using multicriteria analysis: Application in commercial poultry production. *Animals, 11*(12), 3483. https://doi.org/10.3390/ani11123483.

Röös, E., Mie, A., Wivstad, M., Salomon, E., Johansson, B., Gunnarsson, S., Wallenbeck, A., Hoffmann, R., Nilsson, U., Sundberg, C., & Watson, C. A. (2018). Risks and opportunities of increasing yields in organic farming. A review. *Agronomy for Sustainable Development, 38*(2), 14. https://doi.org/10.1007/s13593-018-0489-3.

Savory, J., Percival, D., & Yuill, I. (2002). Influence of perch space allowance on perching behaviour of laying hens. *British Poultry Science, 43*(5), 22–23. https://doi.org/10.1080/00071660276238845 5.

Schmucker, S., Hofmann, T., Sommerfeld, V., Huber, K., Rodehutscord, M., & Stefanski, V. (2021). Immune parameters in two different laying hen strains during five production periods. *Poultry Science, 100*(11), 101408. https://doi.org/10.1016/j.psj.2021.101408.

Schreiter, R., Damme, K., von Borell, E., Vogt, I., Klunker, M., & Freick, M. (2019). Effects of litter and additional enrichment elements on the occurrence of feather pecking in pullets and laying hens – A focused review. *Veterinary Medicine and Science, 5*(4), 500–507. https://doi.org/10.1002/vms3.184.

Schwarzer, A., Plattner, C., Bergmann, S., Rauch, E., Erhard, M., Reese, S., & Louton, H. (2021). Feather pecking in non-beak-trimmed and beak-trimmed laying hens on commercial farms with aviaries. *Animals, 11*(11), 3085. https://doi.org/10.3390/ani11113085.

Schwean-Lardner, K., Fancher, B. I., Gomis, S., Van Kessel, A., Dalal, S., & Classen, H. L. (2013). Effect of day length on cause of mortality, leg health, and ocular health in broilers. *Poultry Science, 92*(1), 1–11. https://doi.org/10.3382/ps.2011-01967.

Schwean-Lardner, K., Fancher, B. I., Laarveld, B., & Classen, H. L. (2014). Effect of day length on flock behavioural patterns and melatonin rhythms in broilers. *British Poultry Science, 55*(1), 21–30. https://doi.org/10.1080/00071668.2013.860211.

Sepeur, S., Spindler, B., Schulze-Bisping, M., Habig, C., Andersson, R., Beyerbach, M., & Kemper, N. (2015). Comparison of plumage condition of laying hens with intact and trimmed beaks kept on commercial farms. *European Poultry Science, 79*, 1–12. https://doi.org/10.1399/eps.2015.116.

Shynkaruk, T., Buchynski, K., & Schwean-Lardner, K. (2022). Lighting programme as a management tool for broilers raised without antibiotics – Impact on productivity and welfare. *British Poultry Science, 63*(6), 761–767. https://doi.org/10.1080/00071668.2022.2083943.

Siegrist, M., & Hartmann, C. (2019). Impact of sustainability perception on consumption of organic meat and meat substitutes. *Appetite, 132*, 196–202. https://doi.org/10.1016/j.appet.2018.09.016.

Simjee, S., & Tice, G. (2023). The risk-benefit balance of resistance to ionophores in *Enterococcus faecium* and *Enterococcus faecalis* for ionophore coccidiostats in broiler chickens. *Journal of Antimicrobial Chemotherapy, 78*(9), 2121–2130. https://doi.org/10.1093/jac/dkad183.

Simm, R., Slettemeås, J. S., Norström, M., Dean, K. R., Kaldhusdal, M., & Urdahl, A. M. (2019). Significant reduction of vancomycin resistant *E. faecium* in the Norwegian broiler population coincided with measures taken by the broiler industry to reduce antimicrobial resistant bacteria. *PLoS One, 14*(12), e0226101. https://doi.org/10.1371/journal.pone.0226101.

Smoluk-Sikorska, J., Śmiglak-Krajewska, M., Rojík, S., & Rojík Fulnečková, P. (2024). Prices of Organic food—The gap between willingness to pay and price premiums in the organic food market in Poland. *Agriculture, 14*(1), 17. https://doi.org/10.3390/agriculture14010017.

Sobotik, E. B., Nelson, J. R., & Archer, G. S. (2020). How does ultraviolet light affect layer production, fear, and stress. *Applied Animal Behaviour Science, 223*, 104926. https://doi.org/10.1016/j.applanim.2019.104926.

Steenfeldt, S., & Nielsen, B. L. (2015). Welfare of organic laying hens kept at different indoor stocking densities in a multi-tier aviary system. I: Egg laying, and use of veranda and outdoor area. *Animal, 9*(9), 1509–1517. https://doi.org/10.1017/s1751731115000713.

Stehr, M., Grashorn, M., Dannenberger, D., Tuchscherer, A., Gauly, M., Metges, C. C., & Daş, G. (2019). Resistance and tolerance to mixed nematode infections in relation to performance level in laying hens. *Veterinary Parasitology, 275*, 108925. https://doi.org/10.1016/j.vetpar.2019.108925.

Stratmann, A., Fröhlich, E. K. F., Gebhardt-Henrich, S. G., Harlander-Matauschek, A., Würbel, H., & Toscano, M. J. (2016). Genetic selection to increase bone strength affects prevalence of keel bone damage and egg parameters in commercially housed laying hens. *Poultry Science, 95*(5), 975–984. https://doi.org/10.3382/ps/pew026.

Sun, Z. W., Yan, L., G, Y. Y., Zhao, J. P., Lin, H., & Guo, Y. M. (2013). Increasing dietary vitamin D3 improves the walking ability and welfare status of broiler chickens reared at high stocking densities. *Poultry Science, 92*(12), 3071–3079. https://doi.org/10.3382/ps.2013-03278.

Tallentire, C. W., Edwards, S. A., Van Limbergen, T., & Kyriazakis, I. (2019). The challenge of incorporating animal welfare in a social life cycle assessment model of European chicken production. *International Journal of Life Cycle Assessment, 24*(6), 1093–1104. https://doi.org/10.1007/s11367-018-1565-2.

Taylor, P. S., Hemsworth, P. H., Groves, P. J., Gebhardt-Henrich, S. G., & Rault, J. L. (2018). Ranging behavior relates to welfare indicators pre- and post-range access in commercial free-range broilers. *Poultry Science, 97*(6), 1861–1871. https://doi.org/10.3382/ps/pey060.

Thøfner, I. C. N., Dahl, J., & Christensen, J. P. (2021). Keel bone fractures in Danish laying hens: Prevalence and risk factors. *PLoS ONE, 16*(8), e0256105. https://doi.org/10.1371/journal.pone.0256105.

Thuy Diep, A., Larsen, H., & Rault, J.-L. (2018). Behavioural repertoire of free-range laying hens indoors and outdoors, and in relation to distance from the shed. *Australian Veterinary Journal, 96*(4), 127–131. https://doi.org/10.1111/avj.12684.

Vaarst, M., Roderick, S., Lund, V., Lockeretz, W., & Hovi, M. (2004). Animal health and welfare in organic agriculture. In M. Vaarst, S. Roderick, V. Lund, & W. Lockeretz (Eds.). *Organic principles and values: The framework for organic animal husbandry*. CAB International.

Vaarst, M., & Alrøe, H. F. (2012). Concepts of animal health and welfare in organic livestock systems. *Journal of Agricultural and Environmental Ethics, 25*(3), 333–347. https://doi.org/10.1007/s10806-011-9314-6.

Vaarst, M., Steenfeldt, S., & Horsted, K. (2015). Sustainable development perspectives of poultry production. *World's Poultry Science Journal, 71*(4), 609–620. https://doi.org/10.1017/s0043933915002433.

Vallance, S., Espig, M., Taylor, A., Brosnahan, C., & McFetridge, A. (2023). Farmer wellbeing and animal welfare: Exploring the connections using the case of facial eczema. *Journal of Rural Studies, 104*, 103140. https://doi.org/10.1016/j.jrurstud.2023.103140.

van der Eijk, J. A. J., Gunnink, H., Melis, S., van Riel, J. W., & de Jong, I. C. (2022). Reducing stocking density benefits behaviour of fast- and slower-growing broilers. *Applied Animal Behaviour Science, 257*, 105754. https://doi.org/10.1016/j.applanim.2022.105754.

van der Eijk, J. A. J., van Harn, J., Gunnink, H., Melis, S., van Riel, J. W., & de Jong, I. C. (2023). Fast- and slower-growing broilers respond similarly to a reduction in stocking density with regard to gait, hock burn, skin lesions, cleanliness, and performance. *Poultry Science, 102*(5), 102603. https://doi.org/10.1016/j.psj.2023.102603.

van Horne, P. (2020). Economics of broiler production systems in the Netherlands: Economic aspects within the Greenwell sustainability assessment model. (Wageningen Economic Research report; No. 2020-027). Wageningen Economic Research. https://doi.org/10.18174/518522.

Van Krimpen, M. M., Kwakkel, R. P., Reuvekamp, B. F. J., Van Der Peet-Schwering, C. M. C., Den Hartog, L. A., & Verstegen, M. W. A. (2005). Impact of feeding management on feather pecking in laying hens. *World's Poultry Science Journal, 61*(4), 663–686. https://doi.org/10.1079/wps200478.

van Krimpen, M. M., Leenstra, F., Maurer, V., & Bestman, M. (2016). How to fulfill EU requirements to feed organic laying hens 100% organic ingredients. *Journal of Applied Poultry Research, 25*(1), 129–138. https://doi.org/10.3382/japr/pfv048.

van Staaveren, N., Ellis, J., Baes, C. F., & Harlander-Matauschek, A. (2021). A meta-analysis on the effect of environmental enrichment on feather pecking and feather damage in laying hens. *Poultry Science, 100*(2), 397–411. https://doi.org/10.1016/j.psj.2020.11.006.

van Wagenberg, C. P. A., de Haas, Y., Hogeveen, H., van Krimpen, M. M., Meuwissen, M. P. M., van Middelaar, C. E., & Rodenburg, T. B. (2017). Animal board invited review: Comparing conventional and organic livestock production systems on different aspects of sustainability. *Animal, 11*(10), 1839–1851. https://doi.org/10.1017/s175173111700115x.

Ventura, B. A., Siewerdt, F., & Estevez, I. (2010). Effects of barrier perches and density on broiler leg health, fear, and performance. *Poultry Science, 89*(8), 1574–1583. https://doi.org/10.3382/ps.2009-00576.

Ventura, B. A., Siewerdt, F., Estevez, I., & Mettke-Hofmann, C. (2012). Access to barrier perches improves behavior repertoire in broilers. *PLoS One, 7*(1), e29826. https://doi.org/10.1371/journal.pone.0029826.

Vezzoli, G., Mullens, B. A., & Mench, J. A. (2015). Relationships between beak condition, preening behavior and ectoparasite infestation levels in laying hens. *Poultry Science, 94*(9), 1997–2007. https://doi.org/10.3382/ps/pev171.

Wei, Y., Zheng, W., Li, B., Tong, Q., Shi, H., & Li, X. (2020). Effects of B-wave ultraviolet supplementation using light-emitting diodes on caged laying hens during the later phase of the laying cycle. *Animals, 10*(1), 15. https://doi.org/10.3390/ani10010015.

Wen, J., Yang, S., Zhu, J., Liu, A., & Rao, Y. (2025). Meta-analysis on the effect of foraging on feather pecking and feather coverage. *Applied Animal Behaviour Science, 285*, 106584. https://doi.org/10.1016/j.applanim.2025.106584.

Wurtz, K. E., Thodberg, K., Berenjian, A., Foldager, L., Tahamtani, F. M., & Riber, A. B. (2022). Commercial layer hybrids kept under organic conditions: A comparison of range use, welfare, and egg production in two layer strains. *Poultry Science, 101*(9), 102005. https://doi.org/10.1016/j.psj.2022.102005.

Zoli, M., Mantovi, P., Ferrari, P., Ferrari, L., & Ferrante, V. (2023). Soil organic matter and nutrient levels in outdoor runs in organic laying farms. *Animals, 13*(3), 401. https://doi.org/10.3390/ani13030401.

Chapter 11

Silvopasture

Leonie Jacobs[1] and Allison Pullin[2]
[1]*School of Animal Sciences, Virginia Tech, Blacksburg, VA, United States,* [2]*Prestage Department of Poultry Science, North Carolina State University, Raleigh, NC, United States*

Introduction

A silvopasture system is an intentional integration of animals, trees, and forage on a single plot of land. The practice is rooted in traditional and Indigenous land management strategies (reviewed by Shi & Conway-Anderson 2022). For example, on a laying hen facility where hens range in an apple orchard or a broiler chicken facility where birds can range in a forested area used for timber, the eggs and apples or the chicken meat and timber both function as sources of income while using a single plot of land.

Access to an outdoor range is generally considered to improve animal welfare. An outdoor range is a more complex environment than just an indoor-only space, offering choice and challenges for birds to experience and providing the birds with the ability to express behaviors that may be limited in an indoor space. Some examples of stimulated behaviors are foraging, scratching, sunbathing, and dustbathing (de Jonge & van Trijp, 2013; Fanatico et al., 2016; Knierim, 2006). Time spent outdoors can positively correlate with other indicators of good animal welfare, including reduced anxiety and improved gait compared to birds that are kept solely indoors (Grigor et al., 1995; Taylor et al., 2018; Stadig et al., 2017a).

There are some inherent risks for animal welfare associated with the outdoors, which include exposure to potential predators (Paneru et al., 2023; Singh et al., 2017), pathogens, and extreme weather conditions (Singh et al., 2017). A silvopasture system, compared to the conventional open-grassland pasture systems commonly used for poultry, may reduce some of these risks, for instance, wild bird disease transmission (Bestman et al., 2018) and parasite exposure (National Center for Appropriate Technology NCAT, 2004). Ranges with trees or tall grass eliminated predation losses entirely, whereas poultry on open grassland experienced predation by raptors and crows, particularly when the birds were young (Dal Bosco et al., 2014).

Silvopasture systems provide opportunities for sustainability beyond any monoculture system. A systematic review of 53 studies from 1983–2021

Integrating Welfare and Sustainability in Poultry Systems. DOI: https://doi.org/10.1016/B978-0-443-40594-5.00013-4

showed consistent benefits, primarily farm income diversification and shade for livestock (Smith et al., 2022). Environmental benefits include improved water quality, soil health, and increased flora and fauna species richness and abundance, based on a review of 267 studies (Shi & Conway-Anderson, 2022).

There is no formal documentation on how many farms incorporate a silvopasture approach for poultry. Since it is not a common system applied in commercial integrated farm operations in the United States or Europe, it is likely that less than 5% of poultry produced commercially has access to such a system. Yet, some evidence does exist that it is applied globally. Small-scale operations may include this practice more frequently, and silvopasture approaches were mentioned to be widely practiced in many countries, including Latin America and the Caribbean (Urrea-Benítez, 2024). When 20 United States farmers were interviewed about their silvopastoral system, six of those produced poultry: four with chickens for meat or eggs, and two with turkeys (Orefice et al., 2017). In the United Kingdom, 286 laying hen operations that supplied McDonald's Restaurants Ltd. were required to plant 5% of their range areas with trees in 2008 (Brigh et al., 2011). Even though the trees may not provide a source of income (which was not explicitly mentioned), the range design mimics a silvopasture system.

In this chapter, we will provide an overview of how silvopasture may impact animal welfare, structuring the impacts through the Three Concepts framework as described by Fraser (2008). In addition, we will consider the sustainability impacts of this production system. We focus on poultry-specific practices over other livestock when possible and focus on silvopasture over free-range or pasture access when possible.

Animal welfare

Biological functioning

Biological functioning encompasses the animal's physical health and productivity. For laying hens, the latter includes egg production characteristics, while for broilers, the emphasis is on feed efficiency and growth.

Hens

Productivity

Egg production and egg quality can be impacted by the housing conditions of the hens, and this includes the type and quality of the range that hens can access. Studies to date have mainly assessed the productivity of brown-feathered hens housed in silvopasture systems.

Hy-Line Brown hens ranged on a medic pasture (a legume) in a crop and pasture rotation system during summer in a Mediterranean climate. While experiencing a heat wave, egg production was lower than the standard performance expected in a cage system, with a rate of lay of 79% versus

93% (the cage standard) between 18 and 40 weeks of age (Glatz et al., 2005). The same research group allowed Hy-Line Brown hens to range in a wheat stubble pasture in spring and observed improved egg production and bird live weight compared to the cage standard, comparable mortality, but lower egg weight (58 vs 61 g) and slightly delayed reproductive maturity (Miao et al., 2005). Egg production of hens ranging in an apple orchard followed expected standards, with >90% egg production from 33 to 46 weeks and a decrease to 72% at 77 weeks of age (Bestman, 2015).

In Italy, egg quality improved when hens of a commercial egg-laying strain (feather color not stated) were raised with access to hazelnut orchards compared to a control (Cosentino et al., 2020). Carbohydrate, lipid, protein, fatty acid, and cholesterol contents differed in eggs from the orchards compared to the control group (Cosentino et al., 2020). The difference was dependent on the site, with some egg characteristics differing between the two orchard sites, too, although this was not formally analyzed. ISA Brown hens in Nigeria showed enhanced performance when raised with access to a legume pasture (*Stylosanthes hamata*) compared to indoor-only or a grass pasture; with heavier ovaries and oviducts at 20 weeks of age, earlier oviposition, and higher hen-day egg production than with access to a grass pasture at mid-lay; and finally, lower feed intake (FI) compared to indoor-only (Oke et al., 2016).

With limited information available, it seems that hen productivity depends on a multitude of factors (for instance, forage type, climate, and weather) and how these are managed; however, silvopasture systems show some promise for improved egg production and quality compared to grassland or indoor-only housing.

Disease and mortality

Disease pressures in flocks with outdoor access can differ from those in indoor-only flocks. Although focused on small- to medium-scale free-range farms rather than silvopastures specifically, up to 10% mortality of hens was reported in a survey with 41 Australian farmers (Singh et al., 2017), with predation (34%), cannibalism (29%), grass impaction (20%), heat stress (24%), and disease outbreaks (10%) reported as the cause (Singh et al., 2017). The most common diseases were fowl cholera (17%), coccidiosis (7.3%), spotty liver disease (7.3%), and infectious laryngotracheitis (4.8%), with other causes of mortality being cold temperatures at night, smothering, vent prolapse, and injury (Singh et al., 2017). Mortality of hens that were ranging on a medic pasture during a heat wave in a Mediterranean summer climate was higher than the cage standard (9.1% vs 1.2%) (Glatz et al., 2005), which was attributed to the heat wave.

Silvopasture may reduce the risk of exposure to some diseases carried by waterfowl, including avian influenza. Risk of avian influenza transmission from wild birds to poultry was theorized to be reduced in heavily vegetated pastures, compared to open grassland in one pilot study (Bestman et al., 2018).

They argued that this was due to the heavily vegetated habitat being avoided by waterfowl, which prefer open grassland for grazing and safety.

Thermal comfort

When hens have access to an outdoor range, they are inherently exposed to weather conditions that may negatively or positively impact their welfare. Yet, the option of staying inside or going outdoors theoretically provides more means to behaviorally respond to these conditions (Bonnefous et al., 2022). To our knowledge, no studies have investigated the associations between thermal stress in poultry and silvopastures. However, there are some inferences that can be made based on free-range studies of laying hens. When artificial shelters are provided with differing gradients of sun-blocking ability, hens mostly preferred shelters with the highest-density shade (Rana et al., 2022). Although thermal comfort was not assessed, outdoor temperature was a key predictor of shelter use (Rana et al., 2022), with approximately 39% to 72% of the variation in shelter use explained by this factor. Fewer hens were using the shelters, so possibly fewer were ranging at higher temperatures. Similarly, in the United Kingdom, the proportion of hens ranging gradually decreased as temperatures increased above 17°C (Richards et al., 2011), supporting the theory that laying hens prefer lower temperatures.

Silvopastures seem to be a promising avenue to provide a cooler microclimate than open grassland pasture. In Brazil, two silvopastoral systems, one with Brazilian native trees and one with eucalyptus trees, had improved microclimate compared to an open grassland pasture, with lower solar radiation transmission and up to three fewer hours of theorized animal thermal stress daily (as calculated using a black globe and humidity index >79) compared to the open grassland (Pezzopane et al., 2019). When lambs were kept on silvopasture, they experienced less radiant heat and lower ground surface temperatures compared to those on open pasture. Consequently, they showed lower rectal temperatures, spent more time grazing and walking, and less time standing at rest, which was interpreted as behavioral indicators of thermal comfort (Vieira et al., 2021).

Footpad and leg health

Although not assessed in a silvopasture system, hens that used their range more had a lower incidence of footpad dermatitis compared to less active rangers. In contrast, hens in the same study that showed more walking indoors had a higher incidence of footpad dermatitis than those less active (Rodriguez-Aurrekoetxea & Estevez, 2016), suggesting that footpad health is not just mediated by activity, but depends on the environment in which the hens are active. When range use is stimulated by a silvopastoral approach (Pacher-Deutsch et al., 2025), this may further lead to improved footpad health, but this has not been studied in hens yet.

Parasites

No parasite infection data from commercial silvopasture farms are available. In a survey across 41 small- to medium-sized semiintensive free-range egg farms in Australia, 32% of respondents reported that their hens had internal and external parasites (Singh et al., 2017). There is some evidence that parasite infestations could be less common in silvopasture systems. A biodiverse environment, compared to a monocultured grass pasture, supports biological control of parasites, and although pests are not eliminated, damage levels to vegetation are low enough to be tolerated in most cases (National Center for Appropriate Technology NCAT, 2004). In systems where poultry and cattle are alternated in the range, there may be additional benefits for the cattle. Poultry may reduce internal and external parasite pressure for cattle as they "disassemble" manure pats, destroying parasite eggs and feeding on larvae (National Center for Appropriate Technology NCAT, 2004).

Broiler chickens

Productivity and meat quality

Broiler production performance is assessed through body weight (BW), feed intake (FI), and feed conversion ratio (FCR), with a lower FCR indicating more efficient feed-to-muscle conversion.

Silvopasture access can improve broiler chicken productivity. Performance of a slow-growing Sasso T451 breed was assessed in indoor-only housing, free-range paddocks containing artificial shelters, and in a silvopasture system with trees used for biomass production (Stadig et al., 2016). Although the indoor-only birds were heaviest at slaughter, FI and FCR did not differ among the treatments. A fast-growing breed (Ross 708) and a slow-growing breed (Sasso) both displayed heavier BW in indoor-only housing compared to silvopasture (He-Demontaron et al., 2025). However, the silvopasture-reared broilers consumed less feed and had a lower FCR at 42 and 56 days of age than indoor-reared broilers (He-Demontaron et al., 2025.). Broilers in silvopasture likely supplemented their diet with forages and insects on the range and consequently consumed less feed (Fig. 11.1), while indoor-housed broilers only ate the feed provided. Similarly, broilers reared with olive trees ingested more forage, followed by birds reared in tall sorghum grass, while birds reared in a standard outdoor paddock ingested the least forage (Dal Bosco et al., 2014). However, when comparing those three different types of outdoor areas, production performance did not differ among the groups (BW, FI, and FCR) (Dal Bosco et al., 2014).

When outdoor paddocks contained one of three legume species, free-range fast-growing Marshall broilers displayed lower FI and lower FCR compared to a grass-only paddock or indoor-only rearing (Oke et al., 2021). Broilers had a higher final BW when their paddocks contained the *S. hamata* legume species specifically, compared to broilers reared on the other two legume species (*Stylosanthes guanensis* or *Mucuna pruriens*), the grass-only paddock, or

FIGURE 11.1 **Foraging can contribute to feed intake and improved productivity.** Slow-growing Sasso broilers forage in tall grasses in a silvopasture plot.
Photo credit: Allison Pullin.

indoor-only (Oke et al., 2021). Taken together, the type of forage differently affects productivity, so silvopasture performance could be altered depending on the type of forage provided on the ground.

For meat-type poultry, meat quality is an important consideration when assessing productivity and production systems. Meat quality attributes include pH, cook loss, water-holding capacity, texture, nutrient composition, shelf life, and color—all of which can affect the tenderness, juiciness, sensory properties, and purchasing decisions by consumers. Impacts of access to a silvopasture on meat quality characteristics vary. A recent review concluded that meat from birds with outdoor access tended to have lower pH, appear darker and tougher, contain more protein and less fat, and have improved flavor and juiciness in sensory panels (Campbell et al., 2025). However, these effects are difficult to attribute solely to the housing system, as they are often confounded with broiler breed, broiler age, time of year, and diet, which also affect meat quality (Campbell et al., 2025). Indeed, one study comparing the meat quality of a fast-growing Ross 708 breed to a slow-growing Sasso breed in indoor-only and silvopasture housing reported that breed had a greater influence than housing system alone (Caceres and Sobreira, 2026). The fast-growing birds had a higher prevalence of muscle myopathies (i.e., woody breast and spaghetti meat), higher levels of moisture and polyunsaturated fatty acids (PUFAs), and less protein, fat, and collagen compared to the slow-growing birds. Regardless of breed, silvopasture-grown meat had lower saturated fatty acids and improved oxidative stability, which were attributed to dietary diversity and increased behavioral activity from the outdoor access, but a sensory panel did not identify taste differences between housing environments (Caceres and Sobreira, 2026.).

Breast meat from slow-growing broilers raised on silvopasture was yellower than meat from birds raised only indoors, which was attributed to the birds ingesting plant material; however, the meat from silvopasture birds was unexpectedly darker than meat from birds raised indoors, despite having similar pH and drip loss (Stadig et al., 2016). Total PUFA were higher for birds reared free-range with artificial shelters compared to indoor-only birds, but only some PUFA were higher for silvopasture-reared birds compared to indoor-only birds, and the total amount did not differ. The researchers attributed different PUFA outcomes in the two outdoor access systems to differences in vegetative diversity between the paddocks. In a sensory panel, the silvopasture meat was scored as less fibrous and more tender than meat from indoor or free-range-reared birds, and juicier than indoor-reared birds (Stadig et al., 2016).

Disease and mortality

In a hedgerow system with a mixture of trees and shrubs, slow-growing Red Ranger broilers displayed stronger immune responses, including higher bacterial-killing activity and reduced hemodilution, compared to birds raised on open pastures (Delgadillo et al., 2021). The researchers suggested that these benefits may stem from birds consuming immune-modulatory agents, such as phytochemicals, willow leaves, and insects while foraging within the hedgerow environment.

Predation is a significant mortality concern in outdoor access production. Silvopasture systems with trees or tall grass prevented predation losses by reducing the visibility of chickens to predators, whereas young, slow-growing broilers on plots of open grassland experienced predation by raptors and crows (Dal Bosco et al., 2014). Two trials with silvopasture-raised fast-growing broilers showed no predator-related mortality in the silvopasture plots, while one predator event was recorded in the open pasture plots, possibly by a raccoon (Paneru et al., 2023). Mortality was twofold higher in a silvopasture system compared to an indoor-only housing system, but predation was not reported as a reason for mortality (He-Demontaron et al., 2025). Instead, the main driver of mortality in the silvopasture system was the aftermath of a severe rainstorm that caused puddling in the uneven terrain under the trees and colder overnight temperatures. Bird mortality was attributed to piling, drowning, and hypothermia (He-Demontaron et al., 2025). Further research is warranted for shelter design and land management (e.g., drainage and terrain leveling) to mitigate the effects of severe weather events on bird welfare.

Feather condition

Feather cleanliness provides insight into management conditions, as dirty feathers are correlated with footpad dermatitis and indicate poor litter condition (Saraiva et al., 2016). Dirty feathers compromise feather structure and consequently could impair birds' thermoregulatory abilities (Welfare Quality (R), 2009). Indoor housing was associated with a higher prevalence of dirty feathers than outdoor housing (Sans et al., 2023), but for silvopasture specifically, data are limited. Slow-growing broiler feather cleanliness did not differ between indoor housing and a silvopasture system (Stadig et al., 2017a). Broiler characteristics also affect feather condition within the same housing environment, suggesting different adaptabilities to outdoor access. In a pasture with olive trees, bushes, and hedges, six broiler genotypes with similar growth rates differed in their feather condition across five body regions ranging from no intact feathers to fully intact feathers (Mancinelli et al., 2020). In an open grassland range, individual fast-growing broilers that utilized the range had cleaner vents compared to birds that never went outside, suggesting either a benefit of outdoor access for cleanliness or that healthier birds go outside more (Taylor et al., 2018).

Thermal comfort

Heat stress in outdoor access production can reduce growth rates, increase inflammatory responses, and lead to mortality (Moyle et al., 2014; Liles et al., 2015; Aksoy et al., 2021). In summer months in an open grassland pasture, heterophil numbers were higher in fast-growing broilers compared to winter months, which was attributed to higher temperatures (Sánchez-Casanova et al., 2022). On sunny days, broilers will stay indoors if no shade is available (Dawkins et al., 2003). Providing shade, particularly through overhead vegetative cover, offers a heat-abatement strategy (see Fig. 11.2). Overhead cover from trees and shrubs in a hedgerow system resulted in faster growth rates in a fast-growing breed, which was attributed to reducing heat stress (Delgadillo et al., 2021). Although thermal comfort was not quantified, a silvopasture system with broilers showed the attenuation of extreme temperatures in the range (Jacobs et al., 2026).

FIGURE 11.2 **Slow-growing broilers shelter (from the sun) under hazelnut bushes.** Silvopastures provide shelter from the sun and other weather conditions.
Photo credit: Leonie Jacobs.

Footpad and leg health

Tree cover may offer a strategy to mitigate ground moisture compared to open grassland pastures by sheltering the ground below from rainfall. Footpad dermatitis was improved in fast-growing Ross 708 broilers housed in a silvopasture system compared to an open grassland pasture system, while gait was excellent in both environments (Paneru et al., 2023). Slow-growing broilers housed in a silvopasture system with olive trees also experienced a lower incidence of footpad dermatitis and breast blisters compared to broilers in an open pasture, which was attributed to more locomotion and less time on high-moisture ground (Dal Bosco et al., 2014). Similarly, slow-growing broilers reared in a silvopasture environment tended to have less hock dermatitis than broilers reared indoors, but did not differ in gait (Stadig et al., 2017a). In Brazil, free-range farms with fewer shelters had caked soil in the outdoor areas, which was associated with a higher risk for footpad dermatitis (Sans et al., 2014). Legume species in paddocks on Nigerian farms reduced lameness compared to indoor-only housing or a free-range paddock with grasses, but this was attributed to more broiler activity in those paddocks rather than ground moisture (Oke et al., 2021).

Severe rainstorms could cause the ground underneath tree canopies to become wet. If there is insufficient drainage, pooling water and higher ground moisture could result, and less sun exposure due to tree cover may prolong the time for wet soil to dry out after a storm. In a silvopasture system that experienced a severe rainstorm event, over 90% of broilers experienced footpad lesions, and one-third were considered severe with a score of two on a three-point scale (He-Demontaron et al., 2025). Gait was not affected, with silvopasture and indoor-housed birds both displaying good walking abilities. Keeping birds inside shelters until ground conditions improve may prevent severe footpad lesions from high ground moisture, provided that the indoor conditions are managed with sufficient floor space allowance and optimal litter and air quality.

Parasites

Similar to laying hens, there is no research on endo- or ecto-parasitic infections in broilers reared in silvopasture environments. Outdoor access in general increases the risk of endo- and ecto-parasitic infections in poultry (Campbell et al., 2025; Jeni et al., 2021; Shifaw et al., 2021). For example, coccidia counts were highest in slow-growing broilers during their first week of pasture access with artificial shelters, at approximately 28 days of age (Terra et al., 2021). Insight from other livestock reared for meat production suggests that silvopasture either reduces the risk or does not worsen the risk of parasitic infection.

In Brazil, beef cattle reared in silvopasture benefited from lower fecal parasite egg counts and higher lymphocyte proliferation compared to cattle in open pasture (Lopes et al., 2022), whereas others reported no differences in fecal egg counts between silvopasture and open pasture (de Mendonca et al., 2014;

Lopes et al., 2020). Another study in Brazil reported a higher prevalence of fecal egg counts from two parasitic species in beef cattle reared in silvopasture compared to open pasture, but the cattle did not differ in packed cell volume or BW gain (Oliveira et al., 2017). Similarly, silvopasture-reared cattle had higher numbers of ectoparasites in the form of ticks during one month of the year than pasture-reared cattle, but no negative effects were found for cattle performance (Bello et al., 2020). In the previous study, fecal egg counts were lower for silvopasture for one month compared to open pasture and did not differ from open pasture for the remainder of the year (Bello et al., 2020).

Taken together, there is a broad diversity of parasitic species across geographic regions and susceptibility to infections at different times of year. Evidence from beef cattle supports that silvopasture environments do not exacerbate parasitic infections, but further research is needed for poultry.

Natural living

Natural living considers both the ability of animals to express their highly motivated natural behaviors and the presence of naturalistic components in the animals' environment. Silvopastures especially fit well within this concept (Fig. 11.3), since a forest habitat is the natural environment of chickens when considering their ancestors, the jungle fowl (Collias & Collias, 1967).

Hens

Range use

A well-established benefit of providing natural cover on the range, such as in a silvopasture system, is that chickens make greater use of the outdoor area. This is likely because the environment more closely resembles the natural habitat of their ancestor, the jungle fowl, which to this day inhabits densely forested areas (Collias & Collias, 1967; Nguyen-Phuc & Berres, 2018). A high percentage of hens using the free-range area can be achieved by providing shelter (Bestman & Wagenaar, 2003; Zeltner & Hirt, 2003), with natural shelter from vegetation being especially preferred (Dal Bosco et al., 2014; Dawkins et al., 2003; Stadig et al., 2017b). On a poultry farm with willows, 75% of the hens were seen outside (Bestman, 2017).

When given a choice between vegetated and open grassland or bare-earth sections, hens seem to prefer the vegetated parts of the range (Larsen et al., 2017; Pacher-Deutsch et al., 2025). Hy-Line Brown hens spent the most time under dense canopy cover compared to areas with less dense or no canopy cover in Australia (Larsen et al., 2017), and Lohmann Brown Classic hens showed a preference for areas with 2-year-old bushes distributed across two formerly uncovered outdoor ranges in Austria, compared to the remaining uncovered sections in those ranges (Pacher-

FIGURE 11.3 **Slow-growers in forested pasture.** A densely forested pasture provides opportunities for natural living, as shown here in a silvopasture system with slow-growing broilers in MN, USA.
Photo credit: Leonie Jacobs.

Deutsch et al., 2025). More Lohmann Brown hens were observed ranging in sections with heavy vegetation cover compared to sections without overhead cover, regardless of temperature, on an Austrian farm, while at another farm, there was no difference between covered and uncovered sections, unless temperatures were high, which resulted in more birds in the covered sections compared to those without vegetation cover (Deutsch et al., 2024).

However, having natural vegetation does not always result in high levels of range use. At a commercial tree nursery in the Netherlands, hens would only leave their mobile house if the caretaker was present, and approximately 20% of the hens were lost due to predation by foxes and birds of prey within the 3 months they were there (Bestman, 2017). This high rate of predation may have resulted in the reluctance to use the range and may in part have been due to limited protection from predators because the fence was not electrified and tree removal led to large open spaces in the range (Bestman, 2017).

Injurious pecking

Free-range access, not exclusively in silvopasture systems, shows promise for the reduction of injurious feather pecking. There seems to be a negative association between the use of a free-range area and the degree of feather-pecking damage, meaning that if more hens in a flock use the free-range area, significantly less feather-pecking damage is observed (Bestman & Wagenaar, 2003; Green et al., 2000; Lambton et al., 2010; Mahboub et al., 2004; Nicol et al., 2003). When more than 50% of the flock used the outdoor area in good weather, a reduced risk of feather pecking was observed (Green et al., 2000). Similarly, a high percentage of hens in the flock using the outdoor run significantly decreased feather-pecking damage at 50 weeks or older (Bestman & Wagenaar, 2003).

Yet, this association is not consistently found, like in ISA Brown hens in an experimental setup, where there was no association between range use and plumage damage (Hartcher et al., 2016). There was also no correlation between the proportion (5% to 90%) of range cover and plumage damage at the end of lay, but plumage damage was negatively correlated with the percentage of canopy cover within tree-planted areas, showing that good canopy cover can reduce plumage damage due to injurious feather pecking (Brigh et al., 2011). When ranging in wheat stubble, there was little evidence of feather pecking and cannibalism (Miao et al., 2005).

With limited data, it does seem that in some cases, especially on United Kingdom laying hen farms, silvopasture-like environments can reduce this abnormal behavior.

Natural behavior

A natural environment (i.e., a forested environment) will likely stimulate a naturalistic behavioral repertoire, since domesticated chickens are descendants of jungle fowl (Collias & Collias, 1967). Indeed, domesticated chickens show strongly motivated behaviors that their ancestors express in the wild (Jacobs, 2026). In a silvopasture system in Brazil (rubber tree plantation), on average, 30% of Lohmann Brown Lite hens were observed foraging or scratching, compared to 12% of their indoor-only counterparts (Camarero et al., 2023).

When Hy-Line Brown hens in Australia could access distinct habitat types, including two types of trees, dense tree saplings, or bare earth without canopy cover, the most common behavior across locations was foraging (33% of observations) (Larsen et al., 2017). Hens foraged more under tree canopies than on bare earth or in the sapling location, depending on time of day (Larsen et al., 2017). Furthermore, comfort behaviors and social behaviors (depending on time of day) were more common in locations with trees than in bare-earth or sapling locations. Overall, the flock in this study provided a clear preference for dense, low canopy cover from trees than any other habitat type, and a greater behavioral repertoire was shown in this highly preferred area (Larsen et al., 2017).

Broiler chickens

Range use

Similar to laying hens, commercial broiler chickens are more likely to utilize outdoor ranges with greater areas of tree cover, as trees protect birds from bright sun and aerial predators (Dawkins et al., 2003). Broilers range less in the winter, which is in part attributed to trees having fewer leaves, and consequently less protection for the birds during this time of year (Dawkins et al., 2003). Lower motivation to range would reduce the behavioral and health benefits of range access. As a result, silvopasture farms may be better suited to geographic regions with less seasonal variability, or to growing flocks only during seasons with sufficient leaf coverage. Regardless, in both winter and summer, more slow-growing broilers were observed outside in silvopasture with olive trees and pastures with tall grasses compared to a standard outdoor paddock (Dal Bosco et al., 2014). More slow-growing broilers were observed outside in silvopasture compared to ranges with artificial shelters 42.8% versus 35.1%, respectively, and twice as many silvopasture broilers ranged further than 5 m from the house than broilers with artificial shelters (10.7% versus 4.1%, respectively; Stadig et al., 2017a). Similarly, 29% of fast-growing broilers were observed ranging in a silvopasture in the mornings (Fig. 11.4), compared to 22% in an open grassland pasture, based on live observations (Jacobs et al., 2026). Silvopasture was highly utilized by fast-growing (46% to 73% of birds) and slow-growing broilers (52% to 78% of birds), depending on age and time of day (He-Demontaron et al., 2025).

Most birds use the silvopasture during the morning compared to the afternoon (Stadig et al., 2017b; Jacobs, 2026; He-Demontaron et al., 2025). Fewer birds use the outdoor area in times of rain, increased radiation, and higher wind speeds, but there were always more silvopasture broilers outside in these weather events than broilers with artificial shelters (Stadig et al., 2017a). Particularly with increasing solar radiation, more silvopasture birds were observed outside compared to birds with artificial shelters, indicating more effective shade from trees (Stadig et al., 2017b).

Broiler breed influences how birds use the range. Breeds classified as slow-growing (<24 g/d) spent the most time on a range with mature trees, hedges, and bushes; fast-growing broilers (>41 g/d) spent the least time on the range, with medium-growing broilers (25 to 40 g/d) being intermediate (Castellini et al., 2016). A similar pattern was observed with distance ranged from the house, with slow- and medium-growing breeds ranging twofold to threefold further than the fast-growing breed (Castellini et al., 2016). Also, in a range with mature trees, hedges, and bushes, a slow-growing breed traveled 10 times the daily distance (1230 m) of a fast-growing breed (125 m; Dal Bosco et al., 2010).

FIGURE 11.4 **Fast-growing broilers in a silvopasture.** Fast-growing broilers in a silvopasture plot will readily range in the morning, although they may not leave the proximity of their coop. *Photo credit: Leonie Jacobs.*

Natural behavior

Silvopasture also promotes the diversity of behaviors displayed by broilers. Slow-growing broilers performed more locomotion in silvopasture with olive trees, followed by pastures with tall grasses, while broilers in a standard outdoor paddock performed the least locomotion (Dal Bosco et al., 2014). Likewise, more fast- and slow-growing silvopasture broilers were walking, standing, sparring, and dustbathing than broilers reared indoors only (He-Demontaron et al., 2025) (see Fig. 11.5).

There were more broilers observed foraging in a silvopasture system compared to those on grasslands with artificial shelters (Stadig et al., 2017b). The type of forage available influences behavioral diversity, with fast-growing broilers displaying more locomotion, dustbathing, and preening when the range contained legume species compared to ranges without legume species or indoor-only housing (Oke et al., 2021) (see Fig. 11.6). The authors attributed these differences to improved foraging opportunities and ground conditions.

FIGURE 11.5 **Silvopastures stimulate the natural behavior of slow-growing broilers.** Slow-growing Sasso broilers roost on tree roots under the shade in a silvopasture plot.
Photo credit: Allison Pullin.

FIGURE 11.6 **Fast-growing broilers show natural behavior in silvopastures.** Fast-growing Ross 708 broilers sit, stand, forage, and eat in a silvopasture plot established in an existing forest border. Broilers can move in and out of shaded areas.
Photo credit: Allison Pullin.

Affective states

Affective states or emotional states refer to the long-term mood states that are impacted by all short-term emotions, experiences, and moods an animal may experience. There is little research on this aspect of animal welfare in a silvopasture context.

Hens

To date, there have been no studies directly focusing on the affective state impacts of silvopasture in hens. However, the expression of certain behaviors can be informative of affective states and thus provide some insights into the emotional welfare of hens. A hen's positive welfare status relates in part to how the hen is inclined to behave under natural conditions (Bracke & Hopster, 2006), being able to show preference, make choices, and fulfill motivational needs. Foraging, sometimes referred to as scratching, is considered a highly motivated natural behavior of chickens (Jacobs, 2026). Higher frequencies of this behavior in silvopastures or silvopasture-like habitats, as reported by Larsen et al. (2017) and Camarero et al., (2023), could indicate improved affective states compared to indoor-only hens or hens in habitats without canopy cover. Furthermore, the reduced prevalence of abnormal behaviors, such as injurious pecking, can also indicate improved affect in those hens compared to hens ranging with less canopy cover (Brigh et al., 2011).

Broiler chickens

Affective state in broilers is typically improved with environmental complexity and novelty, which can be achieved through outdoor access (Jacobs et al., 2023; Riber et al., 2018). Fast-growing broiler chickens reared in silvopasture plots showed reduced fearfulness compared to those reared in open pastures, as measured by tonic immobility tests (Paneru et al., 2023). In two trials comparing indoor-only housing to outdoor access with either artificial shade structures or silvopasture, tonic immobility durations did not differ between environments in either trial, but silvopasture broilers required more tonic immobility inductions in the second trial, indicating less fearfulness (Stadig et al., 2017a). The authors suggested a possible relationship between less fearfulness and more use of the outdoor area, but did not have individual bird ranging data to fully investigate this relationship.

Other studies comparing fearfulness in birds reared with open grassland outdoor access or indoors have also found mixed results when assessing tonic immobility, likely due to the type of outdoor access provided or the breed used. For example, broilers reared with outdoor access did not differ in their fear responses compared to indoor-reared broilers (Abdourhamane & Petek, 2024), and broilers reared on four different types of vegetative species also did not differ in their fear responses (Bashir et al., 2023). Others reported that broilers reared with outdoor access were either more (Zhao et al., 2014) or less fearful (Oke et al., 2021) than indoor-reared broilers. The previous studies all differed in the type of vegetation, breed used, and age tested, and none of the studies reported shelter or tree cover in the outdoor area. Thus it is promising that silvopasture may mitigate fear responses, likely due to tree cover providing protection for the birds.

Corticosterone is a common physiological marker of stress in poultry, but it should be interpreted with caution as the valence of the stressor (i.e., positive or negative), and therefore its relationship with affective state, cannot be determined without additional context and supporting metrics (Koolhaas et al., 2011). Silvopasture-reared broilers displayed higher feather corticosterone concentrations than indoor-reared broilers, but this could be indicative of higher arousal from a stimulating environment rather than distress (He-Demontaron et al., 2025). Finally, play behavior can be an indicator of a positive affective state as it is expressed when birds have more space and environmental complexity (Vas et al., 2023). Sparring (i.e., play fighting) is an example of play behavior in poultry, and there was a higher probability of sparring behavior in a silvopasture environment compared to an indoor environment (He-Demontaron et al., 2025).

Sustainability

Combining tree or produce production (orchards, tree nurseries, willows, and *Miscanthus*) with poultry has advantages for animal welfare (health, natural living, and affect). Furthermore, the integrated land use can benefit sustainability metrics. For one, it reduces the total land use for food production (Bestman, 2017). Other sustainability benefits are well documented, although not often for poultry-specific production systems.

Planet

The soil ecosystem benefits from the presence of trees. Red alder trees (400 stems·ha) in a silvopasture system in the United Kingdom were estimated to fix 31 kg.yr/ha of Nitrogen (N), with a total amount of N that could potentially be added to the soil estimated at almost 41 kg·yr/ha (Teklehaimanot & Mmolotsi, 2007). A combination of Douglas fir (*Pseudotsuga menziesii*), perennial ryegrass (*Lolium perenne*), and subclover (*Trifolium subterraneum*) in Oregon, United States, more efficiently stored carbon (C) than tree plantations (forests) and pasture monocultures, with 740 kg·yr/ha more C than forests and 520 kg·yr/ha more C than pastures in 11 years' time (Sharrow & Ismail, 2004).

The presence of poultry in these systems can benefit the soil ecosystem through nutrient input, but inputs may be concentrated (near the poultry house) and may exceed vegetation (trees, forage, or crop) demand. Hens that ranged a wheat stubble field in Australia did not impact soil pH, but soil nitrate levels were higher (7.8 vs 33.0 mg/L) after access compared to before, with animal droppings likely contributing to the increase (Miao et al., 2005). Similarly, a calculation showed that on a laying hen farm in the Netherlands with an outdoor run, with on average 13% of the hens outside (2500·hens/ha), 340 kg·N and 200 kg phosphorus ha^{-1} (P) would be deposited into the soil annually (Dekker et al., 2012). If that were an apple or pear orchard that annually uses 18–20 kg of P_2O_5 ha^{-1} (Ehlert et al., 2009), P in the soil would gradually increase (Bestman, 2015). This gradual increase risks accumulation in the long term as deposition exceeds demand over time. Thus while moderate nutrient input from poultry may benefit soil health and productivity, excessive inputs may place environmental health at risk.

People

For farm animal management practices to be socially sustainable, they must be accepted by consumers as a means of ethical production (Alonso et al., 2020). Indeed, increasing consumer interest in animal welfare has been met with welfare-friendly product differentiation in the marketplace, corporate social responsibility efforts, and animal welfare legislation (Buller et al., 2018). Silvopasture is currently not differentiated from free-range and pastured poultry product labels (Campbell et al., 2025), but consumer demand for meat from birds reared with outdoor access has increased in the last decade (Rothrock et al., 2019). As existing research demonstrates animal welfare benefits with silvopasture management, it is likely to be socially accepted as an ethical management practice by consumers. Developing silvopasture animal welfare standards is an opportunity to differentiate and certify these products in the future.

Social acceptance by producers is also critical to ensure long-term implementation. Although not poultry-specific, silvopasture adopters most valued the practice for income diversification and shade, with 88% of silvopasture adopters planning to continue the practice into the future (Smith et al., 2022). A survey of silvopasture producers in California found that the primary benefits of the production system were improving soil health, carbon sequestration, savings on labor and fuel with weed management by livestock (albeit the main livestock used in these systems were sheep), shade and forage for livestock, and diversification of revenue streams (Mazaroli, DeLonge, & Carlisle, 2024). Most of the California producers had been involved in silvopasture for at least 5 years. Small-scale silvopasture poultry production may also be an opportunity to improve food security, revenue, and community building in urban and peri-urban "food deserts," but these models have not been empirically evaluated (Conway & Nieman, 2022).

Profit

Considering the economic benefits of pasture-raised poultry in general, rather than silvopasture-specific, producers generally seem positive. Interviews and surveys with 18 pastured poultry producers in California revealed that this production system was directly profitable for 50% of farmers, while 78% reported indirect profits through savings on items such as fertilizer and pest management. In addition, they perceived benefits for marketing appeal (44% of respondents) (Hilimire, 2012). In an economic analysis comparing a large-scale conventional broiler farm to a small-scale pasture-raised broiler farm, the large-scale farm achieved a lower unit operating cost ($0.52/lb) than the pasture-based model ($1.92/lb) (Alves et al., 2026). These differences were mainly due to production volume, where the large-scale system benefits as investment costs are distributed over a much greater annual live-weight output than the pasture-based system. However, premium pricing on pasture-based products allows for significant gains for pastured poultry producers to not only recover their annual operating costs but also generate sustainable profitability under favorable market conditions (Alves et al., 2026). Therefore price and product differentiation can substantially improve the economic outlook of small-scale outdoor systems to become financially viable niche systems (Alves et al., 2026.).

While there are no economic analyses of silvopasture-specific poultry production, mixed enterprises with multiple streams of revenue benefit poultry producers. An investment analysis of a mixed-enterprise farm producing conventional broiler houses and a rotation of hay and pine trees on open cropland in south Alabama suggested a large initial investment cost ($389,166) but found that the returns were greater than most agricultural investments when averaged over 30 years ("Prevatt et al., 2025). The authors concluded that mixed enterprises may make poultry operations more viable, but it is worth noting that they modeled an intensive broiler production farm rather than a silvopasture farm with broilers integrated into the crop area. However, considering that silvopasture producers cite diversified revenue streams as a benefit of silvopasture production (Mazaroli et al., 2024), these results are promising for silvopasture poultry production. Correspondingly, a financial analysis (a 30-year projection) for a silvopasture system with forage, timber, and cattle production showed that it resulted in the highest internal rate of return (6.4%) and net present value per hectare ($1,277) compared to open pasture and two other production systems (Orefice et al., 2019).

Profitability of the trees or produce depends, of course, on the type that is produced and how the stand is managed. For example, willows can produce up to 20 tons of fresh biomass per hectare per year for up to 20 years, which can be used instead of 5500 m^3 of gas (Bestman, 2017). Thirty walnut trees on an organic laying hen farm in the Netherlands produced about 1000 kg of nuts annually, depending on spring weather conditions (Bestman, 2017). *Miscanthus*, or elephant grass, can serve as fuel, bedding material, or feed for cattle. On an organic laying hen farm in the Netherlands, the harvests gradually increased annually, from 6 to 14, and then to 16 tons/ha (Bestman, 2017).

Challenges

Risks of a silvopasture system where vegetation and poultry production are combined include high investment costs needed for planting large surfaces with trees. California silvopasture producers had some access to federal and state cost-share funding programs, but there were still significant gaps between the program funding and the actual cost of planting and maintaining trees (Mazaroli et al., 2024). Poultry producers may lack the skills to manage trees, or tree producers may lack poultry production skills (Bestman, 2017). This can be further exacerbated as vegetation may be difficult to establish and maintain, depending on the climatic conditions and poultry destruction of foliage and soil (de Koning and Kitessa, 2018). To address these challenges in part, producers could identify hardy plants and trees that are resistant to poultry damage and that can do well in poor conditions (de Koning et al., 2019).

Given the scale of global intensive poultry production, a silvopasture production system is anticipated to remain a niche market. The greater land requirement, the increased production costs (Alves et al., 2026), and the sensitivity to seasonality (Jacobs et al., 2026) make this production system not suitable as an alternative to all conventional production. As a niche product, it can be successfully marketed to consumers directly, to restaurants, or via grocery stores.

Future directions

Much is still unknown about silvopasture systems for poultry when considering the scientific literature. Inclusion of evidence from other species provides more insights into this Indigenous and traditional practice of agriculture. Scientific evidence does suggest some benefits to animal welfare and the sustainability of these systems compared to monocultural systems, but some of this evidence is through inferences or theoretical impacts.

Dense canopy cover seems to be preferred by birds and shows some benefits for animal welfare that lower levels of cover, or even artificial shelters, do not. A more controlled assessment of levels of canopy cover on animal welfare, bird production, and soil health metrics would be valuable, especially for broiler chicken production systems (Fig. 11.7), since the evidence is currently biased towards laying hens.

Furthermore, research on parasites and disease, specifically for silvopasture poultry systems, is needed. With some evidence showing reduced predation with silvopasture, tree cover may be an effective tool to reduce interactions between wild birds and poultry, which may have additional benefits of mitigating highly pathogenic avian influenza infections.

Producer training and technical assistance, as well as federal and state research investments and incentive policies, are additional future opportunities to overcome silvopasture challenges with knowledge and funding gaps (Mazaroli et al., 2024).

FIGURE 11.7 **Silvopasture research plot.** A silvopasture research plot established in a forested plot of land on a mixed-enterprise farm, with alternating pens of slow-growing Sasso broilers (brown birds) and fast-growing Ross 708 broilers (white birds).
Photo credit: Allison Pullin.

References

Abdourhamane, İ. M., & Petek, M. (2024). Health-based welfare indicators and fear reaction of slower growing broiler compared to faster growing broiler housed in free range and conventional deep litter housing systems. *Journal of Applied Animal Welfare Science, 27*(3), 442–453. https://doi.org/10.1080/10888705.2022.2100221.

Aksoy, T., Çürek, D.İ., Narinç, D., & Önenç, A. (2021). Effects of season, genotype, and rearing system on broiler chickens raised in different semi-intensive systems: Performance, mortality, and slaughter results. *Tropical Animal Health and Production, 53*(1), 189. https://doi.org/10.1007/s11250-021-02629-y.

Alonso, M. E., González-Montaña, J. R., & Lomillos, J. M. (2020). Consumers' concerns and perceptions of farm animal welfare. *Animals, 10*(3), 385. https://doi.org/10.3390/ani10030385.

Alves, L. K. S., Vukina, T., Bartz, B. M., Eckberg, J. O., & Pullin, A. N. (2026). Pasture-based versus conventional broiler chicken production: Enterprise budget modeling and profitability analysis. *Poultry Science*, 105, 107350. https://doi.org/10.1016/j.psj.2026.107350.

Bashir, N., Şekeroğlu, A., Tainika, B., Gür, F. M., Duman, M., & Şentürk, Y. E. (2023). Effect of different pasture species on welfare and behavior of slow-growing broiler chickens in free-range production system. *Tropical Animal Health and Production, 55*(3), 206. https://doi.org/10.1007/s11250-023-03616-1.

Bello, H. J. S., Gonçalves, J. A., Teixeira, G. S., de Freitas Santos, J. M., do Valle Polycarpo, G., de Almeida, F. A., Amarante, A. F. T., & de Soutello, R. V. G. (2020). Parasitism in Angus x Nellore heifers in a silvopastoral system. *Tropical Animal Health and Production, 52*(4), 1733–1738. https://doi.org/10.1007/s11250-019-02196-3.

Bestman, M. (2015). www.agforward.eu System Report: Agroforestry for Organic Egg Production in the Netherlands

Bestman, M. (2017). Lessons learnt- agroforestry for organic and free-range egg production in the Netherlands. https://library.wur.nl/WebQuery/groenekennis/2227952.

Bestman, M., de Jong, W., Wagenaar, J. P., & Weerts, T. (2018). Presence of avian influenza risk birds in and around poultry free-range areas in relation to range vegetation and openness of surrounding landscape. *Agroforestry Systems, 92*(4), 1001–1008. https://doi.org/10.1007/s10457-017-0117-2.

Bestman, M. W. P., & Wagenaar, J. P. (2003). Farm level factors associated with feather pecking in organic laying hens. *Livestock Production Science, 80*(1-2), 133–140. https://doi.org/10.1016/s0301-6226(02)00314-7.

Bonnefous, C., Collin, A., Guilloteau, L. A., Guesdon, V., Filliat, C., Réhault-Godbert, S., Rodenburg, T. B., Tuyttens, F. A. M., Warin, L., Steenfeldt, S., Baldinger, L., Re, M., Ponzio, R., Zuliani, A., Venezia, P., Väre, M., Parrott, P., Walley, K., Niemi, J. K., & Leterrier, C. (2022). Welfare issues and potential solutions for laying hens in free range and organic production systems: A review based on literature and interviews. *Frontiers in Veterinary Science, 9*, 952922. https://doi.org/10.3389/fvets.2022.952922.

Bracke, M. B. M., & Hopster, H. (2006). Assessing the importance of natural behavior for animal welfare. *Journal of Agricultural and Environmental Ethics, 19*(1), 77–89. https://doi.org/10.1007/s10806-005-4493-7.

Brigh, A., Brass, D., Clachan, J., Drake, K. A., & Joret, A. D. (2011). Canopy cover is correlated with reduced injurious feather pecking in commercial flocks of free-range laying hens. *Animal Welfare, 20*(3), 329–338. https://doi.org/10.1017/s096272860000289x.

Buller, H., Blokhuis, H., Jensen, P., & Keeling, L. (2018). Towards farm animal welfare and sustainability. *Animals, 8*(6), 81. https://doi.org/10.3390/ani8060081.

Caceres, J. C., Sobreira, C. H., Munmun, A. R., Sigmon, C. S., He-DeMontaron, A. D., Pullin, A. N., Grimes, J. L., Walker, L. L., Dean, L., Schilling, M. W., & Campbell, Y. L. (2026). The effect of housing system and broiler breed on meat quality1. *Poultry Science, 105*(8), 107016. https://doi.org/10.1016/j.psj.2026.107016.

Camarero, N. P., Righi, C. A., Ruffini Sabadin, E., Oliveira da Silva, I. J., Menten, J. F. M., & Zarate Couto, H. T. (2023). Freedom for chickens – The influence of the silvopastoral system on poultry behavior. *Brazilian Journal of Agriculture, 98*, 132–140. https://www.revistadeagricultura.com.br/index.php/revistadeagricultura/article/view/4335/4196.

Campbell, Y. L., Walker, L. L., Bartz, B. M., Eckberg, J. O., & Pullin, A. N. (2025). Outdoor access versus conventional broiler chicken production: Updated review of animal welfare, food safety, and meat quality. *Poultry Science, 104*(4), 104906. https://doi.org/10.1016/j.psj.2025.104906.

Castellini, C., Mugnai, C., Moscati, L., Mattioli, S., Amato, M. G., Mancinelli, A. C., & Dal Bosco, A. (2016). Adaptation to organic rearing system of eight different chicken genotypes: Behaviour, welfare and performance. *Italian Journal of Animal Science, 15*(1), 37–46. https://doi.org/10.1080/1828051X.2015.1131893.

Collias, N. E., & Collias, E. C. (1967). A field study of the red jungle fowl in north-central India. *The Condor, 69*(4), 360–386. https://doi.org/10.2307/1366199.

Conway, A., & Nieman, C. (2022). Small-scale silvopasture: Addressing urban and peri-urban livestock challenges in the United States with agroforestry practices. *Urban Agriculture and Regional Food Systems, 7*(1). https://doi.org/10.1002/uar2.20023.

Cosentino, C., Freschi, P., Fascetti, S., Paolino, R., & Musto, M. (2020). Growth control of herbaceous ground cover and egg quality from an integrated poultry-hazelnut orchard system. *Italian Journal of Agronomy, 15*(3), 1594. https://doi.org/10.4081/ija.2020.1594.

Dal Bosco, A., Mugnai, C., Rosati, A., Paoletti, A., Caporali, S., & Castellini, C. (2014). Effect of range enrichment on performance, behavior, and forage intake of free-range chickens. *Journal of Applied Poultry Research, 23*(2), 137–145. https://doi.org/10.3382/japr.2013-00814.

Dal Bosco, A., Mugnai, C., Sirri, F., Zamparini, C., & Castellini, C. (2010). Assessment of a global positioning system to evaluate activities of organic chickens at pasture. *Journal of Applied Poultry Research, 19*(3), 213–218. https://doi.org/10.3382/japr.2010-00153.

Dawkins, M. S., Cook, P. A., Whittingham, M. J., Mansell, K. A., & Harper, A. E. (2003). What makes flee-range broiler chickens range? In situ measurement of habitat preference. *Animal Behaviour, 66*(1), 151–160. https://doi.org/10.1006/anbe.2003.2172.

Dekker, S. E. M., Aarnink, A. J. A., De Boer, I. J. M., & Koerkamp, P. W. G. G. (2012). Total loss and distribution of nitrogen and phosphorus in the outdoor run of organic laying hens. *British Poultry Science, 53*(6), 731–740. https://doi.org/10.1080/00071668.2012.749342.

Delgadillo, E., Glidden, C., Pollak, M., Rysenga, H., Jolles, A., & Beechler, B. (2021). The benefit of hedgerow access on the health and growth rate of pasture raised broiler chickens. *Frontiers in Animal Science, 2*, 649924. https://doi.org/10.3389/fanim.2021.649924.

Deutsch, M., Dalseg, I. K., Kuchling, S., Sefc, K. M., Erregger, B., & Kalcher-Sommersguter, E. (2024). Space use of free-range laying hens on two outdoor ranges with different amounts of vegetation cover. *Applied Animal Behaviour Science, 277*, 106353. https://doi.org/10.1016/j.applanim.2024.106353.

de Jonge, J., & van Trijp, H. C. M. (2013). The impact of broiler production system practices on consumer perceptions of animal welfare. *Poultry Science, 92*(12), 3080–3095. https://doi.org/10.3382/ps.2013-03334.

de Mendonca, R. M. A., Leite, R. C., Lana, A. M. Q., Costa, J. O., & Toth, G. (2014). Parasitic helminth infection in young cattle raised on silvopasture and open-pasture in Southeastern Brazil. *Agroforestry Systems, 88*(1), 53–62. https://doi.org/10.1007/s10457-013-9655-4.

de Koning, C., Kitessa, S. M., Barekatain, R., & Drake, K. (2018). Determination of range enrichment for improved hen welfare on commercial fixed-range free-range layer farms. *Animal Production Science, 59*(7), 1336–1348. https://doi.org/10.1071/AN17757.

de Koning, C., Barekatain, R., Singh, M., & Drake, K. (2019). Saltbush (*Atriplex nummularia* and *A. amnicola*) as potential plants for free-range layer farms: Consequences for layer performance, egg sensory qualities, and excreta moisture. *Poultry Science, 98*(10), 4555–4564. https://doi.org/10.3382/ps/pez294.

Ehlert, P. A. I., Dekker, P. H. M., van der Schoot, J. R., Visschers, R., van Middelkoop, J. C., van der Maas, M. P., Pronk, A. A., & van Dam, A. M. (2009). Fosforgehalten en fosfaatafvoercijfers van landbouwgewassen : eindrapportage. Alterra, Wageningen. https://research.wur.nl/en/publications/fosforgehalten-en-fosfaatafvoercijfers-van-landbouwgewassen-eindr/.

Fanatico, A. C., Mench, J. A., Archer, G. S., Liang, Y., Gunsaulis, V. B. B., Owens, C. M., & Donoghue, A. M. (2016). Effect of outdoor structural enrichments on the performance, use of range area, and behavior of organic meat chickens. *Poultry Science, 95*(9), 1980–1988. https://doi.org/10.3382/ps/pew196.

Fraser, D. (2008). Understanding animal welfare. *Acta Veterinaria Scandinavica, 50*(S1). https://doi.org/10.1186/1751-0147-50-s1-s1.

Glatz, P. C., Ru, Y. J., Miao, Z. H., Wyatt, S. K., & Rodda, B. J. (2005). Integrating poultry into a crop and pasture farming system. *International Journal of Poultry Science, 4*(4), 187–191. https://doi.org/10.3923/ijps.2005.187.191.

Green, L. E., Lewis, K., Kimpton, A., & Nicol, C. J. (2000). Cross-sectional study of the prevalence of feather pecking in laying hens in alternative systems and its associations with management and disease. *Veterinary Record, 147*(9), 233–238. https://doi.org/10.1136/vr.147.9.233.

Grigor, P. N., Hughes, B. O., & Appleby, M. C. (1995). Effects of regular handling and exposure to an outside area on subsequent fearfulness and dispersal in domestic hens. *Applied Animal Behaviour Science, 44*(1), 47–55. https://doi.org/10.1016/0168-1591(95)00576-e.

Hartcher, K. M., Hickey, K. A., Hemsworth, P. H., Cronin, G. M., Wilkinson, S. J., & Singh, M. (2016). Relationships between range access as monitored by radio frequency identification technology, fearfulness, and plumage damage in free-range laying hens. *Animal, 10*(5), 847–853. https://doi.org/10.1017/s1751731115002463.

He-DeMontaron, A. D. 2025. The effects of genetic strain and silvopasture environment on broiler welfare and production performance. Master's Thesis. North Carolina State University, Raleigh. https://www.lib.ncsu.edu/resolver/1840.20/46605.

Hilimire, K. (2012). The grass is greener: Farmers' experiences with pastured poultry. *Renewable Agriculture and Food Systems, 27*(3), 173–179. https://doi.org/10.1017/s1742170511000287.

Jacobs, L., Blatchford, R. A., de Jong, I. C., Erasmus, M. A., Levengood, M., Newberry, R. C., Regmi, P., Riber, A. B., & Weimer, S. L. (2023). Enhancing their quality of life: Environmental enrichment for poultry. *Poultry Science, 102*(1), 102233. https://doi.org/10.1016/j.psj.2022.102233.

Jacobs, L., Paneru, B., Brooks, G., Downing, A., Munsell, J., Fike, J., & Pent, G. (2026). Fast-growing broiler chickens range more in a silvopasture than a grass pasture based on live observations. *Frontiers in Animal Science, 6*, 1721759. https://doi.org/10.3389/fanim.2025.1721759.

Jeni, R. E., Dittoe, D. K., Olson, E. G., Lourenco, J., Seidel, D. S., Ricke, S. C., & Callaway, T. R. (2021). An overview of health challenges in alternative poultry production systems. *Poultry Science, 100*(7), 101173. https://doi.org/10.1016/j.psj.2021.101173.

Koolhaas, J. M., Bartolomucci, A., Buwalda, B., de Boer, S. F., Flügge, G., Korte, S. M., Meerlo, P., Murison, R., Olivier, B., Palanza, P., Richter-Levin, G., Sgoifo, A., Steimer, T., Stiedl, O., van Dijk, G., Wöhr, M., & Fuchs, E. (2011). Stress revisited: A critical evaluation of the stress concept. *Neuroscience & Biobehavioral Reviews, 35*(5), 1291–1301. https://doi.org/10.1016/j.neubiorev.2011.02.003.

Knierim, U. (2006). Animal welfare aspects of outdoor runs for laying hens: A review. *Journal of Life Sciences, 54*(2), 133–145. https://doi.org/10.1016/s1573-5214(06)80017-5.

Jacobs, L. (2026). Understanding the behavioural needs of broilers. In H. van den Brand (Ed.). *Future perspectives in broiler rearing: Incorporating the broiler point of view*. Burleigh Dodds Science Publishing.

Lambton, S. L., Knowles, T. G., Yorke, C., & Nicol, C. J. (2010). The risk factors affecting the development of gentle and severe feather pecking in loose housed laying hens. *Applied Animal Behaviour Science, 123*(1-2), 32–42. https://doi.org/10.1016/j.applanim.2009.12.010.

Larsen, H., Cronin, G., Smith, C. L., Hemsworth, P., & Rault, J.-L. (2017). Behaviour of free-range laying hens in distinct outdoor environments. *Animal Welfare, 26*(3), 255–264. https://doi.org/10.7120/09627286.26.3.255.

Liles, K. M., Bartlett, J. R., & Beckford, R. C. (2015). Comparing the effects of conventional and pastured poultry production systems on the stress levels of broilers. *Professional Agricultural Workers Journal, 2*(7). https://tuspubs.tuskegee.edu/pawj/vol2/iss2/7.

Lopes, L. B., Kamchen, S. G., Gomes, F. J., Natividade, U., Magalhães, L. M. D., de Paula Pimenta, A., & Araujo, R. N. (2022). Influence of silvopastoral systems on gastrointestinal nematode infection and immune response of Nellore heifers under tropical conditions. *Veterinary Parasitology, 309*, 109765. https://doi.org/10.1016/j.vetpar.2022.109765.

Lopes, L. B., Pedreira, B. C., Eckstein, C., dos Santos, L. L., & Peruffo, R. G. (2020). Gastrointestinal nematode egg counts and performance of beef cattle raised on open pastures and silvopastoral systems in Brazil. *Agroforestry Systems, 94*(5), 1693–1700. https://doi.org/10.1007/s10457-020-00486-6.

Mahboub, H. D. H., Müller, J., & Von Borell, E. (2004). Outdoor use, tonic immobility, heterophil/lymphocyte ratio and feather condition in free-range laying hens of different genotype. *British Poultry Science, 45*(6), 738–744. https://doi.org/10.1080/00071660400014267.

Mancinelli, A. C., Mattioli, S., Dal Bosco, A., Aliberti, A., Amato, M. G., & Castellini, C. (2020). Performance, behavior, and welfare status of six different organically reared poultry genotypes. *Animals, 10*(4), 550. https://doi.org/10.3390/ani10040550.

Mazaroli, D. N., DeLonge, M., & Carlisle, L. (2024). The potential of silvopasture in California: Producer perspectives. *Agroecology and Sustainable Food Systems, 48*(10), 1413–1427. https://doi.org/10.1080/21683565.2024.2405886.

Miao, Z. H., Glatz, P. C., Ru, Y. J., Wyatt, S. K., & Rodda, B. J. (2005). Integrating free-range hens into a wheat stubble. *International Journal of Poultry Science, 4*(8), 526–530. https://doi.org/10.3923/ijps.2005.526.530.

Moyle, J. R., Arsi, K., Woo-Ming, A., Arambel, H., Fanatico, A., Blore, P. J., Clark, F. D., Donoghue, D. J., & Donoghue, A. M. (2014). Growth performance of fast-growing broilers reared under different types of production systems with outdoor access: Implications for organic and alternative production systems. *Journal of Applied Poultry Research, 23*(2), 212–220. https://doi.org/10.3382/japr.2013-00882.

National Center for Appropriate Technology (NCAT). (2004). NCAT's organic livestock workbook a guide to sustainable and allowed practices. https://www.calameo.com/books/00071543087e7b23fb13c.

Nguyen-Phuc, H., & Berres, M. E. (2018). Genetic structure in Red Junglefowl (*Gallus gallus*) populations: Strong spatial patterns in the wild ancestors of domestic chickens in a core distribution range. *Ecology and Evolution, 8*(13), 6575–6588. https://doi.org/10.1002/ece3.4139.

Nicol, C. J., Pötzsch, C., Lewis, K., & Green, L. E. (2003). Matched concurrent case-control study of risk factors for feather pecking in hens on free-range commercial farms in the UK. *British Poultry Science, 44*(4), 515–523. https://doi.org/10.1080/00071660310001616255.

Oke, O. E., Akande, J. O., Sodipo, I. D., Ekunseitan, D. A., Rahman, S. A., Smith, O. F., & Onagbesan, O. M. (2021). Evaluation of access to different legume pastures on performance and welfare of broiler chickens during dry season under tropical environment. *Veterinary Medicine and Science, 7*(4), 1359–1368. https://doi.org/10.1002/vms3.461.

Oke, O. E., Ladokun, A. O., & Onagbesan, O. M. (2016). Reproductive performance of layer chickens reared on deep litter system with or without access to grass or legume pasture. *Journal of Animal Physiology and Animal Nutrition, 100*(2), 229–235. https://doi.org/10.1111/jpn.12353.

Oliveira, M. C. S., Nicodemo, M. L. F., Pezzopane, J. R. M., Gusmão, M. R., Chagas, A. C. S., Giglioti, R., Bilhassi, T. B., Santana, C. H., Gonçalves, T. C., Rabelo, M. D., & Néo, T. A. (2017). Gastrointestinal nematode infection in beef cattle raised in silvopastoral and conventional systems in São Paulo state, Brazil. *Agroforestry Systems, 91*(3), 495–507. https://doi.org/10.1007/s10457-016-9950-y.

Orefice, J., Carroll, J., Conroy, D., & Ketner, L. (2017). Silvopasture practices and perspectives in the Northeastern United States. *Agroforestry Systems, 91*(1), 149–160. https://doi.org/10.1007/s10457-016-9916-0.

Orefice, J., Smith, R. G., Carroll, J., Asbjornsen, H., & Howard, T. (2019). Forage productivity and profitability in newly-established open pasture, silvopasture, and thinned forest production systems. *Agroforestry Systems, 93*(1), 51–65. https://doi.org/10.1007/s10457-016-0052-7.

Pacher-Deutsch, M., Meyer, P., Meimberg, H., & Gierus, M. (2025). Enhancing range use in free-range laying hen systems: The impact of vegetation cover over time. *Animals, 15*(9), 1204. https://doi.org/10.3390/ani15091204.

Paneru, B., Pent, G. J., Nastasi, S., Downing, A. K., Munsell, J. F., Fike, J. H., & Jacobs, L. (2023). Effect of silvopasture system on fearfulness and leg health in fast-growing broiler chickens. *PLoS One, 18*(3), e0282923. https://doi.org/10.1371/journal.pone.0282923.

Pezzopane, J. R. M., Nicodemo, M. L. F., Bosi, C., Garcia, A. R., & Lulu, J. (2019). Animal thermal comfort indexes in silvopastoral systems with different tree arrangements. *Journal of Thermal Biology, 79*, 103–111. https://doi.org/10.1016/j.jtherbio.2018.12.015.

Prevatt, W., Yeager, J., Young, G., Simpson, G., & Pepper, H. (2005). An investment analysis of a southern mixed enterprise farm: Broilers and silvopasture. *Journal of the ASFMRA*, 2005, 1–9. https://www.proquest.com/scholarly-journals/investment-analysis-southern-mixed-enterprise/docview/219597503/se-2.

Rana, M. S., Lee, C., Lea, J. M., & Campbell, D. L. M. (2022). Commercial free-range laying hens' preferences for shelters with different sunlight filtering percentages. *Animals, 12*(3), 344. https://doi.org/10.3390/ani12030344.

Riber, A. B., van de Weerd, H. A., de Jong, I. C., & Steenfeldt, S. (2018). Review of environmental enrichment for broiler chickens. *Poultry Science, 97*(2), 378–396. https://doi.org/10.3382/ps/pex344.

Richards, G. J., Wilkins, L. J., Knowles, T. G., Booth, F., Toscano, M. J., Nicol, C. J., & Brown, S. N. (2011). Continuous monitoring of pop hole usage by commercially housed free-range hens throughout the production cycle. *Veterinary Record, 169*(13), 338. https://doi.org/10.1136/vr.d4603.

Rodriguez-Aurrekoetxea, A., & Estevez, I. (2016). Use of space and its impact on the welfare of laying hens in a commercial free-range system. *Poultry Science, 95*(11), 2503–2513. https://doi.org/10.3382/ps/pew238.

Rothrock, M. J., Gibson, K. E., Micciche, A. C., & Ricke, S. C. (2019). Pastured poultry production in the United States: Strategies to balance system sustainability and environmental impact. *Frontiers in Sustainable Food Systems, 3*, 74. https://doi.org/10.3389/fsufs.2019.00074.

sans, E. C. de O, Dahlke, F., Freitas Federici, J., Tuyttens, F. A. M., & Forte Maiolino Molento, C. (2023). Welfare of broiler chickens in Brazilian free-range versus intensive indoor production systems. *Journal of Applied Animal Welfare Science, 26*(4), 505–517. https://doi.org/10.1080/10888705.2021.1992280.

Sans, E. C. O., Federici, J. F., Dahlke, F., & Molento, C. F. M. (2014). Evaluation of free-range broilers using the Welfare Quality® protocol. *Revista Brasileira de Ciencia Avicola / Brazilian Journal of Poultry Science, 16*(3), 297–306. https://doi.org/10.1590/1516-635x1603297-306.

Saraiva, S., Saraiva, C., & Stilwell, G. (2016). Feather conditions and clinical scores as indicators of broilers welfare at the slaughterhouse. *Research in Veterinary Science, 107*, 75–79. https://doi.org/10.1016/j.rvsc.2016.05.005.

Sharrow, S. H., & Ismail, S. (2004). Carbon and nitrogen storage in agroforests, tree plantations, and pastures in western Oregon, USA. *Agroforestry Systems, 60*(2), 123–130. https://doi.org/10.1023/B:AGFO.0000013267.87896.41.

Shi, X., & Conway-Anderson, A. (2022). Assessing the sustainability of silvopasture systems. *CABI Reviews: Perspectives in Agriculture, Veterinary Science, Nutrition and Natural Resources, 2022*. https://doi.org/10.1079/cabireviews202217047.

Shifaw, A., Feyera, T., Walkden-Brown, S. W., Sharpe, B., Elliott, T., & Ruhnke, I. (2021). Global and regional prevalence of helminth infection in chickens over time: A systematic review and meta-analysis. *Poultry Science, 100*(5), 101082. https://doi.org/10.1016/j.psj.2021.101082.

Singh, M., Ruhnke, I., De Koning, C., Drake, K., Skerman, A. G., Hinch, G. N., & Glatz, P. C. (2017). Demographics and practices of semi-intensive free-range farming systems in Australia with an outdoor stocking density of ≤1500 hens/hectare. *PLoS One, 12*(10), e0187057. https://doi.org/10.1371/journal.pone.0187057.

Smith, M. M., Bentrup, G., Kellerman, T., MacFarland, K., Straight, R., Ameyaw, L., & Stein, S. (2022). Silvopasture in the USA: A systematic review of natural resource professional and producer-reported benefits, challenges, and management activities. *Agriculture, Ecosystems & Environment, 326*, 107818. https://doi.org/10.1016/j.agee.2021.107818.

Stadig, L. M., Rodenburg, T. B., Ampe, B., Reubens, B., & Tuyttens, F. A. M. (2017a). Effect of free-range access, shelter type and weather conditions on free-range use and welfare of slow-growing broiler chickens. *Applied Animal Behaviour Science, 192*, 15–23. https://doi.org/10.1016/j.applanim.2016.11.008.

Stadig, L. M., Rodenburg, T. B., Ampe, B., Reubens, B., & Tuyttens, F. A. M. (2017b). Effects of shelter type, early environmental enrichment and weather conditions on free-range behaviour of slow-growing broiler chickens. *Animal, 11*(6), 1046–1053. https://doi.org/10.1017/s1751731116002172.

Stadig, L. M., Rodenburg, T. B., Reubens, B., Aerts, J., Duquenne, B., & Tuyttens, F. A. M. (2016). Effects of free-range access on production parameters and meat quality, composition and taste in slow-growing broiler chickens. *Poultry Science, 95*(12), 2971–2978. https://doi.org/10.3382/ps/pew226.

Sánchez-Casanova, R. E., Sarmiento-Franco, L., & Phillips, C. J. C. (2022). The effects of providing outdoor access to broilers in the tropics on their behaviour and stress responses. *Animals, 12*(15), 1917. https://doi.org/10.3390/ani12151917.

Taylor, P. S., Hemsworth, P. H., Groves, P. J., Gebhardt-Henrich, S. G., & Rault, J. L. (2018). Ranging behavior relates to welfare indicators pre- and post-range access in commercial free-range broilers. *Poultry Science, 97*(6), 1861–1871. https://doi.org/10.3382/ps/pey060.

Teklehaimanot, Z., & Mmolotsi, R. M. (2007). Contribution of red alder to soil nitrogen input in a silvopastoral system. *Biology and Fertility of Soils, 43*(6), 843–848. https://doi.org/10.1007/s00374-006-0163-9.

Terra, M. T. B., Pacheco, W. J., Harrison, M., Mccrea, B. A., & Hauck, R. (2021). A survey of coccidia and nematodes in pastured poultry in the state of georgia. *Avian Diseases, 65*(2), 250–256. https://doi.org/10.1637/aviandiseases-D-20-00120.

Urrea-Benítez J. L. (2024). Tackling food system challenges through silvopasture insights expert workshop Alliance Bioversity International Tackling food system challenges through silvopasture. https://alliancebioversityciat.org/stories/tackling-food-system-challenges-through-silvopasture-insights-expert-workshop

Vas, J., BenSassi, N., Vasdal, G., & Newberry, R. C. (2023). Better welfare for broiler chickens given more types of environmental enrichments and more space to enjoy them. *Applied Animal Behaviour Science, 261*, 105901. https://doi.org/10.1016/j.applanim.2023.105901.

Vieira, F. M. C., Pilatti, J. A., Czekoski, Z. M. W., Fonsêca, V. F. C., Herbut, P., Angrecka, S., de Souza Vismara, E., de Paulo Macedo, V., Dos Santos, M. C. R., & Paśmionka, I. (2021). Effect of the silvopastoral system on the thermal comfort of lambs in a subtropical climate: A preliminary study. *Agriculture, 11*(8), 790. https://doi.org/10.3390/agriculture11080790.

Welfare Quality® (2009). Welfare Quality(R) assessment protocol for poultry (broilers, laying hens). Welfare Quality® Consortium, Lelystad, Netherlands.

Zeltner, E., & Hirt, H. (2003). Effect of artificial structuring on the use of laying hen runs in a free-range system. *British Poultry Science, 44*(4), 533–537. https://doi.org/10.1080/00071660310001616264.
Zhao, Z. G., Li, J. H., Li, X., & Bao, J. (2014). Effects of housing systems on behaviour, performance and welfare of fast-growing broilers. *Asian-Australasian Journal of Animal Sciences, 27*(1), 140–146. https://doi.org/10.5713/ajas.2013.13167.

Chapter 12

Animal welfare and sustainable poultry production in India

Jaydip Rokade[1], Geeta Pipaliya[1], Darshana Bhaisare[2], Manzoor Wani[3], Sai Siva Kumar Katam[1], and Monika Madheswaran[1]

[1]*PHT Division, ICAR- Central Avian Research Institute, Bareilly, Uttar Pradesh, India,* [2]*Poultry Science Division, Nagpur Veterinary Collages, MAFSU, Nagpur, Nagpur, Maharashtra, India,* [3]*ANFT Division ICAR-Central Avian Research Institute, Bareilly, Uttar Pradesh, India*

Introduction

The Indian poultry sector has witnessed incredible growth in recent years, evolving as one of the most vibrant industries, with broiler and layer segments comprising 65.3% and 34.7%, respectively, of poultry production in India. As per the latest figures (Baghel et al., 2025), the country ranks second globally in egg production, following China, with an annual total of 149.11 billion eggs (2024–25). The per capita availability of 106 eggs per annum in 2024–25 marked a 2.91% increase from the previous year (103 eggs per annum), highlighting the industry's growth trajectory. The poultry population has now exceeded 850 million birds, with commercial poultry accounting for a substantial proportion. The sector exhibits a compound annual growth rate of 8%–10%, significantly higher than that of conventional agricultural crops, which typically range from 2%–3%. This rapid expansion has positioned poultry as a critical driver of agricultural diversification and income generation. Poultry production supports the livelihoods of millions of people across the value chain, including farmers, hatchery operators, feed manufacturers, veterinarians, transporters, processors, and marketers. The Indian poultry sector has alone contributed nearly 50% of India's total 10.50 million tonnes of meat production in 2024–25.

Poultry meat and eggs are distinguished for their exceptional nutritive value, imparting protein with high biological value, essential amino acids, and wide-ranging vitamins and minerals, placing them in high demand across diverse populations of different age groups. To meet consumers' demand, producers often face challenges in maintaining high animal welfare and ethical standards. However, in the postpandemic Covid era, consumers have become

Integrating Welfare and Sustainability in Poultry Systems. DOI: https://doi.org/10.1016/B978-0-443-40594-5.00007-9

more insightful, pursuing not only delicacy and taste but also demanding uncompromised standards of health, hygiene, ethics, and animal welfare (Kolluri et al., 2021; Sinclair et al., 2022). The active involvement of NGOs and international bodies has propelled poultry welfare to the center stage, marking a transitional phase for the poultry sector nationwide and worldwide. Sustainability in Indian poultry production is a complex issue that balances economic viability with environmental responsibility and social equity, particularly the welfare of the birds. India should prioritize improving public welfare by addressing key challenges such as the diversion of maize for ethanol production. This demand reduces the availability of maize for animal feed, drives up costs, and forces reliance on lower-quality alternatives. To build a more sustainable system, India needs to invest in water-efficient agricultural practices, promote alternative protein sources like insects and algae, and accelerate the adoption of solar energy. India should also prioritize poultry health, as the tropical climate and widespread use of basic housing management make health challenges a critical concern.

Regulatory frameworks and organizational structure for animal welfare in India

In addition to these regulatory and institutional efforts, public awareness of animal welfare in India is progressively growing, while direct evidence remains limited. Conventionally, societal morals have accentuated sympathy towards animals; still, awareness regarding welfare concerns in intensive poultry production has been relatively low. Indications from livestock stakeholders across Asia, including India, indicate that while financial considerations persist as dominant factors, animal welfare is progressively recognized for its association with livestock health, final product quality, and human health and well-being. In India, consumers exhibit a budding preference for native chicken, specifically linked with perceptions of natural and health-oriented production systems; though, buying decisions remain strongly subjective to price and availability limitations (Kathiravan & Chitrambigai, 2024). Additionally, countrywide initiatives such as the "Eat Right India" program of the Food Safety and Standards Authority of India reveal the growing importance of consumer attentiveness and responsible food choices (Food Safety & Standards Authority of India FSSAI, 2020). India's animal welfare landscape is guided by a combination of regulatory frameworks and institutional structures that define responsibilities across the national, state, and local levels. The Prevention of Cruelty to Animals (PCA) Act, 1960, remains the central legislation, supported by species- and sector-specific rules such as the:

Transport rules: https://awbi.gov.in/uploads/regulations/163309961689TRANSPORT%20OF%20ANIMALS,%20RULES,%201978.pdf,

Slaughter rules: https://awbi.gov.in/uploads/regulations/163309997882(SLAUGHTER%20HOUSE)%20RULES,%202001.pdf
Breeding rules: https://awbi.gov.in/uploads/regulations/163310016236Dog%20Breeding%20and%20Marketing%20English.pdf.

Oversight is shared between the Ministry of Fisheries, Animal Husbandry and Dairying (MoFAHD), the Animal Welfare Board of India (AWBI), and State Animal Welfare Boards (SAWB), with implementation often carried out by district officials, local municipalities, and police departments. Together, this framework sets the legal and administrative foundation for improving welfare standards in farming, research, entertainment, and companion animal sectors across the country.

Publicly-funded organizations

Animal Welfare Board of India

The AWBI was established in 1962 under Section 4 of the PCA Act, 1960 (https://www.indiacode.nic.in/bitstream/123456789/11237/1/the_prevention_of_cruelty_to_animals_act%2C_1960.pdf) and functions under the MoFAHD. The Board, founded under the leadership of the late Smt. Rukmini Devi Arundale serves as the apex advisory body to the Government of India on animal welfare matters. Its major purposes include promoting humane education, funding Animal Welfare Organizations, and guaranteeing implementation of the PCA Act provisions across the country. In the poultry sector, the AWBI issues recommendations related to housing density, handling, transport, and slaughter to minimize stress and suffering in birds. They are regulations (statutory directions) backed by the PCA Act, 1960, but enforcement is typically weak. The AWBI derives its power from Section 38 of the PCA Act, 1960. Orders issued by the AWBI concerning transport, housing, or slaughter are not mere suggestions but statutory directions. Noncompliance is punishable under Section 11 of the PCA Act, where a farmer can be fined and/or imprisoned for up to 3 months. The AWBI Handbook on Animal Welfare is the comprehensive guide detailing space requirements, flooring, ventilation, water trough dimensions, etc., for dairy, poultry, and other livestock (https://awbi.gov.in/uploads/regulations/174073569382IO_HandBook_compressed.pdf). The recommendations are not regularly issued, but if the AWBI does issue a directive, then it becomes the legal minimum standard. The consequence for an individual farmer is currently little, but for commercial entities (slaughterhouses, large dairies), noncompliance can shut down their business via license cancellation.

Committee for the Purpose of Control and Supervision of Experiments on Animals

The Committee for the Purpose of Control and Supervision of Experiments on Animals (CPCSEA) is also operational under the PCA Act, 1960. This committee regulates the use of animals in research, testing, and education. All biomedical and veterinary institutions conducting experiments on poultry or other animals must register with CPCSEA. The Committee ensures that experimental birds are housed and handled under ethical and hygienic conditions, minimizing pain or distress. Institutional Animal Ethics Committees oversee local compliance with these norms.

National Institute of Animal Welfare

The National Institute of Animal Welfare located at Ballabhgarh (Haryana), acts as a training and capacity-building institute under MoFAHD. It promotes awareness, research, and professional education in animal welfare, including welfare standards in poultry housing, transportation, and slaughter.

Organizations under the Ministry of Environment, Forest, and Climate Change

Agencies such as the Wildlife Crime Control Bureau and Central Zoo Authority operate mainly for wild and captive animals but contribute indirectly to national welfare standards and policies that influence ethical frameworks for all species, including poultry.

State and district-level bodies

At the state level, SAWB support the implementation of animal welfare activities and awareness programs. At the district level, Societies for the Prevention of Cruelty to Animals ensure enforcement of PCA provisions and investigate cruelty cases, including those arising from inappropriate poultry handling, overstocking, or inhumane transport.

Private and nongovernmental organizations

People for Animals (PFA), India's largest animal welfare NGO, founded by Smt. Maneka Gandhi, plays a dynamic role in encouraging awareness, policy support, and rescue operations. The organization regularly raises welfare issues related to industrial poultry farming, such as space constraint, beak trimming, transport conditions, and slaughter practices (People for Animals, 2022).

Broiler housing and welfare

India's broiler sector has shifted from small backyard flocks to today's intensive, large-scale production, where chickens are typically raised in deep-litter sheds using fast-growing breeds and short production cycles. Housing is usually at high stocking densities with minimal environmental enrichment and poor litter conditions, which heighten welfare risks such as heat stress and lameness (Campbell et al., 2025). Although India has overarching animal welfare provisions under the PCA framework, along with AWBI guidance, CPCSEA guidelines for research facilities, and food-safety-oriented FSSAI norms, there is no broiler-specific, enforceable national welfare standard comparable to that of the European Union (EU) regulations (EU animal welfare legislation, n.d.). As a result, key welfare issues such as genetics, stocking density (SD) limits, enrichment, humane slaughter, and auditing remain largely unregulated. The Better Chicken Commitment (2026), which addresses these gaps through slower-growing breeds, lower densities, enrichment, and controlled-atmosphere slaughter, is widely adopted in Europe and North America but has no regulatory standing in India and limited voluntary uptake, leaving broiler welfare primarily dependent on industry practices rather than mandated standards. The following sections discuss three management factors identified as key issues to address for improving broiler chicken welfare in India, namely transport, lighting, and SD.

Welfare issues related to transport stress in broilers

The majority of the commercial broilers are reared in industrial farms with limited floor space per bird and are transported in poor conditions as live birds to the market (Sasidhar, 2017). The broilers' welfare is thus compromized to a greater extent during rearing and transport than in countries with stricter regulations, such as within the EU. During the journey, the birds endure noise, speed, high/low temperature, mixing with unfamiliar groups, restricted space, withdrawal of feed and water, and improper handling, which can all lead to stress as well as body weight losses (Mohan et al., 2025). The major welfare-related issues of broilers include leg problems, lameness, metabolic disorders, breast burn, and transport stress (Rokade et al., 2025). Transportation of broilers compromises their welfare to a greater level in Indian conditions when compared with nations that are more tightly regulated. In India, broilers are transported by different vehicles such as bicycles, motorcycles, and trucks while contained in plastic or metal crates. On a smaller scale, birds are also transported without crates/containers in two-wheeler vehicles. The birds are typically transported during the daytime, without shade, in the hotter part of the day. This is further complicated by withdrawal of feed/water, vibrations, and noises from the vehicle, compounding the stress experienced by the bird (Mitchell & Kettlewell, 2008). This transport stress can cause fearfulness and lead to changes in the bird's normal physiology, resulting in increased mortalities and weight loss during/after transport (Sarkar et al., 2013; Siddharth et al., 2024). Prakash & Prabakaran (2020) compared shrinkage percentage loss across different transport distances/durations and seasons. While the seasonal weather did not have a significant impact on shrinkage, there was a positive linear relationship between transport distance/duration and percentage loss (Prakash & Prabakaran, 2020). Physiologically, transport causes depletion of muscle glycogen reserves and affects the muscle pH to a greater extent. Thus, transport stress causes physiological changes that impact bird welfare as well as the carcass quality of the birds during slaughter (Siddharth et al., 2024; Utomo et al., 2019).

Consequently, broiler welfare indicators during transport in India have to be identified and analyzed to check whether the birds are in a good welfare condition or not. Mohan et al. (2025) found that a transport duration of more than 4 hours under Indian conditions caused leg weakness and impaired the walking ability of the birds. Preslaughter rest time after transport in a cooler environment could alleviate some of the stress experienced by the birds (Rokade et al., 2025). Mohan et al. (2025) found that there was no significant difference between the lairage period of 2 or 4 hours for stress alleviation, and moreover, a longer lairage period with feed withdrawal resulted in more body weight loss (Mohan et al., 2025). So, it can be concluded that a transport time of more than 4 hours is not recommended, and a 1-hour rest period should be mandatory to avoid stress and production loss. As another method of reducing transport stress, Nagesh (2024) reported that an antistressor formulation of liposome supplementation at 25 ml/L had a significant positive impact by reducing shrinkage loss, enhancing overall bird welfare, and yielding cost savings of up to three Indian rupees per bird.

Lighting management

A central factor influencing the welfare of broiler chickens in indoor housing of tropical regions is the management, where open sheds with cross ventilation are common. Within these open-sided systems, paramount management concerns include the floor space per bird and lighting. Light is a key environmental component with color or wavelength, intensity, and duration as predominant elements (Olanrewaju et al., 2006). The color of the light source has a beneficial effect on broilers by influencing numerous behavioral and physiological mechanisms (Pal et al., 2019; Soliman & El-Sabrout, 2020). Chickens have a distinctive visual system and can visualize in the range of 315–750 nm wavelength. Most of their behavior is mediated by vision, where the color of the light plays a critical role in both poultry performance and welfare (Prescott et al., 2003). Research conducted specifically on commercial farms in India showed that the lighting schedule impacted bird mortality, with birds raised under near continuous light showing higher mortality compared to birds provided 6, 8, or 10 hours of darkness each day (Khutal et al., 2022). However, there was no impact of the lighting schedule detected on other welfare measures such as body weight, gait, and heterophil/lymphocyte ratio. Another study in India compared the behavior of broilers in the evening while exposed to blue, green, red, or standard incandescent light (Khaliq et al., 2018). More resting and comfort behaviors were observed under blue light, more locomotion under the standard incandescent light, and more aggression under the red and standard lights (Khaliq et al., 2018). In contrast, Ravi et al. (2026) found that light color (white, blue, green) in broiler open-sided houses had no significant effect on performance or welfare. Kumar et al. (2019) compared the behavior of broilers in open-sided sheds while raised under different light wavelengths (white, green, blue) using LED and incandescent bulbs (Kumar et al., 2019). Feeding significantly increased under the LED lighting compared with the incandescent on some weeks, as did dozing behavior. While most behaviors did not vary, the authors concluded that LED bulbs could increase feeding while not compromising bird welfare and could be more energy efficient for poultry producers (Kumar et al., 2019). Thus lighting research is key for understanding what types of changes could be implemented in Indian poultry systems to improve bird welfare, while ensuring farmers are able to meet their production needs.

Stocking density management

The SD, defined as the number of housed birds (or live weight) per m^2 of floor space at the end of the growth phase (or depopulation), is another factor of importance in management. Increasing the SD in broiler houses is usually carried out to maximize outcome from a production standpoint, but this objective may not always be achieved and might compromise the birds' welfare. The limited space availability under high SD conditions limits chickens' natural behaviors and movements, and might worsen contact dermatitis, such as hock burn and footpad dermatitis (HB and FPD) (Guinebretière et al., 2024). In India, high SD has traditionally been viewed as a direct path to higher economic sustainability, but now this paradigm is shifting. The "cost" of bird welfare is increasingly becoming a direct economic cost, making the initial assumption less stable. High SD is often pursued on economic grounds, but the resulting compromise in bird welfare incurs significant economic costs. These include elevated morbidity and mortality, increased medication use, higher culling rates, processing losses, and reduced feed conversion efficiency—all of which undermine the profitability the model seeks to achieve. Understanding optimal stocking densities for improved bird welfare that align with production imperatives in India is necessary for coping with the challenges posed in this global region. High SD can therefore be viewed as a tactic for greater economic efficiency, but it frequently comes at the expense of bird welfare, increased risk of disease, and possible environmental load per bird, all of which could jeopardize long-term sustainability. Research on SD by Ravi et al. (2026) showed that SD did not influence most production parameters in open-sided houses, though lower density improved male body weight. Welfare benefits were limited to better gait scores and litter quality at low and medium densities (Ravi et al., 2026). In India, most commercial broiler operations use higher densities than 12 kg/m^2, typically in the range of 16–20 kg/m^2, to meet rising demand for affordable poultry meat with constrained land and fixed facility costs (EFSA AHAW Panel et al., 2023). Therefore under Indian open-house conditions, research is needed to determine the SD that will minimize bird welfare compromise, while still meeting consumption needs. The sustainability impacts of lower bird density remain a challenge on a global scale (see Chapter 3 for more discussion on indoor broiler housing systems).

Layer housing and welfare

Layer housing in India is dominated by conventional battery cage systems, where hens are confined in wire enclosures that severely restrict movement and natural behaviors such as perching, nesting, and dust-bathing. While this system has long been the industry norm due to perceived efficiency, it has drawn increasing scrutiny from veterinarians, policymakers, and civil society for its welfare costs. In recent years, cage-free housing models have begun to emerge, driven by consumer awareness, global corporate commitments, and improved farm management practices, but adoption remains uneven across states. The AWBI, PCA (Egg Laying Hens) Rules, 2023 (https://awbi.gov.in/Document/rules), and several High Court rulings are engaged in deciding the future prospect of layer welfare, which is an increasingly important part of India's broader conversation on agricultural reform and humane food systems.

Cage-free versus caged housing: understanding the welfare gap

Cage-free housing aligns far more closely with the Five Freedoms because it allows hens to move, spread their wings, nest, forage, perch, and dust bathe, all of which are essential natural behaviors. In contrast, cages confine birds to cramped spaces that restrict movement, prevent exercise, and contribute to bone weakness, foot injuries, frustration, and aggression (Hartcher & Jones, 2017). Cage-free systems provide nest boxes that support hens' strong instinct to lay eggs in a private space, deep litter that enables natural foraging and reduces harmful redirected pecking, dust-bathing areas that maintain plumage health and reduce stress, and perches that improve comfort, bone strength, and social stability (see Chapter 2 for more discussion on laying hens in indoor systems). Because conventional cages prevent exhibition of these behaviors, they create some of the greatest welfare challenges for hens, while cage-free environments managed properly offer significantly better physical and psychological well-being (Hartcher & Jones, 2017).

An Initiative for cage-free egg production in India

With progressing knowledge and awareness of animal ethics and welfare in consumers, the demand for cage-free eggs in India has increased (Rokade et al., 2024). This demand-driven production has obliged producers to settle on cage-free production methods, demanding time and expertise to implement it effectively (https://www.dw.com/en/cage-free-indias-movement-for-eco-friendly-egg-farming/video-72637078). For this change, in 2019, a Singapore-based consulting company, namely Global Food Partners (GFP), formed an exclusive partnership with the Cage-free and Free-range Poultry Producers Association (CFFRPPA) of India (https://globalfoodpartners.com/wp-content/uploads/2023/02/GFP-CFFRPPA-MOU-press-release.pdf). The partnership strives towards the capacity building of

Indian egg producers through online resources, in-person sessions, and training workshops covering the aspects associated with birds' health, welfare, and best practices in cage-free production systems (https://www.cffrppa.com/). The partnership consists of hypothetical as well as applied (practical) training sessions, including farm visits, bird assessments, handling, and collective learning at GFP's cage-free innovation and welfare hub in Indonesia, with active involvement from Indian farmers and trainers. With this initiative, GFP and CFFRPPA are also developing buyer engagement to help food and hospitality businesses procure cage-free eggs and, thus, strengthen the domestic market for better welfare production. In collaboration with international universities, GFP offers an online academy with e-learning courses in several Asian languages to teach farmers the basics of cage-free egg production and better flock management (https://www.thepoultrysite.com/articles/gfp-partners-with-prestigious-universities-to-create-online-courses-to-prepare-asian-egg-industry-for-cage-free-transition). This association ascends at a significant moment, as many prominent food and hospitality enterprises, including Nestlé, Unilever, Marriott, and many more have sworn to use totally cage-free eggs by the year 2025, although only now in some global regions. Thus the Indian poultry egg producers are working to upgrade their systems and implement progressive welfare practices.

Another major breakthrough was created in the year 2024 by the passing of an MoU between the Central Avian Research Institute and People for Animals Public Policy Foundation (PFA-PPF) for founding India's first model cage-free egg production and training center in Uttar Pradesh (Fig. 12.1) to set a nation-wide illustration and offer a platform for knowledge exchange and market access for cage-free producers. Efforts are in progress to develop such training centers in northern India, presenting a broader approach to develop capacity across regions and match the increasing demand for cage-free eggs. PFA-PPF, in partnership with

FIGURE 12.1 India's first collaborative cage-free poultry training facility at Trichy, Tamil Nadu, India.

TANSIM, established another cage-free training facility in Tamil Nadu to strengthen on-ground capacity building and demonstrate scalable cage-free production systems (PTI-News Voir, 2025; Certified Humane Asia, 2025, https://www.ptinews.com/press-release/inauguration-of-india-s-first-collaborative-cage-free-training-facility-in-trichy-tamil-nadu/2759991 inauguration of India's first collaborative cage-free training facility). Alongside this, the organization actively engaged over 25 global food companies to develop India-specific cage-free transition roadmaps, ensuring their global commitments translate meaningfully within the Indian supply chain. PFA also represented India in Mercy for Animals' Egg-quity Index, bringing visibility to the country's progress and gaps on corporate cage-free transitions (https://mercyforanimals.org/eggquity-index/). To further support companies, PFA collaborated with EY to publish a corporate cage-free transition guide (Fig. 12.2), offering practical implementation pathways tailored to the Indian market. In September 2024, the team convened a high-impact cage-free buyer-seller meet in Jaipur, bringing together leading hospitality brands and producers to unlock procurement linkages and accelerate demand for cage-free eggs. On the supply side, PFA's team conducted over 150 farm visits and 10 structured training sessions, collectively reaching more than 31,000 farmers and stakeholders across six states (https://www.jaipurkiran.com/2024/09/rajasthans-hospitality-industry-takes.html). These engagements focused on technical training, business viability, and animal welfare best practices. At the policy and district level, PFA is working closely with the district administration of Rudraprayag to develop India's first cage-free district model, creating a replicable blueprint for regional transitions. Additionally, the organization has signed an MoU with the Compass Foundation to establish another model cage-free farm in Andhra Pradesh, further expanding demonstration infrastructure and farmer adoption pathways Fig. 12.1–12.2.

FIGURE 12.2 Inauguration day picture of first collaborative cage-free poultry training facility at Trichy, Tamil Nadu, India.

Conclusion

Poultry housing and management have progressed significantly worldwide, reflecting rising scientific understanding, ethical attention, and public requests for better animal welfare. Laying hens are commonly kept in free-range, cage, or cage-free systems, while broilers are primarily reared in deep-litter, slatted, or free-range systems, each having unique trade-offs in welfare, production, and biosecurity. Broiler production methods present unique challenges in balancing bird welfare, production efficiency, and biosecurity. Internationally, enriched cages, aviaries, and free-range systems are gradually selected for encouraging natural behaviors, health, and positive emotional states in laying hens; however, cautious management is essential to alleviate disease, injury, and environmental stress. In the Indian poultry industry, sustainability is not just about being "green", it is about creating a system that is economically viable for farmers, responsible to the environment, and humane for the animals. Prioritizing the welfare of the birds is a practical strategy that serves all these goals. To achieve ecological sustainability, high-performance systems must conserve resources, optimize feed and water use, and reduce ammonia and greenhouse gas emissions through enhanced housing design and management. Consolidation and regular updating of regulatory frameworks, consumer awareness, and technical capacity will be decisive to complement productivity with birds' well-being, safeguarding sustainable, ethical, and resilient poultry production systems across diverse regions.

AI disclosure

During the preparation of this work, the author(s) used DeepSeek in order to search the latest developments. After using this tool/service, the author(s) reviewed and edited the content as needed and take(s) full responsibility for the content of the publication.

References

Baghel, S. P., S. G. Kurian & Gangwar, S. N. P. (2025). Government of India Ministry of Fisheries, Animal Husbandry & dairying Department of Animal Husbandry & Dairying (DAHD). Release of basic animal husbandry statistics. https://dahd.gov.in/sites/default/files/2025-11/BAHS2025Brochure.pdf.

Campbell, Y. L., Walker, L. L., Bartz, B. M., Eckberg, J. O., & Pullin, A. N. (2025). Outdoor access versus conventional broiler chicken production: Updated review of animal welfare, food safety, and meat quality. *Poultry Science, 104*(4), 104906. https://doi.org/10.1016/j.psj.2025.104906.

EFSA AHAW Panel, et al. (2023). Welfare of broilers on farm. *EFSA Journal, 21*, e07788. https://doi.org/10.2903/j.efsa.2023.7788.

EU animal welfare legislation. (n.d.). https://food.ec.europa.eu/animals/animal-welfare/eu-animal-welfare-legislation_en.

Food Safety and Standards Authority of India (FSSAI). 2020. Eat Right India: A movement for safe, healthy and sustainable food. https://eatrightindia.gov.in/.

Guinebretière, M., Warin, L., Moysan, J. P., Méda, B., Mocz, F., Le Bihan-Duval, E., Thomas, R., Keita, A., & Mignon-Grasteau, S. (2024). Effects of strain and stocking density on leg health, activity, and use of enrichments in conventional broiler chicken production. *Poultry Science, 103*(9), 103993. https://doi.org/10.1016/j.psj.2024.103993.

Hartcher, K. M., & Jones, B. (2017). The welfare of layer hens in cage and cage-free housing systems. *World's Poultry Science Journal, 73*(4), 767–782. https://doi.org/10.1017/s0043933917000812.

Kathiravan, G., & Chitrambigai, K. (2024). Consumer preferences for native chicken meat in India: implications for sustainable production and household dynamics. *Current Research in Nutrition and Food Science Journal, 12*(1), 166–180. https://doi.org/10.12944/crnfsj.12.1.14.

Khaliq, T., Khan, A. A., Dar, P. A., Nazir, T., Bilal, M., Tarique, P., Afzal, I., Khaliq, T., Khan, A. A., Dar, P. A., Nazir, T., Afzal, I., Bilal, M., & Tarique, P. (2018). Behavioral study of broilers reared under different colours of light in the evening hours. *Journal of Entomology and Zoology Studies, 6*(4), 1624–1627.

Khutal, G. P., Kadam, A. S., Dhumal, M. V., Lonkar, V. D., Patodkar, V. R., Mhase, P. P., & Dhaygude, V. S. (2022). Effect of lighting schedule on the performance of broilers. *Indian Journal of Veterinary Sciences and Biotechnology, 18*(3), 84–87. https://doi.org/10.21887/ijvsbt.18.3.19.

Kolluri, G., Tyagi, J. S., & Sasidhar, P. V. K. (2021). Research Note: Indian poultry industry vis-à-vis coronavirus disease 2019: A situation analysis report. *Poultry Science, 100*(3), 100828. https://doi.org/10.1016/j.psj.2020.11.011.

Kumar, S., Gupta, R. K., Singh, Y., Sharma, A., & Kashyap, N. (2019). Behaviour of broiler birds reared under coloured light emitting diodes vis-a-vis incandescent light supplemental lighting programme. *International Journal of Current Microbiology and Applied Sciences, 8*(07), 423–430. https://doi.org/10.20546/ijcmas.2019.807.052.

Mitchell, M. A., & Kettlewell, P. J. (2008). Engineering and design of vehicles for long distance road transport of livestock (ruminants, pigs and poultry). *Veterinaria Italiana, 44*(1), 201–213.

Mohan, S., Madheshwaran, M., Kumar, A., Mariappan, G., Biswas, A. K., Rokade, J. J., Nagesh, S., Tiwari, A. K., & Singh, Y. (2025). Evaluating the impact of pre-slaughter transport stress on broiler welfare and meat quality in India. *Scientific Reports, 15*(1), 39725. https://doi.org/10.1038/s41598-025-23436-7.

Nagesh, S. S. (2024). *Strategies to ameliorate post harvest losses in chicken meat and eggs* (PhD thesis). ICAR–Indian Veterinary Research Institute, Uttar Pradesh, India.

Olanrewaju, H. A., Thaxton, J. P., Dozier, W. A., Purswell, J., Roush, W. B., & Branton, S. L. (2006). A review of lighting programs for broiler production. *International Journal of Poultry Science, 5*(4), 301–308. https://doi.org/10.3923/ijps.2006.301.308.

Pal, P., Dey, D., Sharma, B., Choudhary, S., Sahu, J., Kumar, S., & Ghosh, S. (2019). Effect of light management in broiler production: A review. *Journal of Entomology and Zoology Studies, 7*(3), 437–441.

People for Animals. 2022. Harsh realities of poultry farms in India. https://peopleforanimalsindia.org/articles/detals/197.

Prakash, D., & Prabakaran, R. (2020). Factors affecting live shrink loss in commercial broilers. *Journal of Entomology and Zoology Studies, 8*(6), 1770–1773. https://doi.org/10.22271/j.ento.2020.v8.i6x.8074.

Prescott, N. B., Wathes, C. M., & Jarvis, J. R. (2003). Light, vision and the welfare of poultry. *Animal Welfare, 12*(2), 269–288. https://doi.org/10.1017/s0962728600025689.

Ravi, S., Rokade, J. J., Bhanja, S. K., Kolluri, G., Marappan, G., Tiwari, A. K., Shameeh, M., Madheshwaran, M., & Srivastava, V. (2026). Effect of light and stocking density regimen on welfare and zoo-technical performance of broilers under tropical conditions. *Animal Science Journal, 97*(1), e70165. https://doi.org/10.1111/asj.70165.

Rokade, J., Champati, A., Sonale, N., Wadajkar, P., Madheshwaran, M., Bhaisare, D., & Tiwari, A. K. (2024). The cage-free egg sector: Perspectives of Indian poultry producers. *Frontiers in Veterinary Science, 11*, 1442580. https://doi.org/10.3389/fvets.2024.1442580.

Rokade, J., Krishna Gopal, S., Khulape, S., Sonale, N., Patel, P., Champati, A., Kolluri, G., & Madheswaran, M. (2025). Effects of transport stress on gene expression and biomarkers in broilers with and without an anti-stress supplement. *Animal Production Science, 65*(9), AN24372. https://doi.org/10.1071/AN24372.

Sarkar, P., Ghosh, S., Batabyal, S., & Chatterjee, S. (2013). Biochemical stress responses of broiler chickens during transport. *Indian Journal of Animal Research, 47*(1), 29–34. http://www.indianjournals.com/ijor.aspx?target=ijor:ijar1&volume=47&issue=1&article=005&type=pdf.

Sasidhar, V. K. (2017). Poultry welfare education in India. In S. K. Bhanja, J. J. Rokade, G. Kolluri, & M. Gopi (Eds.). *Stress and welfare: Concepts and strategies for addressing current challenges in poultry production, 2017* (pp. 232–240). ISBN no. 978-93- 5279-729-5.

Siddharth, M., Rokade, J. J., Bhanja, S. K., Tyagi, J. S., Monika, M., Pearlin, B. V., Kumar, A., & Gopi, M. (2024). Transportation stress: Impact on behaviour and welfare in meat-type chickens under indian scenario. *Heliyon, 10*(5), e27129. https://doi.org/10.1016/j.heliyon.2024.e27129.

Sinclair, M., Lee, N. Y. P., Hötzel, M. J., de Luna, M. C. T., Sharma, A., Idris, M., Islam, M. A., Iyasere, O. S., Navarro, G., Ahmed, A. A., Curry, M., Burns, G. L., & Marchant, J. N. (2022). Consumer attitudes towards egg production systems and hen welfare across the world. *Frontiers in Animal Science, 3*, e27129. https://doi.org/10.3389/fanim.2022.995430.

Soliman, F. N. K., & El-Sabrout, K. (2020). Light wavelengths/colors: Future prospects for broiler behavior and production. *Journal of Veterinary Behavior, 36*, 34–39. https://doi.org/10.1016/j.jveb.2019.10.014.

The Better Chicken Commitment. (2026). https://betterchickencommitment.com/us/.

Utomo, M. C. C., Sarjana, T. A., Suprijatna, E., Kismiati, S., Nasoetion, M. H., Atmomarsono, U., & Windriasari, E. (2019). The influence of transportation distance during the dry season on broiler chickens' carcas traits. *IOP Conference Series: Earth and Environmental Science, 247*(1), 012036. https://doi.org/10.1088/1755-1315/247/1/012036.

Chapter 13

Sustaining natural resources and building resiliency

Anne C. Fanatico
Sustainable Development Department, Appalachian State University, Boone, NC, United States

Need for conservation and resiliency

We depend on natural resources for our food and for life support systems for the earth, but unfortunately, they are being used up. In agriculture, soil is being eroded, aquifers are depleted, and biodiversity is lost. In addition, there are many environmental and societal challenges, including climate change, emerging diseases, and hunger. Ironically, while agriculture can be harsh on the environment and cause some of the problems, agriculture and poultry production are also important parts of the solutions, depending on the methods used. Intensive agriculture can be very efficient. Poultry have been raised intensively since the 1970s for high-quality protein production (Cheeke, 2003). Chickens, turkeys, and ducks efficiently convert feed to meat or eggs, and are more efficient than other farm animals. In poultry systems, efficiency, maximum production, high yield, profits for companies, and food for a growing world population have been emphasized. The importance of genetics, diet, and high-tech housing is often highlighted for high-yield poultry production. However, with the many global challenges, we need more than efficiency; we also need resilience (Fig. 13.1).

The poultry supply chain for intensive production is vulnerable to challenges, which were particularly evident during the coronavirus pandemic in 2020–21 as the world and workforce were significantly disrupted. Outside of unprecedented global events, large-scale agriculture uses intensive tillage, synthetic chemicals, and irrigation, which have contributed to the degradation of the environment, including soil erosion, water depletion, and pollution of the environment, compromising future production (Gliessman et al., 2022). Thus, agriculture has become dependent on external inputs, such as fossil fuels, further increasing system vulnerability as well as environmental impact. This accompanies a suite of specific problems related to intensive poultry production, including:

- Antibiotic resistance: Routine use in feeding is now being phased out.
- Emerging diseases, such as highly pathogenic avian influenza.

Integrating Welfare and Sustainability in Poultry Systems. DOI: https://doi.org/10.1016/B978-0-443-40594-5.00011-0

Large-scale, Intensive ↔ Small-scale, Extensive

Feature		
Scale	World-wide	Bioregional production (local and regional)
Housing	Energy-efficient housing	Open housing
Environment	Controlled indoor environment	Access to outdoors
Genetics	Fast-growing, high-yielding genetics	Differentiated genetics
Feed	High performance feed	Alternative feeds and feeding
Health care	Targeted health care	Holistic health
Slaughter	Centralized slaughter at large facilities	De-centralized slaughter at small facilities
Focus	Economies of scale	Resilience, robust
Change	Continuous improvement	Transformative change

Efficient ↔ Robust

FIGURE 13.1 **Relationship between robustness and efficiency in poultry production.** Robustness and efficiency are on a continuum and impacted by scale.

- Loss of diversity: Natural lands have been cleared for crops, agrichemicals applied, and habitat lost.
- Loss of biodiversity: Many of our crops and food animals have uniform genetics and are raised in monocultures or specialized units, and are vulnerable to pathogen outbreaks.
- Excess nutrients: Poultry manure or litter is often applied to pastures or crop fields. It may be applied for the nitrogen content, but the phosphorus content may be too high and cause eutrophication of waterways.
- Depletion of fresh water and aquifers: Water is used intensively in feed crop and livestock production.
- Use of fossil fuels in production, processing, distribution, and consumption: Greenhouse gas emissions cause global warming.

There are also many social issues in intensive agriculture. In addition to producing food, we also produce income with agriculture. In fact, the creation of wealth for companies may often be a focus. Many poultry welfare issues are related to high-yield production, including fitness, livability, and the ability to resist disease. Without a focus on equity and welfare principles, people and animals can be marginalized and even exploited. Therefore, it is important to foster methods of production that ensure agriculture and food animal production do not consume their resource bases. There is a critical need to maintain natural resources and include goals of ecology and equity, along with production and income. Agriculture is a unique sector because it can provide ecosystem services, including building soil, sequestering carbon, and increasing biodiversity, depending on the methods used. Poultry and livestock can produce high-protein food, generate income, and provide ecosystem services, without depleting the base of natural and human/animal capital.

Solutions and benefits

Natural resources can be maintained in agriculture by using ecological methods. Basic ecological processes include the flow of energy from the sun and capture by plants as glucose, nutrient cycling of minerals, such as phosphorus, among soil, crops, and animals; and biodiversity and mutualisms among plants, animals, and trees. Ecological processes rely on inputs that are internal to the ecosystem instead of fossil fuels and pharmaceuticals, which are external to the system. Developing alternative energy sources and circularity of materials and value are key components of bio-economies. With the exception of solar radiation, most of our natural resources are nonrenewable, and we need to focus on conservation rather than extraction. In fact, resources need to be more than conserved, they need to be regenerated. Key natural resources sustain life on the planet and can also drive human endeavors and business. They include:

- Healthy atmosphere, with sequestration of excess carbon in plants and the soil
- Fertile soil
- Abundant fresh water
- Rich biodiversity and ecological diversity
- Healthy ecosystems like forests, grasslands, and healthy populations and communities of soil microbes, trees, and animals.

Human resources are a special type of natural resource because we are ultimately part of ecosystems. Human resources are maintained with building community, social equity, human rights, and international development. To build social capital and ensure social distributive justice for humans, it is important to build relationships with stakeholders and community. Aligning with nurtured human resources, it is important to also ensure health and well-being for poultry. Healthy environments are tied to healthy birds. In fact, animal health is often a first indication of poor environmental health. Animal welfare guidelines are typically in place to facilitate the physical and mental well-being of poultry where understanding relationships among sustainability aspects, environment, poultry welfare, and human rights can help us set goals. For example, the relationship between agricultural productivity versus robustness is a continuum. To be productive and maintain resources at the same time, a balance is needed, and trade-offs are considered. A focus solely on profits can alter these considerations and cause an imbalance. Environmental accounting and even alternative economic systems can help balance raising poultry for food and income, while ensuring ecosystem and social benefits. A greater focus on robustness can mean less productivity in the short-term, but health and sustained resources in the long-term. The relationship of welfare and sustainability is an overlap rather than a continuum, because welfare is

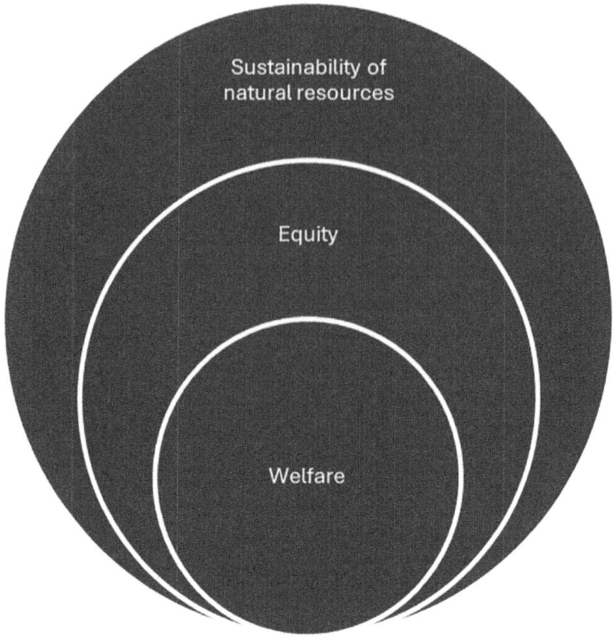

FIGURE 13.2 **Relationship between welfare and sustainability in poultry production.** Welfare is intrinsic to sustainability.

intrinsic to sustainability. There is no sustainability without justice for people and welfare for birds (Fig. 13.2).

Ecological poultry production that follows nature and justice has many benefits. It can maintain natural resources and provide ecological services, as well as the following:

- Provide high-quality, concentrated animal protein
- Contribute to livelihoods for farmers
- Bring income to producers and companies
- Create equity and social justice; build social resilience and community development
- Reduce hunger and support international development.

A well-developed agriculture provides food security where fertile land is a key asset for communities and nations. Poultry can have a strong impact because of the large scale of production, worldwide popularity, and healthy protein from meat and eggs. Human populations are growing in many areas of the world, and meat production is also increasing. According to FAO (2026), poultry is the most consumed meat on a global basis. Sustainable poultry production can help the human population to be more resilient for the future.

Basic ecological principles in poultry production

Ecological agriculture is site-specific and adapted to the needs of local environments. There is no specific recipe for ecological poultry production, but rather basic principles, including overall ecosystem functioning and structure (biotic and abiotic).

Soil fertility and organic matter

Feed crops, such as corn and soybeans, are often grown geographically distant from the poultry houses and may not be a visible part of production. However, the impact of feed crop production is an integral part of the sustainability of poultry production. Agricultural basics include crop production for poultry feeds. Fertile soil, organic matter, and microbes are the basis of regenerative agriculture and, indirectly, animal production. Plants capture solar energy from the sun and create biomass and food from carbon dioxide in the atmosphere and water via photosynthesis. The main photosynthate, glucose, is used for fuel by the plants and also for plant structure. Animals (and poultry) eat plants and release energy and nutrients stored in plant biomass by consumption, digestion, excretion, and decay. Poultry and other animals enable the flow of energy from the chemical bonds of glucose from crops to animals to humans (Gliessman et al., 2022). Energy leaves the farm ecosystem as animal products are marketed and consumed by people (Fig. 13.3).

Some of the plant biomass remains in the farm ecosystem as crop residues, building organic matter and supporting soil microbes. Soil biota break down crop residues, dead roots, animal waste, etc., into smaller compounds. This organic matter has many benefits, including (1) increasing nutrient retention in the soil and improving fertility, (2) increasing water-holding capacity so that less water is needed for feed crop production, (3) helping build soil structure to reduce erosion, and (4) providing food for soil microbes. Soil microbes, such as mycorrhizal fungi, provide plants with soil-derived nutrients, such as phosphorus. Soil microbes can also transform organic matter into complex polymers (humus). Being a relatively stable form of carbon (Gliessman et al., 2022), this synthesis of humus sequesters carbon in the soil for long periods, removing it from the atmosphere. While energy flows through ecosystems, nutrients/minerals are cycled. Nutrients, like nitrogen and phosphorus, are cycled through inorganic forms (reservoirs in the atmosphere or soil) and organic forms (plants, animals, debris). With their roots, plants bring up nutrients from deep layers in the soil. Minerals in the soil are a key natural resource, and nutrient cycling maintains fertility. For example, phosphorus is mined for agricultural crops from coastal deposits and transported as fertilizer to crop-growing areas. Phosphorus is limited in supply and is important to conserve. Unfortunately, run-off often occurs during heavy rain events, and phosphorus then contaminates surface water, causing eutrophication. Phosphorus can end up in the ocean and cause

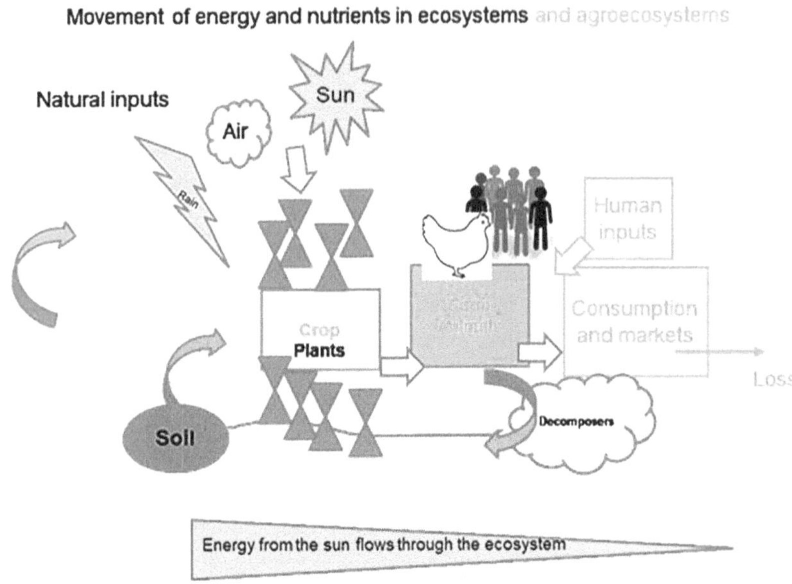

FIGURE 13.3 **Energy and nutrient movement in agroecosystems.** Energy flows and nutrients cycle.
Modified from Gliessman, S.R., Méndez, V.E., Izzo, V.M., & Engles, E.W. (2022). Agroecology: Leading the transformation to a just and sustainable food system. CRC Press. http://doi.org/10.1201/9781003304043.

dead zones in estuaries. However, instead of mining, phosphorus can be cycled via compost, crop residues, etc. See Science and Technology for Phosphorus Sustainability (Steps, 2025). Another common agronomic practice for corn, soybean, and sorghum crops raised for livestock and poultry feed in the United States is intensive tillage. This practice can degrade the soil through erosion and compaction, opening up the soil and exposing the soil microbes to air. The oxygen then increases the rate at which the microbes consume organic matter and can deplete the soil (Gliessman et al., 2022). In contrast, reduced tillage helps maintain organic matter in the soil. Reduced tillage methods use cover crops, crop residues, etc. for soil cover, and limit tillage to only the seed bed instead of the whole field. Animal manure and compost are added to build organic matter and increase nitrogen in the soil. Leguminous cover crops, such as alfalfa and clover, also add nitrogen. In ecological poultry production, feed crops are produced with reduced tillage, cover crops, and soil amendments as part of regenerative agriculture.

Poultry production systems

Living conditions for poultry impact natural resources and poultry welfare. In commercial production, poultry are generally raised in environmentally controlled sheds with higher stocking densities. In contrast, ecological production usually includes more space, windows, enrichments, and outdoor access. These features are common in organic poultry production. Organic agricultural production is meant to be ecological. Many countries have organic certification programs that adhere to the principles of the International Federation of Organic Agriculture Movements (2026), but unfortunately, not all certification programs adhere well to ecological principles in practice. Studies have been conducted to determine the impact of low stocking density, natural light, outdoor access, physical activity, enrichments, agroforestry, small flock size, and alternative genetics on poultry well-being (Bonnefous et al., 2022). Guinebretière et al. (2025) found that complex environments can promote natural behavior, reduce emotional reactivity, and alter gut microbes, improving broiler welfare. Birds were less fearful and more inquisitive. However, there are welfare challenges in any production system, whether it is intensive or extensive, confined or outdoor access, industrial or ecological/organic. Outdoor access can lead to better welfare outcomes (i.e., footpad dermatitis, mortality, and lameness) (Campbell et al., 2025), but there are trade-offs. For example, birds with outdoor access may eat more in cold weather to stay warm. However, according to Sarmiento-García et al. (2021), they may experience less heat stress during warm months. In another example of trade-offs, less feed efficiency means raising birds for longer periods. This can also equate to fewer birds, which may mean less food security for people. For more discussion on organic poultry systems, see Chapter 10. For pastured systems, see Chapter 9, and for silvopasture systems, see Chapter 11.

Feeding poultry sustainably

Feeding poultry is closely related to ecological sustainability. It is energetically more efficient to feed grains directly to people. However, we raise livestock because meat and animal-derived food products are high-quality proteins for human food. Also, meat holds an important place in culture and can increase the quality of life for people. As countries become wealthier, most populations eat more animal products. Gross domestic product and urbanization are drivers of meat consumption (Jia et al., 2024). However, if resources (arable land, soil-based minerals like phosphorus, and fresh water) continue to be used up, grains may need to be used for human food instead of livestock feed. Fortunately, poultry are omnivorous and can eat many feeds that are not edible to humans, including crop by-products and culls (Cheeke, 2003). Circularity in feeding reduces waste. For example, roots and tubers are important sources of energy, particularly when sourced as culls, although some cull crops may need to be fermented or processed first. For example, in North Carolina, a potential regional source of energy feed is culled sweet potatoes. Fanatico and Smith (2014) fermented sweet potatoes to destroy antinutritive factors in the skin and found that fermented sweet potato could replace half the corn in an alternative chicken production system with outdoor access. Other possibilities include beet pulp, fruit pomace, and pumpkin pulp. Innovative sources of protein feeds for poultry include insect larvae. Insects can bioconvert crop by-products and food scraps to high-quality protein via their larvae. For example, black soldier fly larvae (*Hermetia illucens*) can be reared on crops and food residues, and other organic material, and then the larvae harvested as feed (Fanatico et al., 2018); (Fig. 13.4). For more discussion on sustainable poultry feeding, see Chapter 14.

Alternative feeding methods, such as cafeteria (different ingredients are separated for hens to consume varying foods as needed) and high-protein concentrate plus grain, can be used to incorporate ecological feeds like cull crops and insect larvae into diets (Fanatico et al., 2013, 2016). Filho et al. (2020) found that, in cafeteria feeding, broilers had a preference for insect larvae. Studies often examine the impact of alternative feeds on bird

FIGURE 13.4 **Black soldier fly production.** Larvae and flies can be raised in a tub with a ramp for self-harvest; a solar dehydrator dries larvae for use as a poultry feed.

performance and health (Siwach et al., 2025), but there is little research on the impact of feeding methods and self-selection on welfare. While cafeteria feeding generally results in lower carcass yield and breast yield, feed efficiency can be improved (Fanatico et al., 2013). However, many research questions remain: Is bird welfare increased by having a choice of feeds? By self-selection of nutrients? Interestingly, the Soil Association (2026) organic standards for livestock include "straights" or uncompounded feed. Low-protein diets are also sometimes used in ecological production to lower environmental impact, but birds should have adequate amino acids for well-being. Poultry production does have a smaller carbon footprint than other meats as poultry do not contribute significant enteric methane to the atmosphere. Thus, poultry, as monogastrics, have a unique potential to be carbon neutral, depending on how they are raised.

Biodiversity

Biodiversity is an important natural resource in ecological poultry production. Biodiversity includes both genetic diversity and the number of species in a community. Genetics are raw materials for poultry breeding, and diversity enables mutualistic interactions. Many types of domestic poultry species are raised around the world (chickens, turkeys, ducks, geese, guineafowl, quail, etc.). Varieties of poultry breeds are valuable resources. Unfortunately, breeds of farm animals become extinct on a regular basis. In the United States, the Livestock Conservancy helps protect endangered breeds of livestock and poultry (Livestock Conservancy, 2026). Nature itself has high genetic diversity due to varied environments. Environments determine which of an organism's qualities are advantageous. The environment acts on genomes so that offspring survive in larger numbers (Gliessman et al., 2022). In poultry production, humans construct the environment and decide which poultry genetics are advantageous. In large-scale broiler production, fast-growing, high-yielding hybrid genetics are used to provide a large breast meat yield. In egg production, hens have a high rate of lay. However, high yield must be balanced with overall fitness and the ability to resist disease. In addition, high-yielding genetics are extremely uniform and therefore vulnerable to pathogenic outbreaks. In ecological poultry production, more genetic variety is used, including medium- and slow-growing hybrids for meat chickens. These genetics are often suited for outdoor production with a high level of activity for foraging. For more discussion on slow-growing meat chickens, see Chapter 3. Native, standard breeds are also used. Standard breeds are low-yielding, but have ties to geographic areas and are well-adapted to a particular environment. It is important to conserve genes that can help mitigate climate challenges, emerging diseases, and improve welfare in the future (Gliessman et al., 2022). For more discussion on specialty poultry breeds, see Chapter 8. Living organisms are a unique type of natural resource due to their biological interactions with each other and the environment. In ecological agriculture, avian species are often raised in association with crops, grazing animals, pasture forages, and trees. For more discussion on silvopasture poultry systems, see Chapter 11. Interactions among individuals and species can help regulate populations, including pests or pathogens, instead of using pesticides or other external inputs (Gliessman et al., 2022). Poultry have been used in many agricultural systems over time, with multiple examples of beneficial interactions, including:

- In nature, egrets and ruminants are a well-known mutualism. Cattle stir up insects with grazing, and the egrets eat external parasites that plague cattle (Fig. 13.5). Domestic chickens can also play this role instead of egrets.
- Chickens can scratch and till small plots for vegetable crop production.
- Poultry can be raised over aquaculture ponds, and their manure provides feed for fish.
- Poultry can provide biological control of crop pests. Chickens eat pests and fallen fruit in orchards, and birds benefit from the shade and shelter of the trees.
- An important mutualism is to raise livestock and crops on the same farm. The manure from the livestock can fertilize the feed crops. It allows the reduction of external inputs, such as synthetic fertilizers.
- Free-range poultry require extensive land. However, raising chickens on the same land as other enterprises, such as cattle production, allows complementary enterprises to be combined and use less land (Fig. 13.6).

FIGURE 13.5 **Mutualism with cattle and egrets.** Cattle stir up insects that egrets eat, and egrets help control cattle parasites.

FIGURE 13.6 **Raising cattle and layers on the same land.** Mixed production uses less land.

Mutualisms are a type of biological interaction that play a key role in ecological agriculture, including microbial interactions. As mentioned earlier, soil microbes provide phosphorus and other minerals to crop plants. In exchange, plants provide glucose to soil bacteria and fungi. Gut microbes provide another example of mutualisms. The bird's gastrointestinal tract (GIT) provides nutrients and a warm environment, while the microbes help with digestion. The GIT and gut microbes also assist in immune function and defense against pathogens (Ruiz, 2019). The GIT not only absorbs nutrients; it also acts as an interface between the outside environment and the bird's internal environment. Deregulation of the microbial-gut-brain axis can lead to aggressive pecking and feather pecking in layers. Modifying gut microbes can reduce injurious behavior (Fu et al., 2024). Jiang et al. (2022) found that the probiotic *Bacillus subtilis* has the potential to reduce aggressive behavior in laying hens. In intensive poultry production, antibiotics and pharmaceuticals are often used to prevent illness. Antimicrobials need to be used responsibly, avoiding those that are medically important to humans. When using biological interactions and mutualisms, inputs that are external to the ecosystem may not be needed.

Ecological diversity and health

As mentioned, the rich diversity of life is a key natural resource. However, diversity is much more than the biodiversity of genetics and the number of species in a community. Nature is also diverse spatially, temporally, and functionally. Perennial components in ecosystems are particularly important for diversity. In annual cropping of grains like corn, the soil is repeatedly disturbed, and the agricultural system stays in an immature, early successional stage. Immature systems lack perennial qualities that can provide fertility, self-regulation, and other ecosystem services. Ecological agriculture incorporates many perennial crop plants, animals, poultry, bushes, trees, and other mature ecosystem components. In mature agroecosystems, disturbances such as tillage of the soil and intensive pest control are reduced. According to the intermediate disturbance theory, if a disturbance is not too frequent, wide-spread, or intense, then both productivity and fertility remain high (Gliessman et al., 2022). Therefore, "patchy," small-scale disturbances are used in ecological agriculture, instead of a clear cut or swath over a large area (Fig. 13.7).

Mature ecosystems and agroecosystems can provide ecological services, such as water filtration and carbon sequestration. Forests can buffer challenges like floods. However, humans have altered natural landscapes. We have cleared forests and replaced them with concrete, cities, and simplified environments, such as monoculture crops. Wild landscapes have been lost to urban sprawl. This loss of habitat reduces diversity and in many areas of the world, there is immense pressure on our natural resources. Dense populations of humans and livestock may come in contact with wildlife and lead to

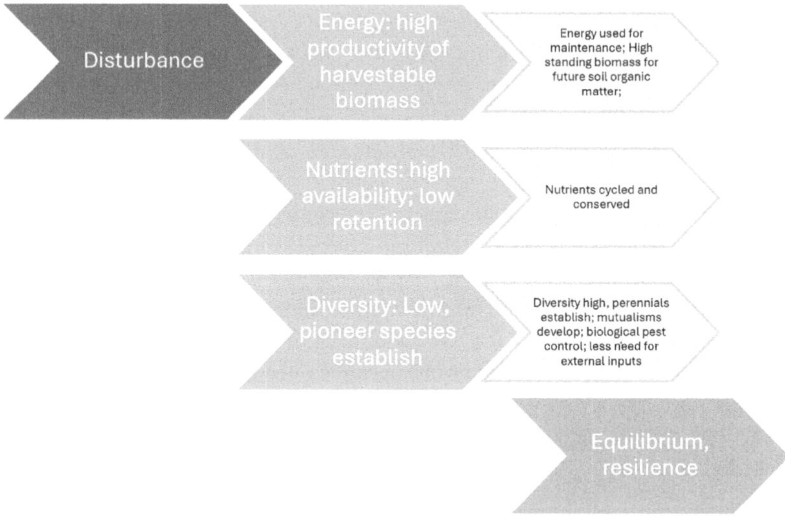

FIGURE 13.7 **Recovery from a disturbance in an agroecosystem.** Perennial components increase ecosystem services and resilience.
Modified from Gliessman, S.R., Méndez, V.E., Izzo, V.M., & Engles, E.W. (2022). Agroecology: Leading the transformation to a just and sustainable food system. CRC Press. http://doi.org/10.1201/9781003304043.

pathogen spill over. Diversity is needed and health is a natural outcome. Diversity can provide an ecological firebreak, especially in the case of emerging diseases, and it can function as an immune system for nature (Gliessman et al., 2022). The resilience of multifunctional landscapes is a key ecosystem service. A One Health approach recognizes that the health of animals, people, and the environment is interconnected (Zinsstag et al., 2015). Ecological production should lead to better welfare. However, there are many research questions. For example, does biodiversity mean more exposure to wildlife and pathogen spill over, including avian influenza?

Circularity and zero waste

Circularity of materials is part of maintaining natural resources in ecological poultry production. Ecosystems cycle many materials, including elemental nutrients with atmospheric reservoirs (i.e., carbon, nitrogen, oxygen) and mineral nutrients with soil or rock reservoirs (i.e., phosphorus, calcium). Circular economies uncover new value in by-products, and "waste" can be reduced to zero. Poultry manure is an important example of circularity. In intensive agriculture, where animals and crops are raised separately in specialized units, manure has become a liability. While poultry litter is a healthy fertilizer for pastures and crops, too much of it can lead to nitrates in groundwater or eutrophication of surface waters due to phosphorus run-off. If manure is managed as a liquid in lagoons, methane, a powerful greenhouse gas, may be emitted. If liquid manure is sprayed in spray fields, odor can be an issue in nearby communities. In contrast, in ecological poultry production, animals and crops are grown on the same farm or region, and nutrients are cycled. Manure is an important production stream as a fertilizer for crop plants. Poultry manure is higher in nitrogen and phosphorus than many other manures. Manure is a slow-release fertilizer and releases nitrogen at a natural rate for soil microbes to incorporate in soil for uptake by plants. Avoiding the use of fast-release nitrogen fertilizers can reduce the emission of nitrous oxide. Animal manure can be cycled in many ways. Rich organic waste, such as liquid manure or viscera from slaughter plants, can be managed with biodigesters. Waste slurries can be separated into solid and liquid. The liquid can be piped into a bag or biodigester, and biogas captured for energy during fermentation. The biogas (mainly methane) then can be converted into electricity to power a building or equipment. The solid manure and the effluent from biogas production are used as fertilizers for crops (Fig. 13.8).

Poultry litter from birds raised in floor systems has the advantage of being dry. It can be composted in place in the poultry house and used for multiple flocks. Additionally, composted manure and litter are useful in organic crop production, especially when it has been heat-treated or composted to 55°C to destroy pathogens (USDA Organic Agriculture Standards/ § 205.203 Soil fertility and crop nutrient management practice standard, 2025). Compost is an

FIGURE 13.8 **Biodigester in Costa Rica.** Dairy manure is converted to biogas to power dairy equipment.

excellent soil amendment for crop production. Compost retains water and nutrients and provides organic matter for soil microbes. Some poultry litter may have a high amount of lignin from wood chips and has been used at times for fuel. However, it may also be useful as a substrate for biochar production. Biochar is a carbon-negative process that can stabilize carbon. It is used as a soil amendment, filter, and for other uses (Graves et al., 2022). It has a porous nature and increases fertility, helps hold nutrients in the soil, and increases organic matter (Khan et al., 2024). It is also important to cycle by-products from slaughter or egg processing. Organic waste that ends up in landfills contributes to methane production. Feathers and bonemeal are commonly used as crop fertilizers to provide nitrogen and phosphorus. Many "waste" materials are actually key resources for building the soil, carbon sequestration, feed crop production, and renewable energy. Zero waste is part of a circular economy, which keeps the value of products circulating (Cappellozza et al., 2019).

Renewable energy flows

Energy flows in poultry production have a large impact on natural resources. Solar energy and other renewable energy sources are needed. In the past, farms were operated on solar and biological energy. Poultry production jumped to become an industrial process in the 1970s (Cheeke, 2003). However, the further from nature the system moves, the more energy that is required to power it. In the United States, agriculture and poultry production currently depend heavily on fossil fuels, including fuel for machinery, irrigation pumps, petroleum-based fertilizer, and pesticides. Electricity (from coal-powered plants) is used to cool and ventilate poultry houses, operate processing plants, equipment, freezers, etc., and propane heats poultry houses. Transportation of raw materials and products uses gasoline and diesel. Fossil fuels are plant carbon that was removed from the atmosphere millions of years ago, and extracting it from the earth for combustion releases excess carbon dioxide into our atmosphere. Carbon dioxide is a greenhouse gas and contributes to climate change. Energy use can be reduced in poultry production. In ecological poultry production, housing is often open so that birds have outdoor access, and windows provide natural lighting. Natural ventilation can make use of cross currents of air. With outdoor access, birds have fresh air, sunlight, and generous space/enrichments. They also have access to vegetation rooted in the soil, and their manure can be incorporated into the soil and pasture in an aerobic process that does not produce methane. Energy can be reduced across the entire poultry meat and egg supply chains (inputs, production, processing, distribution, marketing, and consumption) to achieve carbon neutrality. Energy from fossil fuels can be reduced by using renewable energy, sustainable practices, and local production. Many poultry companies have carbon management strategies. In processing plants, many use renewable energy and reduce packaging.

Water and hydrological cycling

Fresh water is a limited natural resource and is cycled between the earth and the atmosphere (Gliessman et al., 2022). The production of meat and eggs requires a significant investment of water. The use of water in poultry production is largely related to feed crop production. In effective systems, plants bring up water from the soil, which exits to the atmosphere via transpiration from plant stomata. Additionally, organic matter in the soil can retain water (Gliessman et al., 2022). Poultry, like other animals, are largely made up of water and require direct water inputs in the form of drinking water to maintain their biomass and cool themselves by panting. Interestingly, poultry actually need less water than mammalian farm animals because they eliminate excess nitrogen via uric acid instead of urea (Cheeke, 2003). Poultry slaughter and processing plants require a large amount of water; many plants are increasingly efficient in their cycling of water.

Human socio-economic equity and animal welfare in poultry production

In addition to the environment, social equity and animal welfare are key parts of sustainability. As part of nature, humans provide important resources and create human capital. Our knowledge, education, technology, labor, health, and relationships generate a large amount of resources. For example, indigenous

FIGURE 13.9 **Mayan farmers interested in sustainable agriculture.** Indigenous knowledge in turkey production is a valuable resource.

knowledge about chickens in Southeast Asia and turkeys in Central America is invaluable for ecological poultry production (Fig. 13.9).

We belong to ecosystems and are dependent ultimately on solar energy, air, soil, water, seeds, and animals. Equity is related to power and wealth. Social power serves to meet one's needs and is often used to control resources. Unfortunately, power, wealth, and food are unevenly distributed. There is a particularly large disparity of wealth around the world. Social justice is a form of societal health and is critically needed across the globe.

Intensive agriculture and food systems

Large-scale poultry production is generally based on a capitalistic economic system, including private ownership and competition in the marketplace. The typical focus is maximum yield for profits. There are social impacts of a food system that is driven by the market. Nature's resources are often used without replenishment. Many social and environmental costs are externalized and pushed off onto the environment, public health, and marginalized people. The labor of marginalized people without power to negotiate may become exploited (Gliessman et al., 2022). Large poultry companies often consolidate, and power ends up in the hands of a few. In the United States, a gap has grown between producers and consumers, and consumers may not understand the impact of their food choices on the environment and on people (Wunderlich & Smoller, 2019). However, capitalistic economic systems can be designed to value social good, as well as profit. Special markets develop to address concerns (e.g., carbon market, biodiversity market). There are also government incentives or subsidies that reward farmers for environmental care. Participatory approaches and relationships that support human rights are a powerful social resource to increase equity. Engagement among companies, consumers, health professionals, educators, and other stakeholders can promote social resilience. Poultry companies can engage with consumers and stakeholders based on shared values, such as sustainability, animal welfare, and environmental care. For example, Perdue Foods Inc holds an annual animal care summit where viewpoints of many stakeholders are exchanged (Animal Care Summits, 2025).

Community-based food systems

A different socio-economic approach is localizing agriculture and bringing people together in a bioregional "foodshed" (Gliessman et al., 2022). Small farms are well-positioned to take advantage of local feed resources, cycle nutrients locally, and reduce waste. Producing food close to the site of consumption reduces the transportation costs of raw materials. Local resources can meet local needs (Schumacher, 1973). Local production also decreases the distance that poultry must travel to a processing plant, which is important for welfare. Local producers can build relationships with local consumers and other stakeholders. Exchanges are not limited to economic exchanges; people can build relationships centered around food. Many consumers want to know their farmers and how their food is produced. Some consumers even share risks and rewards with farmers in models such as "community-supported agriculture." In this model, customers subscribe to a regular share of a farm's harvest, sometimes paying at the beginning of the production season for a share (Rossi & Woods, 2024). Farmers have an important role in protecting the environment while producing food; however, they need the support of other stakeholders. The small scale of local agriculture keeps it tied to the environment. Instead of scaling on a national level, it is replicated across communities. As mentioned, ecological agriculture is site-specific and is equitable agriculture. The approach depends on the needs of human communities. Sites differ widely, and a diversity of solutions is needed that may include both intensive poultry production on a large scale or small-scale local agriculture (Fig. 13.10).

An increasing amount of the world's population lives in cities. The intensity of poultry production is often related to feed and other available resources and to the density of the human population. Intensive poultry production may be needed where there are large human populations, and extensive production is more suited for peri-urban or rural areas.

International development

A well-developed agriculture, land, and healthy soil contribute to food security for countries and may help reduce political conflicts. Global

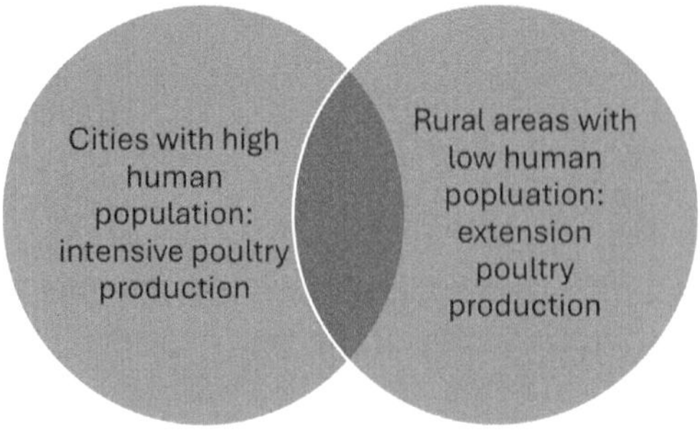

FIGURE 13.10 **Different approaches based on human population.** More intensive human population may require more intensive poultry production.

security is related to food and impacts us all. Much of hunger around the world is due to poverty and inequitable distribution of food, rather than a lack of food (Holt-Gimenez, 2012). Locally based food economies may be more effective than the global market in addressing hunger and poverty, and can ensure that resource-poor farmers have the land and resources they need for sustainable livelihoods. By 2050, there will be almost 10 billion people on the earth, and strategies are needed to ensure healthy food for all.

Goals

Setting goals and measuring progress can help balance environmental, human socio-economic, and animal needs. Goals in dimensions of poultry production may include (Fig. 13.11):

- Environment: Ecology
- Human: Well-being, equity
- Socio-economic: Livelihood, profit
- Bird: Welfare

The US Roundtable for Sustainable Poultry and Eggs uses a framework of "planet, people, poultry" for sustainability (US Roundtable for Sustainable Poultry and Eggs, 2026). On a global level, the United Nations has a System for Environmental and Economic Accounts (System of Environmental Economic Accounting, 2025) that assigns monetary value to natural capital and ecosystems. This can also be used in agriculture and is helpful in decision-making and policy. In the past, GDP has been the main way to track economic progress. However, with natural capital accounting, we can also value clean air and water, healthy soil, etc. (World Bank, 2025).

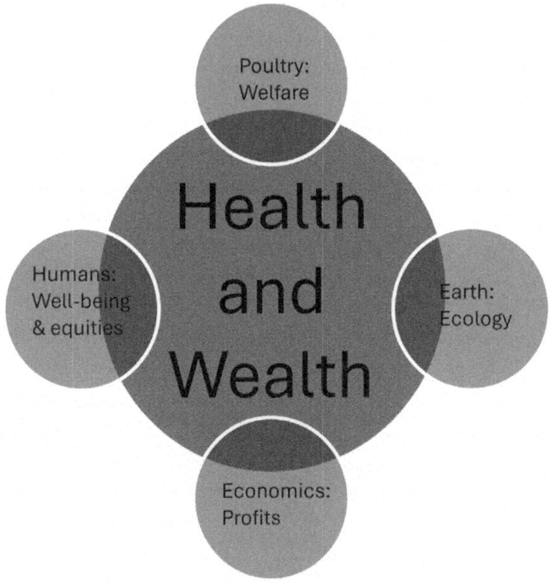

FIGURE 13.11 **Goals in poultry production.** Goals of environment, equity, profit, and welfare overlap.

Conclusion

Poultry production has great potential to maintain natural resources and social resilience while increasing poultry welfare. But there may be a significant disparity between intensive systems designed to produce protein at scale compared with more extensive systems that are integrated into the environment and society but produce protein at a much smaller scale. Finding a balance between these system types and their respective implications for sustainability, as well as poultry welfare, will help us face environmental and social challenges in the future.

References

Animal Care Summits Perdue Farms, Inc Animal Care Summits. (2025). https://corporate.perduefarms.com/responsibility/animal-care/commitments/.

Bonnefous, C., Collin, A., Guilloteau, L. A., Guesdon, V., Filliat, C., Réhault-Godbert, S., Rodenburg, T. B., Tuyttens, F. A. M., Warin, L., Steenfeldt, S., Baldinger, L., Re, M., Ponzio, R., Zuliani, A., Venezia, P., Väre, M., Parrott, P., Walley, K., Niemi, J. K., & Leterrier, C. (2022). Welfare issues and potential solutions for laying hens in free range and organic production systems: A review based on literature and interviews. *Frontiers in Veterinary Science, 9*, 952922. https://doi.org/10.3389/fvets.2022.952922, journal.frontiersin.org/journal/veterinary-science.

Campbell, Y. L., Walker, L. L., Bartz, B. M., Eckberg, J. O., & Pullin, A. N. (2025). Outdoor access versus conventional broiler chicken production: Updated review of animal welfare, food safety, and meat quality. *Poultry Science, 104*(4), 104906. https://doi.org/10.1016/j.psj.2025.104906.

Cappellozza, S., Leonardi, M. G., Savoldelli, S., Carminati, D., Rizzolo, A., Cortellino, G., Terova, G., Moretto, E., Badaile, A., Concheri, G., Saviane, A., Bruno, D., Bonelli, M., Caccia, S., Casartelli, M., & Tettamanti, G. (2019). A first attempt to produce proteins from insects by means of a circular economy. *Animals, 9*(5), 278. https://doi.org/10.3390/ani9050278.

Cheeke, P. (2003). *Contemporary issues in animal agriculture*. Pearson Publishing.

Fanatico, A. C., Arsi, K., Upadhyaya, I., Morales Ramos, J., Donoghue, D., & Donoghue, A. M. (2018). Sustainable fish and invertebrate meals for methionine and protein feeds in organic poultry production. *Journal of Applied Poultry Research, 27*(4), 437–448. https://doi.org/10.3382/japr/pfy037.

Fanatico, A. C., Brewer, V. B., Owens-Hanning, C. M., Donoghue, D. J., & Donoghue, A. M. (2013). Free-choice feeding of free-range meat chickens. *Journal of Applied Poultry Research, 22*(4), 750–758. https://doi.org/10.3382/japr.2012-00687.

Fanatico, A. C., Owens-Hanning, C. M., Brewer Gunsaulis, V., & Donoghue, A. M. (2016). Choice feeding of protein concentrate and grain to organic meat chickens. *Journal of Applied Poultry Research, 25*(2), 156–164. https://doi.org/10.3382/japr/pfv076.

Fanatico, A. C., Smith, D. P., Boyette, M., & Coffey, C. (2014). Performance of free-range meat chickens fed sweet potatoes as a farm-raised energy source. *Poultry Science, 93*.

FAO. (2026). https://openknowledge.fao.org/items/40b80ef2-47b2-47f2-99e5-cf2190b364fa.

Filho, M. A. N., Pereira, R. T., de Oliveira, A. B. S., Suckeveris, D., Burin Junior, A. M., Mastrangelo, T. D. A., da Costa, D. V., & Menten, J. F. M. (2020). Cafeteria-type feeding of chickens indicates a preference for insect (*Tenebrio molitor*) larvae meal. *Animals, 10*(4), 627. https://doi.org/10.3390/ani10040627.

Fu, Y., Hu, J., Zhang, H., Erasmus, M. A., Johnson, T. A., & Cheng, H. W. (2024). The impact of early-life cecal microbiota transplantation on social stress and injurious behaviors in egg-laying chickens. *Microorganisms, 12*(3), 471. https://doi.org/10.3390/microorganisms12030471.

Gliessman, S. R., Méndez, V. E., Izzo, V. M., & Engles, E. W. (2022). *Agroecology: Leading the transformation to a just and sustainable food system*. CRC Press. http://doi.org/10.1201/9781003304043.

Graves, C., Kolar, P., Shah, S., Grimes, J., & Sharara, M. (2022). Can biochar improve the sustainability of animal production? *Applied Sciences, 12*(10), 5042. https://doi.org/10.3390/app12105042.

Guinebretière, M., Guyard-Nicodème, M., Mocz, F., Calandreau, L., Scheubel, A., Moysan, J. P., Chemaly, M., Keita, A., & Warin, L. (2025). Structuring and enriching the rearing environment in conventional broiler chicken production: effects on behavioral indicators, emotional states, and cecal microbiota composition. *Poultry Science, 104*(11), 105663. https://doi.org/10.1016/j.psj.2025.105663.

Holt-Giménez, E., Shattuck, A., Altieri, M., Herren, H., & Gliessman, S. (2012). We already grow enough food for 10 billion people … and still can't end hunger. *Journal of Sustainable Agriculture, 36*(6), 595 -598. https://doi.org/10.1080/10440046.2012.695331.

International Federation of Organic Agriculture Movements. (2026). https://www.ifoam.bio/.

Jia, J., Wu, F., Yu, H., Chou, J., Han, Q., & Cui, X. (2024). Global meat consumption driver analysis with machine learning methods. *Food Security, 16*(4), 829–843. https://doi.org/10.1007/s12571-024-01455-y.

Jiang, S., Hu, J. Y., & Cheng, H. W. (2022). The impact of probiotic bacillus subtilis on injurious behavior in laying hens. *Animals, 12*(7), 870. https://doi.org/10.3390/ani12070870.

Khan, M. A., Basir, A., Shah, S. T., Rehman, M. U., Hassan, M. U., Zheng, H., Basit, A., Székely, Á., Jamal, A., Radicetti, E., Dewir, Y. H., & Mancinelli, R. (2024). Sustainable soil management in alkaline soils: The role of biochar and organic nitrogen in enhancing soil fertility. *Land, 13*(11), 1869. https://doi.org/10.3390/land13111869.

Livestock Conservancy. (2026). https://livestockconservancy.org/.

Rossi, J., & Woods, T. (2024). Placing community supported agriculture in local food systems. *International Journal of Agricultural Sustainability, 22*(1), 2318936. https://doi.org/10.1080/14735903.2024.2318936.

Ruiz, B. (2019). Gut health at ESPN: Beyond microorganisms. WattAgNet.

Sarmiento-García, A., Revilla, I., Abecia, J.-A., & Palacios, C. (2021). Performance evaluation of two slow-medium growing chicken strains maintained under organic production system during different seasons. *Animals, 11*(4), 1090. https://doi.org/10.3390/ani11041090.

Schumacher, E. F. (1973). *Small is beautiful: Economics as if people mattered.* Harper & Row.

Siwach, A., Saini, S., Giri, A., Khatri, P., Kuhad, R. C., & Kumar, A. (2025). Sustainable poultry feed formulations from fruit and vegetable residues for advancing animal health. *Bioresource Technology Reports, 30*, 102159. https://doi.org/10.1016/j.biteb.2025.102159.

Soil Association Organic Standards for Great Britain Farming and growing. (2026). Version 1.7. Soil Association https://www.soilassociation.org/what-we-do/our-standards/soil-association-organic-standards/organic-standards-for-great-britain/, https://www.soilassociation.org/media/28350/sa-gb-farming-growing.pdf.

Steps. (2025). STEPS Center https://steps-center.org/.

System of Environmental Economic Accounting United Nations. (2025). 7–25. https://seea.un.org/home/Natural-Capital-Accounting-Project.

US Roundtable for Sustainable Poultry and Eggs. (2026). https://www.us-rspe.org/.

USDA Organic Agriculture Standards/ § 205.203 Soil fertility and crop nutrient management practice standard USDA National Organic Program. (2025). https://www.ecfr.gov/current/title-7/subtitle-B/chapter-I/subchapter-M/part-205/subpart-C.

World Bank. (2025). https://www.worldbank.org/en/topic/natural-capital.

Wunderlich, & Smoller (2019). Consumer awareness and knowledge about food sources and possible environmental impact. *International Journal of Environmental Impacts, 2*(1), 85–96. https://doi.org/10.2495/EI-V2-N1-85-96.

Zinsstag, J., Schelling, E., Waltner-Toews, D., & Tanner, M. (2015). *One health: The theory and practice of integrated health approaches*. CABI.

Chapter 14

Sustainable feeding

Clive J.C. Phillips[1,2]
[1]*Curtin University Sustainability Policy (CUSP) Institute, Curtin University, Bentley, WA, Australia,* [2]*Institute of Veterinary Medicine and Animal Science, Estonia University of Life Sciences, Tartu, Estonia*

Introduction

Worldwide, about 77 billion chickens are slaughtered annually, increasing by about 4% each year—twice as fast as the other popular farm animals for meat, pork, and beef (PoultryPro, 2025). The highest concentrations of chickens are found in eastern China, eastern United States, south-east Brazil, India, Pakistan, and western Europe (Robinson et al., 2014). Globally, chicken consumption has increased from 3 kg/person/year in 1961 to 65 kg/person/year today. Continued growth is expected until at least 2050 (Mottet et al., 2017). Along with pork, chicken is now the most popular meat across the planet. To achieve this, about four chickens have to be maintained for every person on the planet at any one time. For every chicken reared for egg production, eight chickens are reared for meat.

Feed accounts for 60%–70% of production costs for chickens. They eat primarily maize (corn), wheat, and soybeans, 465 million tonnes per year, grown on between 1 and 2 million ha of land (FAO Food & Agriculture Organisation, 2018; Govoni et al., 2021). Maize is primarily used as an energy source in the United States and Asia; wheat in Europe, Australia, and New Zealand. Maize is also a common basic food for people in developing countries, and it is in increasingly short supply. In 2025 alone, maize export prices rose by 10% due to climate-driven crop shortages.

Because of the rapid growth of poultry consumption, 8 out of the top 10 producers in the world now have to import large quantities of these raw ingredients. Only the United States and Brazil are largely self-sufficient. Mostly, large quantities of feed are sent from the Americas and Australia to Asia. The biggest trade is soybeans sent from the United States to China, with the latter importing 65% of the world's traded soybeans as they are only 9% self-sufficient (Govoni et al., 2021). Most countries in the world (74%) rely on imported soybeans to feed their poultry population. In Brazil, industrial-scale

Integrating Welfare and Sustainability in Poultry Systems. DOI: https://doi.org/10.1016/B978-0-443-40594-5.00008-0

soybean cultivation has been pursued by big multinational companies with the aid of subsidies from the government (Garrett & Rausch, 2016). This could undermine the small-scale farmers producing local crops for local consumption.

Transport costs have been rising rapidly through increases in fuel costs. Transport and concentrate feeds can together make up the majority of carbon dioxide contributions to greenhouse gas emissions (Taylor et al., 2014). Soy from Brazil has a very high greenhouse gas footprint, partly due to land-use change (Carbon Trust, 2010). Reducing the transport distance is of growing importance, both in relation to the increasing cost and greenhouse gas emissions.

Many countries in sub-Saharan Africa and SE Asia cannot produce enough feed for their chickens; some make it up by imports, but this then increases retail prices reducing the affordability of meat and eggs (Poultry & Livestock Review Africa, 2025). The feed shortfall largely comes from scraps and swill in backyard feeding systems. The poorest countries have high levels of malnourishment of their human populations, which would be exacerbated by the use of their own staple foods to feed chickens. They are also vulnerable to low-value chicken products being dumped at low cost to undermine the sustainability of their own chicken farmers (Hedman et al., 2021).

Whilst this system of feeding chickens is sustainable in the short term, in the long term, it has significant risks. The resources required to grow chicken feed and transport it from the Americas to Asia are significant. There is effectively a large-scale transfer of land, fertilizers, and agronomy expertise to Asia, as well as large quantities of water, due to the widespread reliance of maize and soybean crops in Brazil and the United States on irrigation. In Brazil, the deforestation of the Amazon region to produce these crops is causing local pollution and having a damaging impact on both the local and global climate (Prasad, 2016). Drought, unheard of in previous times, is now featuring regularly in the Amazonian rainforest, as well as hitting the US Midwest and Argentina hard, impacting soy production. Freshwater biodiversity is also threatened by intensification of agricultural crop production (Albert et al., 2021).

With considerable uncertainty surrounding the availability of water for irrigating crops for chicken feed in the future, growing groundwater deficits (Partridge et al., 2023), and the high energy (and climate) cost of transporting chicken feed around the world, the sustainability of current chicken feed production must be questioned. Furthermore, there are significant disease risks that can impact poultry supply. Outbreaks such as avian influenza disrupt product availability and cost across supply chains, highlighting their vulnerability to such disruptions (Zamani et al., 2024). In the United States, avian flu has spread to dairy herds, reducing milk production and risking the disease spreading to humans in contact with cows or consuming unprocessed dairy products (Hossain et al., 2024). The other major disease risk to poultry, and ultimately humans, comes from antibiotic resistance. With antibiotics still being used in flocks in the United States and Brazil, albeit at a lower rate than previously (Da Silva et al., 2023; Wallinga et al., 2022), the inevitability of transmission of antibiotic resistance to humans will pose a major threat to human health in the future, unless antibiotics in feeds are removed immediately, which is very unlikely.

It is clear that humans have reached or are rapidly approaching the limit of sustainable growth of chicken production systems. The big three crops used in intensive chicken production worldwide, maize, wheat, and soya, are all increasing rapidly in cost. Alternative grain crops have disadvantages of high fiber content; poor protein, vitamin, and mineral bioavailability; and the presence of antinutritional factors. Plant-based meat alternatives are growing in popularity with continual advancements in their composition and production (Chen et al., 2026). Lab-produced cultured chicken is feasible now, but it remains expensive.

The social licence for intensive chicken production is increasingly questioned by consumers (Parker et al., 2018), on account of the welfare concerns for birds in both egg-laying and meat production units. It has been widely documented that consumers are willing to pay extra for better animal welfare and even that they prioritize animal over worker welfare (Kaminski et al., 2024). They particularly value chickens' freedom of movement in free-range systems, believing that it will reduce the stress on the chickens and improve the quality of their products (tastier, healthier meat), although it is acknowledged that production costs and the cost of the meat increase (Yunes et al., 2017). Many members of the public are poorly informed about the details of intensive chicken production systems (Erian & Phillips, 2017), and this has some bearing on their attitude to the animals' welfare (Erian et al., 2019). Empathy and knowledge of production systems may be linked, but further research is needed in this area to determine whether more knowledgeable people have higher levels of concern for chicken welfare.

In the United Kingdom at least, there is strong support for government legislation that improves layer hen welfare. As well as choosing to consume alternatives to poultry products, concerned members of the public can engage in a variety of antagonistic behaviors, such as influencing friends and family, writing to politicians, donating money or time to a welfare organization, and voicing their opinions publicly. The media may become involved in exposing the cruel treatment of chickens, aided by concerned members of the public. Improving chicken welfare will not only assuage public concerns, but it may also improve product quality and reduce costs of production.

Feeding chickens in a new climate

Climate change is creating more frequent droughts, higher temperatures globally, and more variable precipitation. Drought-resistant crops are urgently needed. If the anticipated 2°C increase in global temperature by the end of this century eventuates, cereal production is anticipated to decline by 20% (Fatima et al., 2020).

Plant species that use the C_4 photosynthetic pathway cope with high temperatures and drought better than maize, wheat, and soybeans, which are C_3 photosynthesizers. A potentially valuable C_4 photosynthesizer is the pseudocereal amaranth. This plant has been cultivated in the tropics and subtropics of Latin and South America for centuries. Nowadays, it is increasingly being cultivated in China as animal feed. It is more resistant to drought and high temperatures than traditional cereal grains grown for chickens, and it confers health benefits due to its high antioxidant content (Alizadeh-Ghamsari et al., 2021; Orczewska-Dudek et al., 2018). Amaranth increases n-3 polyunsaturated fatty acids in chickens, and its high content of antioxidants prevents lipid oxidation. Amaranth grain is a good source of digestible fiber, and performance is comparable to that of chickens fed on corn-soybean diets, provided it is heat-treated for antinutritive compounds if included at high concentrations in the diet (Janmohammadi et al., 2022). It can be used as a protein supplement due to its high protein content and excellent amino acid composition (Janmohammadi et al., 2022), and it is a rich source of antioxidants (vitamins A, C, E, β-carotene, and calcium), iron, manganese, and bioactive compounds (carotenoids, flavonoids, polyphenols, and phytosterols) (Hosseintabar-Ghasemabad et al., 2024). Amaranth leaves can also be used for up to 15% of the diet without jeopardizing growth (Orczewska-Dudek et al., 2018).

Quinoa is another drought-tolerant South American grain crop that has the potential to replace nutrients from traditional grains in chicken diets (Zagarin et al., 2025). It is rich in fiber and protein, but contains antinutritional saponins, which reduce protein availability unless proteases are added. However, saponins also have health benefits, being antioxidants, immunostimulants, and acting as antimicrobial and antiviral agents. Treating the quinoa to remove saponins eliminates these health benefits. Feed conversion ratios with quinoa are good, since feed intake is reduced and growth is similar to that of chickens on conventional diets (Zagarin et al., 2025).

Millet was one of the ancient food grains cultivated in Asia for millennia (Qaisrani et al., 2019). It is the most drought-tolerant of all cereals, more than sorghum, is pest-resistant and it grows well with little fertilizer application (Ravindran et al., 2023). However, several varieties have less metabolizable energy than maize, although the crude protein (CP) is often higher. The high content of nonstarch polysaccharides reduces nutrient digestion, and millet also has antinutritive tannins in its hull. It is therefore recommended that the maximum inclusion rate of pearl millet in the diet of poultry is 40%, even though theoretically it can completely replace maize (Ravindran et al., 2023). Although millet is a commonly grown cereal in India, sub-Saharan Africa, and China, little is currently fed to chickens because of its value as a human food in these regions. If the land devoted to millet continues to increase, pearl millet can replace maize in the diet of some chickens, with beneficial effects in reducing fertilizer and water use (Qaisrani et al., 2019; Rao et al., 2004). Millet has the added advantage in sub-Saharan Africa that processing is not required, the whole grain being small enough to feed to chickens.

Sorghum originated in Africa, but today it is also commonly grown in Asia, Australia, and the United States. It is often used in livestock feed in these regions and is valued for its drought and heat tolerance (Deauville et al., 2023). The grains are starch-rich, and the energy content is between that of barley and maize. Sorghum grains have a protein content that is similar to wheat but higher than maize. Like millet, sorghum grains contain tannins and phytate that bind with proteins in the gut and reduce protein digestibility and uptake of amino acids. Phytate-degrading enzymes can be added to enhance protein digestibility. However, even if the grain is low tannin and phytase supplemented, the content of another antinutritive agent, kafirin, will limit broiler performance (Selle et al., 2010). Lysine is commonly deficient in sorghum grains. Processing is effective in improving digestibility, but may be too expensive. However, promising results are being obtained with transgenic sorghum, which has a higher protein content and a modification to the kafirin protein bodies (Macelline et al., 2024). Transgenic sorghum could replace up to 20% of soybean meal, without loss of performance, and confirms that sorghum, rather than wheat, is better for chickens if the CP content of the diet is low (Wang et al., 2025).

Oilseed residues can replace soybean meal, such as canola/rapeseed meal and sunflower meal, but their high level of fiber, low energy concentration, and deficiency in several amino acids, especially lysine, threonine, and tryptophan, limit their use in chicken diets. If used for chickens, oilseed residues are more suited to growing pullets, layers, and breeders than broilers (Blair et al., 2023). Rapeseed meal also contains antinutritive substances, erucic and eicosanoic acids, glucosinolates, phytates, and nonstarch polysaccharides, which reduce feed intake and growth (Arney et al., 2023).

Edible insects have long been part of the human diet and are a natural feed for chickens; hence, the potential exists for them to replace some of the proteinaceous feed in poultry. Their conversion of feed to edible matter is efficient, about 50%, their growth is rapid, and they have a high protein content (about 60% on a DM basis) with a good amino acid balance (Khan et al., 2016). The most variable component is fat, which in part reflects the substrate on which the insects are fed, but typically the fat content exceeds that of soybean meal. Some processing is desirable to separate the high-protein, high-energy component from the chitinous byproduct (Hawkey & Hall, 2023). The latter has antimicrobial activity when included in the diet of broilers. The fatty acid profile of chickens' breast muscle may be improved by feeding insect oils, with more n-3 and n-6 fatty acids (Benzertiha et al., 2019). Up until recently, feeding insects to chickens was not allowed in several jurisdictions, but notably, since 2021, the European Union has allowed it (Hawkey & Hall, 2023).

The most suitable insects to industrially farm are black soldier flies, yellow mealworms, silkworms, grasshoppers, termites, and a variety of larvae (Margerison & Bogueva, 2023). The black soldier fly has been included at up to 15% of the diet of broilers, with the best chicken growth at 10% inclusion (Zhang et al., 2025). Silkworms are already farmed commercially in the production of silk, and they have an output of about 8 kg pupae/kg silk (Zhang et al., 2025). Most are treated as waste and could be fed to chickens. As well as silkworms, other worms are suitable as chicken feed—maggot meal, earthworms, and mealworms. Like insects, worms are a natural feed for chickens. Earthworms and mealworms have over 60% CP and good amino acid composition (Khan et al., 2016).

The use of insects in the diet of chickens has significant benefits for the reduction of land and water resources required and greenhouse gas emissions. Insects can utilize environmental waste, such as food waste, slaughterhouse waste, crop byproducts, and livestock manure, including that from chickens. Currently, although the sector is growing, the supply chain lacks automation, since insects require killing and drying, and it has high costs associated with the small-scale production units currently in use. There is also increasing concern about the welfare of insects (Boppré and Vane-Wright, 2019), whereas welfare concerns in the past were largely confined to vertebrate animals. In contrast to insects, agriculture and forestry byproducts are rarely suitable for chicken feeding because of their high fiber content and low digestibility. However, forest trimming residues, known as slash, can be processed into chicken feed, whereas they currently present a fire hazard. The value of slash depends on the proportion of foliage (Ateş et al., 2007).

Replacing antibiotics with immunostimulatory plants

Feeding lipid-rich diets to chickens results in oxidative stress, which damages the birds' DNA, lipid biomembranes, and proteins, as well as tissue damage. Immune stress is common in modern, intensive broiler production (Hofmann et al., 2020). Immune stress results in metabolic disturbance, reduced growth, and cytokine release, as well as reducing the integrity of the intestinal barrier and the efficiency of digestive enzymes (Hofmann et al., 2020). There are systemic inflammatory responses that are treated with routine antibiotic inclusion in feed. A thinner-walled gut and a reduction in bacterial activity are posited as some of the benefits of antibiotic growth promoters (Fernández Miyakawa et al., 2024). However, antibiotic inclusion in feed is banned in many countries because it leads to the emergence of resistant bacteria and a lack of functionality in antibiotics that may be of value in human medicine. Worldwide, antibiotics need to be replaced by alternative feed additives, particularly those containing flavonoids that are available in large quantities for the poultry industry, such as naringin present in the peel of citrus fruits (Shilpa et al., 2023).

Green tea, a popular beverage that contains polyphenols, known as catechins, has activity against tumors, inflammation, oxidative stress, bacteria, viruses, and parasites (Jang et al., 2007). It is also effective as a coccidiostat in chickens (Jelveh et al., 2022). Similar functionality is claimed for other plant products, such as turnip extract (Eghbaldost-Jadid et al., 2021). The extracted compounds from brassicas, such as turnips, are both water soluble (phenols, flavonoids, and ascorbic acid) and lipid soluble (tocopherols and carotenoids). However, the extraction costs are significant, and the extract is more valued for human drugs and cosmetics. More feasible for small-scale producers are green tea infusion residues, which also have beneficial properties, or loquat tea byproducts (Matsunaga et al., 2024). These can be incorporated at low levels into chicken feed. Berries also have many antioxidants; for example, sumac is a plant that grows by the roadside in the United States, and the berries contain numerous antioxidants that have the potential to improve the health of chickens (Ahmadian et al., 2020). Feasibility for large-scale production must be considered, even if the plant is efficacious for chicken health improvement. Olive leaf extract is one of the most feasible products to replace antibiotics in the diet of chickens (Phillips et al., 2023).

Algae are photosynthetic aquatic plants with numerous bioactive compounds, including carotenoids, tocopherols, and omega-3 fatty acids, as well as high protein contents and readily digestible energy (Abdel-Wareth et al., 2024). Included at low concentrations (1–4 g/kg) in the diet of both layers and broilers, either processed or raw, algae have antioxidant and antiinflammatory functions from their carotenoids and omega-3 fatty acids, respectively. They also have antiviral and antibacterial properties and increase the concentration of docosahexaenoic acid, an omega-3 fatty acid, in poultry products. The antibacterial properties come from prebiotic effects, and further microbiome benefits are derived from a range of seaweed algae. There is wide variation in composition between species, and *Spirulina* and *Chlorella* are two that might be useful in poultry feeds. However, algae can be more valuable elsewhere, such as for producing plant-based meats, and it must be noted that some algal blooms in the chickens' water supply create significant risks due to their neurotoxins (Abdel-Wareth et al., 2024).

Towards better use of available land

Growing crops and providing pasture for chickens uses 156 million ha of the world's agricultural land (Mottet et al., 2017). Nearly all feed utilized in chicken production systems is suitable for human food. Specifically, layers and broilers require 2.9 and 5.1 kg of protein from human-edible food/kg protein product, respectively (Mottet et al., 2017). This translates to 17 and 25 kg of human-edible food/kg protein product. As the human population increases from its current 8 billion towards 10–11 billion by the end of the century, competition for food resources will increase. More effective use of land to produce chicken feed will be required. The use of insects to turn unusable wastes into food, discussed above, is one such way; however, it requires considerable investment in new and efficient technologies to produce feed on an industrial scale. Another way is to use higher output crops, such as sugar cane and aquatic plants, so that carbon capture is increased and carbon neutrality can ultimately be achieved for chicken production. Crushed sugar cane can be included, but performance declines with inclusion rate because of the high fiber content (Teixeira et al., 2021). In ground form, sugar cane can be included at up to 15% of the diet without loss of performance.

Aquatic plants

Floating aquatic plants, such as duckweed, typically grow faster than terrestrial plants because of the optimization of the resources required for growth—temperature, water availability, and light (López-Pozo et al., 2023). Continuous contact of half the plant with water buffers the plant against changes in environmental temperature. Lemnaceae (duckweed) living in ideal conditions produce a new daughter plant every day or two. By contrast, terrestrial plants only have flushes of green growth after transfer to ideal conditions at timed intervals, from 1 /week to once or twice a year. Duckweed maintains high photosynthetic ability even in low light levels and requires minimal vascular infrastructure for water and sugar transport. Like some terrestrial plants, duckweed uses nitrogen-fixing micro-organisms to secure nitrogen in deficient environments. Being aquatic, duckweed has little need for structural carbohydrates, which render the stems of tall terrestrial plants of low digestibility for animal species. Most of the energy in duckweed is stored as starch, which is readily digestible, especially if enzymes such as phytase are added.

These advantages allow duckweed to grow at rates of 100 tonnes DM/ha/year under ideal conditions, over 25 times that of conventional chicken feed crops. More conservative estimates, of 10–30 tonnes/year, are still much higher than the typical 3.3 tonnes/year for wheat and 2 tonnes/year for soybeans (Baghban-Kanani et al., 2023). Duckweed's high rate of carbon capture can help to advance animal production towards carbon neutrality, including by releasing land to be used for carbon capture by trees. Azolla (an aquatic fern) is another suitable alternative, with very rapid growth rates. Inclusion of duckweed in the diet of chickens up to 15% had no detrimental effect on egg production, and it improved yolk color and apparently had a hepatoprotective effect (Baghban-Kanani et al., 2023). At higher concentrations, some reduction in feed intake and performance has been detected (Haustein et al., 1994). The high concentration of antioxidants in duckweed would be of benefit to intensively housed chickens.

In the future, some of the large areas of land previously considered unsuitable for crop production (Mottet et al., 2017) could be used for the construction of ponds for the growing and harvesting of duckweed. For example, most of the Australian interior was deemed unsuitable for cropping by Mottet et al. (2017), but with the high temperatures for most of the year, and extremely low productivity of the land currently, largely used for extensive beef and sheep production, this alternative use warrants further research.

Towards better welfare

Free-range systems are not necessarily more efficient users of land than the aforementioned crops; however, by allowing birds freedom to roam on an extensive area of land, welfare will usually be improved (see Chapter 9 for further discussion on free-range laying hen systems). Nevertheless, there are risks of predation by wildlife, particularly of young broiler birds, and disease spread from wildlife. The birds may need protection from wildlife by elaborate fences, netting, roofing, motion-activated lights, and/or a guardian animal(s), such as a dog (McKellar et al., 2023). In the tropics, there is the risk of heat stress, but providing outdoor access still increases birds' activity and strengthens their legs, which is likely to be beneficial in avoiding lameness (Sanchez-Casanova et al., 2019, 2022).

Free-range systems don't have to rely on pasture grasses; chickens can be offered a range of plants, including kale, Swiss chard, spinach, lettuce, cabbage, broccoli, greens of beet crops, dandelion, comfrey, alfalfa, chickweed, clover, sunflower, pumpkins, zucchini, and borage. Making these plants available is likely to be easier in small- and medium-scale enterprises, with large-scale enterprises tending to rely on pasture grass, which is easier to manage. In the tropics and subtropics, suitable crops to take advantage of high temperatures include a variety of trees, shrubs, and perennials to provide shelter, protection, and nutrition. Legumes such as cowpea (*Vigna unguiculata)* can usefully reduce nitrogen applications but contain trypsin inhibitors that retard digestion. Tropical grasses offer little in the way of nutrition because of their high fiber content. Grasses also often accommodate snakes, which prey on hens' eggs or young birds.

An increasingly promising type of free-range system for chickens is silvopastoral systems, in which pasture is combined with trees (for more detail on this topic, see Chapter 11). These are particularly valuable in the tropics, where high temperatures may cause heat stress in birds on the open range without cover. Shade and forage may usefully be provided by nitrogen-fixing *Acacia* species, such as the black wattle, tree spinach *Cnidoscolus aconitifolius, Acacia auriculiformis, Albizia falcataria*, or the neem tree, *Azadirachta indica*, *Gliricidia sepium, Leucaena leucocephala*, and cassava *Manihot esculenta.* In dry conditions, *Moringa oleífera* from northern India is particularly appropriate. However, antinutritive compounds and high fiber content may detract from the nutritive value of the forages to chickens (Sánchez-Casanova et al., 2020). Furthermore, the type of pasture is constrained by the trees' shading, with legumes hard to establish and maintain, compared with open pasture in which they can reliably substitute for some nitrogen fertilizer (Balocchi & Phillips, 1997).

Free-range systems are easier to manage with layers than broilers because, in their short lives, broilers may not learn to use the range sufficiently well. They are likely to stay close to the sheds, particularly if there is no shade or cover to protect them if they move to the outer regions of the range (Göransson et al., 2021). Ideally, birds should be stocked at a low rate indoors and have access to an outdoor area, even if not too large, which will bring benefits in faster growth (Sanchez-Casanova et al., 2021). There is considerable opportunity to improve chicken welfare with free-range systems in the tropics, using knowledge gained from the increasingly widespread adoption of free-range systems in developed countries (Sánchez-Casanova et al., 2020). The birds are likely to forage for 60% of the day, diversifying their diet and usefully occupying their time.

Another method of diversifying a chicken's diet is free-choice supplements (Fanatico et al., 2013), which allow them to enrich their diet. This is based on the principle that chickens know better than we do what plants to select to provide the best balance of nutrients. However, managing free choice supplementation can be difficult, as some unpreferred feed may spoil, and there may be aggression to access preferred feeds.

Conclusions

When concerns about the growing human population are combined with the anticipated reduction in cereal production with climate change, it is likely that the shortfall in cereal production per person will be approximately 50% by the end of the century. Even considering that chickens are relatively more efficient than ruminant animals at converting cereals to human food, there is still a major reduction in efficiency by using an animal to digest and metabolize the food. This animal is only made up of 50% meat, the rest being waste. Given the increased demand by humans for dwindling cereal stocks and the inevitable complete removal of antibiotics from the diet of chickens, which improved food conversion efficiency by 3% (El-Fateh et al., 2024), it is clear that there must be progress away from cereal-based diets for chickens. This chapter has described several promising novel methods of feeding chickens: using insects fed on waste products, duckweed or algae grown in ponds, drought- and heat-tolerant crops, and free-range chickens under trees that are offered more attractive feed to eat than grass. However, considerable development is required before any of these more efficient ways of using the earth's limited resources to feed chickens can be implemented on the industrial scale necessary to maintain chicken product supply at the current level. Some of the techniques require considerable infrastructural investment, for example, building ponds. There is growing popularity of plant-based chicken alternatives, largely for health and environmental concerns, which is projected to have a compound annual growth rate of 18% between 2025 and 2035 (Anon, 2024). It is therefore likely that investment banks will be cautious in their support for expensive infrastructural developments. This may force the chicken industry to consider low-cost solutions for alternative feeds. Some of these have been developed for insect farming—polytunnels or retrofitted containers with flyboxes for converting organic waste into insect larvae (Ibitoye et al., 2025). The same degree of infrastructural development has not yet happened for feeding aquatic plants to chickens. Simple tanks are in use, but extracting and drying the product is labor intensive. The adoption of alternatives will also depend on the willingness of the big multinational companies to abandon their current model for intensive broiler and layer production and invest in novel infrastructure. The labor required to manage the system is likely to be significant, with limited availability in many regions. Automated systems are being developed, for example, for robotic harvesting of insects, which are likely to prove attractive to big companies. By the middle of this century, chicken farming may be quite different from the industry we know today.

References

Abdel-Wareth, A. A. A., Williams, A. N., Salahuddin, M., Gadekar, S., & Lohakare, J. (2024). Algae as an alternative source of protein in poultry diets for sustainable production and disease resistance: Present status and future considerations. *Frontiers in Veterinary Science, 11*, 1382163. https://doi.org/10.3389/fvets.2024.1382163.

Ahmadian A., Seidavi A., Phillips C. J. C. (2020). Growth, Carcass Composition, Haematology and Immunity of Broilers Supplemented with Sumac Berries *(Rhus coriaria L.)* and Thyme *(Thymus Vulgaris)* Animals. 10, 513. https://doi.org/10.3390/ani10030513.

Albert, J. S., Destouni, G., Duke-Sylvester, S. M., Magurran, A. E., Oberdorff, T., Reis, R. E., Winemiller, K. O., & Ripple, W. J. (2021). Scientists' warning to humanity on the freshwater biodiversity crisis. *Ambio, 50*(1), 85–94. https://doi.org/10.1007/s13280-020-01318-8.

Alizadeh-Ghamsari, A. H., Hosseini, S. A., Soleymani, M. R., & Nahavandi, R. (2021). Performance, intestinal histomorphology, and blood variables of broilers fed amaranth grain in pellet diet. *Tropical Animal Science Journal, 44*(1), 71–78. https://doi.org/10.5398/tasj.2021.44.1.71.

Anon. (2024). Plant-based chicken market. Retrieved November 17, 2025. www.pmarketresearch.com/hc/plant-based-chicken-market/

Arney, D. (2023). Rape. In C. J. C. Phillips, (Ed.). *The encyclopaedia of animal nutrition* (pp. 570–571). Wallingford, UK: CABI. https://doi.org/10.1079/9781789247282.0018.

Ateş, S., Akyıldız, M. H., Vurdu, H., & Akgül, M. (2007). The utilization of slash in Turkey. *Kastamonu University Journal of Forestry Faculty, 7*(1), 93–103. https://doi.org/10.5555/20093170356.

Baghban-Kanani, P., Oteri, M., Hosseintabar-Ghasemabad, B., Azimi-Youvalari, S., Di Rosa, A. R., Chiofalo, B., Seidavi, A., & Phillips, C. J. C. (2023). The effects of replacing wheat and soyabean meal with duckweed (*Lemna minor*) and including enzymes in the diet of laying hens on the yield and quality of eggs, biochemical parameters, and their antioxidant status. *Animal Science Journal, 94*(1). https://doi.org/10.1111/asj.13888.

Balocchi, O. A., & Phillips, C. J. C. (1997). Grazing and fertilizer management for establishment of *Lotus uliginosus* and *Trifolium subterraneum* under Pinus radiata in southern Chile. *Agroforestry Systems, 37*(1), 1–14. https://doi.org/10.1023/a:1005833528619.

Benzertiha, A., Kierończyk, B., Rawski, M., Kołodziejski, P., Bryszak, M., & Józefiak, D. (2019). Insect oil as an alternative to palm oil and poultry fat in broiler chicken nutrition. *Animals, 9*(3), 116. https://doi.org/10.3390/ani9030116.

Blair, R. (2023). Sunflower (*Helianthus annus*). In C. J. C. Phillips, (Ed.). *The encyclopaedia of animal nutrition* (pp. 667–669). Wallingford, UK: CABI. https://doi.org/10.1079/9781789247282.0019.

Boppré, M., & Vane-Wright, R. I. (2019). Welfare Dilemmas Created by Keeping Insects in Captivity. In Carere, C., & Mather, J. The Welfare of Invertebrate Animals. Cham: Springer Verlag. pp. 23–67. https://doi.org/10.1007/978-3-030-13947-6_3

Carbon Trust. (2010). *Carbon Trust FPX database version 3.2 using data from: Renewable transport fuels obligation (RTFO)*. The Carbon Trust London, Omnitech International, DEFRA.

Chen, L., Lin, X., & Xia, Y. (2026). Plant-based meat alternatives: Advancements in ingredients, processing, consumer perceptions, and future directions. *Journal of Future Foods*, *6*(3), 374–388. https://doi.org/10.1016/j.jfutfo.2025.04.005.

Da Silva, R. A., Arenas, N. E., Luiza, V. L., Bermudez, J. A. Z., & Clarke, S. E. (2023). Regulations on the use of antibiotics in livestock production in South America: A comparative literature analysis. *Antibiotics, 12*(8), 1303. https://doi.org/10.3390/antibiotics12081303.

Deauville, E. (2023). Sorghum (*Sorghum bicolor*). In C. J. C. Phillips, (Ed.). *The encyclopaedia of animal nutrition* (pp. 645–647). Wallingford, UK: CABI. https://doi.org/10.1079/9781789247282.0019.

Eghbaldost-Jadid, R., Nosrati, M., Rasouli, B., Seidavi, A., & Phillips, C. J. C. (2021). The effects of turnip (*Brassica rapa*) extract on the growth performance and health of broilers. *Animals, 11*(3), 1–12. https://doi.org/10.3390/ani11030867.

El-Fateh, M., Bilal, M., & Zhao, X. (2024). Effect of antibiotic growth promoters (AGPs) on feed conversion ratio (FCR) of broiler chickens: A meta-analysis. *Poultry Science, 103*(12), 104472. https://doi.org/10.1016/j.psj.2024.104472.

Erian, I., & Phillips, C. J. C. (2017). Public understanding and attitudes towards meat chicken production and relations to consumption. *Animals, 7*(3), 20. https://doi.org/10.3390/ani7030020.

Erian, I., Sinclair, M., & Phillips, C. J. C. (2019). Knowledge of stakeholders in the livestock industries of east and southeast asia about welfare during transport and slaughter and its relation to their attitudes to improving animal welfare. *Animals, 9*(3), 99. https://doi.org/10.3390/ani9030099.

Fanatico, A. C., Brewer, V. B., Owens-Hanning, C. M., Donoghue, D. J., & Donoghue, A. M. (2013). Free-choice feeding of free-range meat chickens. *Journal of Applied Poultry Research, 22*(4), 750–758. https://doi.org/10.3382/japr.2012-00687.

Fatima, Z., Ahmed, M., Hussain, M., Abbas, G., Ul-Allah, S., Ahmad, S., Ahmed, N., Ali, M.A., Sarwar, G., ul Haque, E., Iqbal, P., & Hussain, S. (2020). The fingerprints of climate warming on cereal crops phenology and adaptation options. *Scientific Reports*, *10*, 18013. https://doi.org/10.1038/s41598-020-74740-3.

FAO (Food and Agriculture Organisation). (2018). *Food outlook: Biannual report on global food markets*. FAO Rome. https://doi.org/10.4060/cc2864en.

Fernández Miyakawa, M. E., Casanova, N. A., & Kogut, M. H. (2024). How did antibiotic growth promoters increase growth and feed efficiency in poultry? *Poultry Science, 103*(2), 103278. https://doi.org/10.1016/j.psj.2023.103278.

Garrett, R. D., & Rausch, L. L. (2016). Green for gold: Social and ecological tradeoffs influencing the sustainability of the Brazilian soy industry. *Journal of Peasant Studies, 43*(2), 461–493. https://doi.org/10.1080/03066150.2015.1010077.

Göransson, L., Gunnarsson, S., Wallenbeck, A., & Yngvesson, J. (2021). Behaviour in slower-growing broilers and free-range access on organic farms In Sweden. *Animals, 11*(10), 2967. https://doi.org/10.3390/ani11102967.

Govoni, C., Chiarelli, D. D., Luciano, A., Ottoboni, M., Perpelek, S. N., Pinotti, L., & Rulli, M. C. (2021). Global assessment of natural resources for chicken production. *Advances in Water Resources, 154*, 103987. https://doi.org/10.1016/j.advwatres.2021.103987.

Haustein, A. T., Gilman, R. H., Skillicorn, P. W., Hannan, H., Díaz, F., Guevara, V., Vergara, V., Gastañaduy, A., & Gilman, J. B. (1994). Performance of broiler chickens fed diets containing duckweed (*Lemna gibba). The Journal of Agricultural Science, 122*(2), 285–289. https://doi.org/10.1017/s0021859600087475.

Hawkey, K., & Hall, H. (2023). Insects as animal feeds. In C. J. C. Phillips, (Ed.). *The encyclopaedia of animal nutrition* (pp. 372–376). Wallingford, UK: CABI. https://doi.org/10.1079/9781789247282.0009.

Hedman, H. D., Zhang, L., Butt, B., Papias, P., Trostle, J. A., & Eisenberg, J. N. S. (2021). "Chicken dumping": Motivations and perceptions in shifting poultry production practices. *One Health, 13*, 100296. https://doi.org/10.1016/j.onehlt.2021.100296.

Hofmann, T., Schmucker, S. S., Bessei, W., Grashorn, M., & Stefanski, V. (2020). Impact of housing environment on the immune system in chickens: A review. *Animals, 10*(7), 1–26. https://doi.org/10.3390/ani10071138.

Hossain, M. A., Monem, M. A., & Rina, A. A. (2024). The growing risk of avian influenza (H5N1) in dairy cows: Clinical, economic, and public health consequences. *New Microbes and New Infections, 62*, 101531. https://doi.org/10.1016/j.nmni.2024.101531.

Hosseintabar-Ghasemabad, B., Di Rosa, A. R., Janmohammadi, H., Slozhenkina, M. I., Gorlov, I. F., Mosolov, A. A., Seidavi, A., & Phillips, C. J. C. (2024). The potential of amaranth grain for feeding to poultry. *World's Poultry Science Journal, 80*(2), 481–509. https://doi.org/10.1080/00439339.2024.2308228.

Ibitoye, O., Ayeni, O., Ayanniyi, O., Wealth, A., Kolejo, O., Adenika, O. A., Murtala, M., Oyedijii, O., Aremu, A., & Muritala, D. (2025). Advancing urban insect farming: Integrating automation, vertical farming, and sustainable waste management systems. *Discover Agriculture, 3*(1), 37. https://doi.org/10.1007/s44279-025-00194-8.

Jang, S. I., Jun, M. H., Lillehoj, H. S., Dalloul, R. A., Kong, I. K., Kim, S., & Min, W. (2007). Anticoccidial effect of green tea-based diets against *Eimeria maxima. Veterinary Parasitology, 144*(1-2), 172–175. https://doi.org/10.1016/j.vetpar.2006.09.005.

Janmohammadi, H., Hosseintabar-Ghasemabad, B., Amirdahri, S., Gorlov, I. F., Vladimirovna, K. E., Slozhenkina, M. I., Bilal, R. M., Seidavi, A., & Phillips, C. J. C. (2022). The energy value for broiler chickens of heat-treated and untreated amaranth grain, with and without enzyme addition. *Agriculture, 12*(11), 1810. https://doi.org/10.3390/agriculture12111810.

Jelveh, K., Rasouli, B., Kadim, I. T., Slozhenkina, M. I., Gorlov, I. F., Seidavi, A., & Phillips, C. J. C. (2022). The effects of green tea in the diet of broilers challenged with coccidiosis on their performance, carcass characteristics, intestinal mucosal morphology, blood constituents and ceca microflora. *Veterinary Medicine and Science, 8*(6), 2511–2520. https://doi.org/10.1002/vms3.923.

Kaminski, D. M., Caputo, V., & McKendree, M. G. S. (2024). The US public's attitudes on animal and worker welfare in the dairy and poultry industries. *Journal of Agricultural and Resource Economics, 49*(1), 23–38. https://doi.org/10.22004/ag.econ.330845.

Khan, S., Naz, S., Sultan, A., Alhidary, I. A., Abdelrahman, M. M., Khan, R. U., Khan, N. A., Khan, M. A., & Ahmad, S. (2016). Worm meal: A potential source of alternative protein in poultry feed. *World's Poultry Science Journal, 72*(1), 93–102. https://doi.org/10.1017/s0043933915002627.

López-Pozo, M., Adams, W. W., Polutchko, S. K., & Demmig-Adams, B. (2023). Terrestrial and floating aquatic plants differ in acclimation to light environment. *Plants, 12*(10), 1928. https://doi.org/10.3390/plants12101928.

Macelline, S. P., Godwin, I. D., Liu, G., Restall, J., Cantor, D. I., McInerney, B. V., Toghyani, M., Chrystal, P. V., Selle, P. H., & Liu, S. Y. (2024). Transgenic, high-protein sorghums display promise in poultry diets in an initial comparison. *Poultry Science, 103*(6), 103698. https://doi.org/10.1016/j.psj.2024.103698.

Margerison, J. K., & Bogueva, D. B. (2023). Insects as human food. In C. J. C. Phillips, (Ed.). *The encyclopaedia of animal nutrition* (pp. 376–378). Wallingford, UK: CABI. https://doi.org/10.1079/9781789247282.0010.

Matsunaga, S., Fukagawa, S., Nakamura, K., Ohtsuka, A., & Ijiri, D. (2024). Effect of a mixed fermented loquat leaf tea by-product on thgrowth performance and meat quality of tsushima-jidori crossbred chicken. *Journal of Poultry Science, 61*, 2024024. https://doi.org/10.2141/jpsa.2024024.

McKellar, R. A., Kreplins, T. L., & Fleming, P. A. (2023). Chicken's best friend- Livestock guardian dog bonding with free-ranging chickens. *Translational Animal Science, 7*(1). https://doi.org/10.1093/tas/txad014.

Mottet, A., de Haan, C., Falcucci, A., Tempio, G., Opio, C., & Gerber, P. (2017). Livestock: On our plates or eating at our table? A new analysis of the feed/food debate. *Global Food Security, 14*, 1–8. https://doi.org/10.1016/j.gfs.2017.01.001.

Orczewska-Dudek, S., Pietras, M., & Nowak, J. (2018). The effect of amaranth seeds, sea buckthorn pomace and black chokeberry pomace in feed mixtures for broiler chickens on productive performance, carcass characteristics and selected indicators of meat quality. *Annals of Animal Science, 18*(2), 501–523. https://doi.org/10.2478/aoas-2018-0002.

Parker, C., Carey, R., & Scrinis, G. (2018). The meat in the sandwich: Welfare labelling and the governance of meat-chicken production in Australia. *Journal of Law and Society, 45*(3), 341–369. https://doi.org/10.1111/jols.12119.

Partridge, T., Winter, J., Kendall, A., Basso, B., Pei, L., & Hyndman, D. (2023). Irrigation benefits outweigh costs in more US croplands by mid-century. *Communications Earth and Environment, 4*(1), 274. https://doi.org/10.1038/s43247-023-00889-0.

Phillips, C. J. C., Hosseintabar-Ghasemabad, B., Gorlov, I. F., Slozhenkina, M. I., Mosolov, A. A., & Seidavi, A. (2023). Immunomodulatory effects of natural feed additives for meat chickens. *Life, 13*(6), 1287. https://doi.org/10.3390/life13061287.

Prasad, R. K. (2016). Impacts of human farm activities on tropical deforestation and climate change: Interactive statistical models. *International Journal of Global Warming, 9*(1), 81. https://doi.org/10.1504/IJGW.2016.074309.

PoultryPro, 2025. Understanding Global Chicken Population Trends and Statistics. www.featheredfarmlife.com/chicken-population-trends-statistics/.

Poultry & Livestock Review Africa (2025). Soaring feed costs in African poultry: Can subsidies save the day? https://poultryandlivestockafrica.com/soaring-feed-costs-in-african-poultry-can-subsidies-save-the-day/. Accessed July 8 2026.

Qaisrani, S. N., Murtaza, S., Khan, A. H., Bibi, F., Iqbal, S. M. J., Azam, F., Hussain, I., & Pasha, T. N. (2019). Variability in millet: Factors influencing its nutritional profile and zootechnical performance in poultry. *Journal of Applied Poultry Research, 28*(2), 242–252. https://doi.org/10.3382/japr/pfy073.

Rao, S. V. R., Raju, M. V. L. N., Reddy, M. R., & Panda, A. K. (2004). Replacement of yellow maize with pearl millet (*Pennisetum typhoides*), foxtail millet (*Setaria italica*) or finger millet (*Eleusine coracana*) in broiler chicken diets containing supplemental enzymes. *Asian-Australasian Journal of Animal Sciences, 17*(6), 836–842. https://doi.org/10.5713/ajas.2004.836.

Ravindran, V. (2023). Millets. In C. J. C. Phillips, (Ed.). *The encyclopaedia of animal nutrition* (pp. 456). Wallingford, UK. https://doi.org/10.1079/9781789247282.0013.

Robinson, T. P., Wint, G. R. W., Conchedda, G., Van Boeckel, T. P., Ercoli, V., Palamara, E., Cinardi, G., D'Aietti, L., Hay, S. I., & Gilbert, M. (2014). Mapping the global distribution of livestock. *PLoS ONE, 9*(5), e96084. https://doi.org/10.1371/journal.pone.0096084.

Sanchez-Casanova, R., Sarmiento-Franco, L., & Phillips, C. (2021). The effects of outdoor access and stocking density on the performance of broilers reared under tropical conditions. *British Poultry Science, 62*(5), 632–637. https://doi.org/10.1080/00071668.2021.1918633.

Sánchez-Casanova, R., Sarmiento-Franco, L., Phillips, C. J. C., & Zulkifli, I. (2020). Do free-range systems have potential to improve broiler welfare in the tropics? *World's Poultry Science Journal, 76*(1), 34–48. https://doi.org/10.1080/00439339.2020.1707389.

Sanchez-Casanova, R., Sarmiento-Franco, L., Phillips, C. J. C., & Segura-Correa, J. (2019). Effects of outdoor access and indoor stocking density on behaviour and stress in broilers in the subhumid tropics. *Animals, 9*(12), 1016. https://doi.org/10.3390/ani9121016.

Sánchez-Casanova, R. E., Sarmiento-Franco, L., & Phillips, C. J. C. (2022). The effects of providing outdoor access to broilers in the tropics on their behaviour and stress responses. *Animals, 12*(15), 1917. https://doi.org/10.3390/ani12151917.

Selle, P. H., Cadogan, D. J., Li, X., & Bryden, W. L. (2010). Implications of sorghum in broiler chicken nutrition. *Animal Feed Science and Technology, 156*(3-4), 57–74. https://doi.org/10.1016/j.anifeedsci.2010.01.004.

Shilpa, V. S., Shams, R., Dash, K. K., Pandey, V. K., Dar, A. H., Ayaz Mukarram, S., Harsányi, E., & Kovács, B. (2023). Phytochemical properties, extraction, and pharmacological benefits of naringin: A review. *Molecules, 28*(15), 5623. https://doi.org/10.3390/molecules28155623.

Taylor, R. C., Omed, H., & Edwards-Jones, G. (2014). The greenhouse emissions footprint of free-range eggs. *Poultry Science, 93*(1), 231–237. https://doi.org/10.3382/ps.2013-03489.

Teixeira, A.de O., Resende, V. A.de, Moreira, L. M., Corassa, A., Brighenti, C. R. G., Gaya, L.de G., Teixeira Valente Júnior, D., Teixeira, H. Z., & Reis, R.de S. (2021). Effect of crushed sugarcane on performance, bone characteristics and intestinal morphology of free-range chicken lines. *Semina: Ciências Agrárias, 42*(5), 2893–2906. https://doi.org/10.5433/1679-0359.2021v42n5p2893.

Wallinga, D., Smit, L. A. M., Davis, M. F., Casey, J. A., & Nachman, K. E. (2022). A review of the effectiveness of current US policies on antimicrobial use in meat and poultry production. *Current Environmental Health Reports, 9*(2), 339–354. https://doi.org/10.1007/s40572-022-00351-x.

Wang, M., Toghyani, M., Macelline, S. P., Lemme, A., Holmes, A. J., Selle, P. H., & Liu, S. Y. (2025). Sorghum surpasses wheat as a feed grain for broiler chickens following dietary crude protein reductions. *Journal of Animal Science and Biotechnology, 16*(1), 20. https://doi.org/10.1186/s40104-024-01147-9.

Yunes, M. C., Von Keyserlingk, M. A. G., & Hötzel, M. J. (2017). Brazilian citizens' opinions and attitudes about farm animal production systems. *Animals, 7*(10), 75. https://doi.org/10.3390/ani7100075.

Zagarin, A. Y., Shitikova, A. V., Selionova, M. I., Akchurin, S. V., & Gladkikh, M. Y. (2025). Comparative evaluation of the effectiveness of using quinoa grain *(Chenopodium quinoa* Willd.) with high and low saponin content in broiler chicken feeding. *Animals, 15*(17), 2574. https://doi.org/10.3390/ani15172574.

Zamani, O., Bittmann, T., & Ortega, D. L. (2024). The effect of avian influenza outbreaks on retail price premiums in the United States poultry market. *Poultry Science, 103*(10), 104102. https://doi.org/10.1016/j.psj.2024.104102.

Zhang, J. Y., Bormon, C. C., Ataher, M. S., Hossain, M., Dutta, A., & Mahfuz, S. (2025). Insects as sustainable source of feed for poultry. *JSFA reports, 5*(9), 370–383. https://doi.org/10.1002/jsf2.70015.

Chapter 15

Life cycle assessment

Giacomo Falcone[1] and Jacopo Bacenetti[2]
[1]*Department of Agricultural Sciences, Università degli Studi Mediterranea di Reggio Calabria, Reggio Calabria, Italy,* [2]*Department of Environmental and Policy Science, Università degli Studi di Milano, Milan, Italy*

Introduction

Global poultry meat consumption has increased dramatically over the past few decades, with a projected 21% increase in consumption projected by 2034 (OECD/FAO, 2025). Consumers are reducing consumption of red meat in favor of more poultry, with almost 22 kg per person consumed annually in high-income countries across 2022–24 (OECD/FAO, 2025). Aligning with these trends, in 2023 global hen egg production reached 91 million tonnes (*FAO Food Outlook: Global output of key food commodity crops on course for new records*, 2025). Poultry products play a major role in human nutrition, especially in developing countries, due to their affordability, wide availability, freedom from religious restrictions, and high nutritional value. From an environmental perspective, poultry products are considered among the most efficient among major livestock production chains, particularly concerning carbon footprint and resource depletion (e.g., land and energy use). Nevertheless, they fall into the medium-high impact range of products when considering a wider basket of edible fresh foods. Therefore continuous improvement of the sustainability of poultry supply chains is essential.

The environmental implications of poultry production are multifaceted, involving various stages of the agricultural process—from feed production and poultry rearing to processing, transportation, and waste management. As the demand for poultry products continues to rise, it is imperative to assess and mitigate their ecological footprint to promote sustainability within the sector. Among the tools available, life cycle assessment (LCA) has emerged as a key methodology, providing a systematic framework to evaluate environmental impacts across the entire life cycle of poultry products. LCA is a holistic and standardized approach (based on ISO 14040 and 14044 standards) that quantifies environmental impacts associated with all stages of a product's life—from raw material extraction to disposal. Within LCA two different approaches can be used to evaluate the environmental impacts of products,

Integrating Welfare and Sustainability in Poultry Systems. DOI: https://doi.org/10.1016/B978-0-443-40594-5.00016-X

systems, or services. Attributional LCA (aLCA) quantifies the environmental burdens associated with producing and using a product within a defined system boundary. It describes the current or average environmental profile of a product and typically uses average data. It answers: "What are the impacts of this product?" Consequential LCA (cLCA) estimates how total environmental impacts change as a consequence of a decision (e.g., increasing demand, introducing a policy, or switching technologies). It models market responses and substitutions using marginal data. It answers the question: "What are the environmental consequences of this change?"

LCA was originally developed for industrial processes, where the monitoring of energy and material flows is easier; however, in the past two decades, LCA has been increasingly applied to agri-food sectors as a well-established decision-making tool. The deployment of LCA in the poultry sector has been instrumental in identifying hotspots, comparing different rearing and management practices, and informing policymakers and industry stakeholders about potential pathways toward sustainability. Its comprehensive analysis enables a nuanced understanding that could not previously be achieved through single-impact assessments or compartmentalized studies.

LCA also finds application in Type III certification programs (UNI, 2010) for creating Environmental Product Declarations (EPD), which are increasingly adopted in the agri-food sector for transparency, marketing, and eco-labeling. The International EPD® System provides product category rules for fresh eggs (Birds' eggs in shell, fresh. Product category classification: UN CPC 023, 2011:15. Version 4.0, valid until 2026-05-18) and poultry meat (meat of poultry [fresh, frozen or chilled]; Product category classification: UN CPC 2112, 2114, 2117, 2010:13, Version 3.0, valid until: 2026-04-02.) However, there are currently no specific Product Environmental Footprint (PEF) Category Rules drawn up by the European Commission for these products. The Food and Agriculture Organization of the United Nations, through the Livestock Environmental Assessment and Performance Partnership (LEAP), has published a guideline document to harmonize the international methodology for the environmental assessment of greenhouse gas (GHG) emissions and energy use in poultry supply chains (FAO, 2016). In the context of poultry, LCA evaluates impacts related to GHG emissions, land use and transformation, water consumption, eutrophication, acidification, resource depletion, and other environmental categories.

Although LCA reviews have been conducted in specific livestock sectors such as milk, pig, and beef production systems (Baldini et al., 2017; de Vries et al., 2015; Goldáraz-Salamero et al., 2025), as well as poultry (mainly in conference proceedings), most of these reviews are quite outdated. Furthermore, previous reviews often analyzed a broader concept of "environmental impact" or "sustainability." For the poultry sector, the most recent review on LCA studies was published by Costantini et al. (2021), with the scope of literature analysis up until 2020. Given the crucial importance of the poultry sector and its projected future growth, this chapter aims to review current knowledge on the environmental performance of poultry industries, focusing attention on the last 5 years. The main objectives of this review chapter are:

- To summarize the purpose and application of LCA in the poultry sector in the last 5 years.
- To discuss strengths, weaknesses, and limitations regarding the application of the LCA approach to the poultry sector.
- To systematically compare different LCA studies of poultry products.
- To identify the main environmental hotspots (i.e., the subprocesses as well as the main inputs or emission sources) primarily responsible for the environmental effects of the poultry supply chain.
- Based on the results of the literature review, identify possible mitigation strategies for the environmental impacts of poultry production processes and discuss their effectiveness.

Literature review methodology

This systematic review focused on LCA studies published in the last 5 years within the poultry industry using a similar approach to Costantini et al. (2021), who reviewed studies prior to 2021. The STARR-LCA framework, as outlined by Zumsteg, Cooper, & Noon (2012), was followed, where the methodology is structured into nine key stages.

Each identified study was analyzed for its LCA application, paying particular attention to the different steps of LCA as well as the main methodological choices carried out by the analysts. More specifically, the following aspects were considered in the LCA steps:

- selection of the functional unit (FU),
- definition of the system boundary,
- solving the multifunctionality (allocation or system expansion),
- inventory data, paying particular attention to the collection of primary data (directly collected and/or measured) and to the estimation of secondary data (coming from databases and literature, estimated by specific models, etc.),
- selection of the characterization methods,
- methodological choices such as modeling of specific aspects of the poultry supply chain (e.g., modeling of emissions from animal rearing, manure management, crop cultivation, etc.), selection of allocation criteria, scenario analysis, and sensitivity and uncertainty analyses.

STARR-LCA framework stages:

1. Title, keywords, and abstract: The review emphasizes the importance of clear and informative titles and keywords to enhance discoverability. Abstracts should incorporate essential elements such as context, aims, data sources, eligibility criteria, and other domain-specific details.

2. Justification for the review: The rationale must be grounded in existing literature, demonstrating how the review contributes to broader interdisciplinary understanding.
3. Research questions and objectives: The review formulates precise and answerable questions using structured formats such as PIFT, which define the product category, impact types, contributing flows or sectors, and the LCA type/approach used.
4. Protocol description: A comprehensive protocol outlines the review process, including search strategies, inclusion and exclusion criteria, data collection methods, and any planned metaanalytical techniques.
5. Study characteristics and findings: Each selected study is examined for its strengths, limitations, and key insights. Visual tools such as tables and charts are used to support the analysis.
6. Bias evaluation: The review assesses potential biases both within individual studies and across the dataset, including publication bias and discipline-specific sources of distortion.
7. Evidence synthesis: Both qualitative and quantitative synthesis methods are applied. Any modifications in metaanalyzes are explained, along with their influence on variability.
8. Review limitations: Constraints related to time, geography, and technology are discussed. The authors also reflect on the relevance of the findings and propose directions for future research.
9. Summary and conclusions: The review concludes by summarizing the main evidence and interpreting it within the context of existing knowledge, highlighting implications for practice and policy.

The literature search was conducted using Scopus and Web of Science, targeting publications from 2021 to 2025. Advanced search queries were tailored to each database, focusing on titles, abstracts, and keywords (see Table 15.1). Irrelevant topics falling under the same "LCA" acronym, such as "latent class analysis" and "longitudinal chromatic aberration," were excluded.

After removing duplicates, 292 records were identified. A two-step screening process—first by title and then by content—led to the selection of 281 relevant studies. Each article was downloaded and assigned a unique ID for cataloging.

Studies were excluded if they focused on by-product valorization (e.g., chemical extraction from poultry waste) or did not apply LCA methodology. The final selection was based on a detailed qualitative and quantitative assessment of each study, considering factors such as:

- product type,
- study design and methodology,
- goals and scope,
- system boundaries,
- data sources and quality,
- impact assessment methods and categories,

- emission models and sensitivity analyzes, and
- key findings.

TABLE 15.1 Parameters used in different search engines.

Search engines	Search fields: TITLE-ABS-KEY
SCOPUS	TITLE-ABS-KEY ((SLCA OR "Social Life Cycle Assessment" OR LCC OR "Life Cycle Costing" OR LCA OR "Life Cycle Assessment" AND eggs) OR (SLCA OR "Social Life Cycle Assessment" OR LCC OR "Life Cycle Costing" OR LCA OR "Life Cycle Assessment" AND poultry) OR (SLCA OR "Social Life Cycle Assessment" OR LCC OR "Life Cycle Costing" OR LCA OR "Life Cycle Assessment" AND chicken) OR (SLCA OR "Social Life Cycle Assessment" OR LCC OR "Life Cycle Costing" OR LCA OR "Life Cycle Assessment" AND broiler)) AND PUBYEAR > 2020 AND PUBYEAR < 2026 AND NOT "latent class analysis" AND NOT "longitudinal chromatic aberration" AND (EXCLUDE (DOCTYPE, "er"))
Web of Science	TS = ((SLCA OR "Social Life Cycle Assessment" OR LCC OR "Life Cycle Costing" OR LCA OR "Life Cycle Assessment") AND (eggs OR poultry OR chicken OR broiler)) AND PY = (2021–2025) NOT TS = ("latent class analysis" OR "longitudinal chromatic aberration") NOT DT = ("Early Access Online")

These 55 studies formed the basis for the literature analysis, with the main parameters of products analyzed, FUs, system boundaries, and solving of multifunctionality, impact assessment methods, and data sources as detailed in Table 15.2–15.4.

Results of the literature review

Goal and scope

As outlined in the previous section, a total of 55 studies were thoroughly analyzed. The literature review shows a clear distinction between studies focusing on poultry meat production, egg production, and both systems. Most studies concerned poultry meat production (32 studies), with a particular focus on broiler chickens, which have been investigated in terms of energy and environmental sustainability in different geographical contexts such as China (Cheng et al., 2023), Brazil (Araújo dos Santos et al., 2023), Finland (Usva et al., 2023), and South Korea (Kang et al., 2025), and under different management strategies, such as the use of low-protein diets (Azmi et al., 2023), the integration

TABLE 15.2 Main methodological aspects: product and functional unit (FU).

Authors and year	Journal	Products analyzed	Functional/declared unit
Alves et al. (2023)	Journal of Cleaner Production	Chicken meat; scenarios with viscera meal	1 kg of chicken meat with 0.22 kg of protein
Araújo dos Santos et al. (2023)	Journal of Environmental Management	Chicken meat, poultry litter, biogas, and biofertilizers	1 kg of chicken meat ready for packaging
Arrigoni et al. (2023)	Science of the Total Environment	Beef, pork, processed pork, and poultry consumed in Italy	DU1: 100 g of product ingested (for comparison); DU2: daily meat consumption by the omnivorous Italian population for 1 year (2018)
Arulnathan et al. (2022)	Journal of Environmental Management	Egg production	1 tonne of eggs
Arulnathan et al. (2025)	Journal of Cleaner Production	Egg production	1 tonne of eggs
Azmi et al. (2023)	Tropical Animal Science Journal	Fried chicken meat consumed	1 kg of fried chicken consumed
Beal et al. (2023)	Poultry Science	Broiler production (chicken)	1 kg live weight
Behboodi et al. (2025)	Journal of Agricultural Engineering	Ostrich meat and eggs	1 tonne of ostrich meat and eggs
Behboodi et al. (2023)	Environmental Resources Research	Ostrich meat and egg production	1 tonne of ostrich meat and egg production
Berton et al. (2025)	Animal	Animal-based dietary proteins (dairy cattle, beef cattle, pigs, poultry)	1 kg of crude protein in animal feed
Bui et al. (2023)	Animals	Pig and broiler chicken productions	1 kg of live animal feed with (vs without) CEFA; one 25 kg bag of CEFA

(Continued)

TABLE 15.2 (Continued)

Authors and year	Journal	Products analyzed	Functional/ declared unit
Chen and Qi (2025)	Environmental Technology and Innovation	Animal-derived foods (pork, chicken, beef, mutton, eggs, milk)	1 kg carcass weight (CW) (for raw meat)
Cheng et al. (2023)	Journal of Cleaner Production	Chicken meat production	1 broiler chicken (2.5 kg live weight, 1.85 kg bone-in chicken meat) (referred to as 1 p)
Ghinea et al. (2024)	2024 12th E-Health and Bioengineering Conference (EHB 2024)	Chicken meat (slaughtering and preparation)	1 kg of chicken meat (for slaughterhouse); 1 kg of chicken meat products (for meat processing plant)
Giusti et al. (2022)	Sustainable Production and Consumption	Brazilian chilled chicken meat	1 kg of chilled Cobb breed chicken
Guðjónsdóttir et al. (2025)	Poultry Science	Broiler chicken meat and hen eggs	1 kg of broiler chicken (CW) (FU1); 1 kg of eggs (FU2). For scenario analysis: 100 g of edible protein
Guillaume et al. (2022)	Agriculture	Egg production	1 kg of shelled eggs
Harasztiné Hargitai et al. (2024)	Journal of Water Process Engineering	Poultry processing plant; wastewater reuse options	1 kg of chicken-based product or 1 m^3 of treated wastewater
Heines et al. (2022)	Resources	Egg and broiler chicken production, with and without black soldier fly larvae in feed	FU1: 1 tonne of feed protein consumed by birds; FU2: 1 kg of protein in end products for human consumption

(Continued)

TABLE 15.2 (Continued)

Authors and year	Journal	Products analyzed	Functional/ declared unit
Kang et al. (2025)	Food and Bioproducts Processing	Chicken meat production	1 kg of chicken meat
Kheiralipour (2021)	Interdisciplinary Applications of the Life Cycle Assessment (LCA) Tool	Broiler, turkey, and ostrich production systems	1 kg of product
Kolawole et al. (2025)	Environmental Impact Assessment Review	Broiler production system (with and without mycotoxin-contaminated feeds)	1 kg of live chicken meat at slaughter weight
Kyttä et al. (2023)	International Journal of LCA	Protein-rich foods (meatballs made from beef, pork, chicken, trout, perch, chickpeas, soy mince, shredded oats)	Specific nutritional indices per product group (nFU) calculated per 100 g of cooked product
Leinonen (2025)	Animal	Broiler genotypes (fast- and slow-growing)	1 kg of eviscerated chicken carcass
Li et al. (2025)	Animals	Laying hens	One box of eggs (18 kg)
Li et al. (2024)	Applied Geography	Household meat consumption (beef, mutton, pork, poultry, seafood)	1 kg protein; 1 kg weight
Li et al. (2021)	Journal of Cleaner Production	Egg production	1 tonne of eggs
Lu et al. (2024)	Journal of Cleaner Production	Beef, pork, chicken, eggs, milk	1 kg CW for meat, 1 kg for eggs, 1 kg FPCM (milk corrected for fat and protein)
Maciel et al. (2024)	Animals	Battery cage egg production systems and associated products	1 tonne of eggs produced; in addition, various UF for the different stages

(Continued)

TABLE 15.2 (Continued)

Authors and year	Journal	Products analyzed	Functional/declared unit
			(feed production: 1 tonne; pullet production: 1000 pullets; egg grading: 1 tonne; liquid egg production: 1 tonne)
Maciel et al. (2023)	Animals	Egg production	1 tonne of eggs per unit of time
Mata et al. (2023)	Environmental Science and Engineering	Chicken slaughter and preparation process	1 kg of chicken live weight at the company gate
Matton et al. (2025)	Animals	Broiler chickens	1 tonne of feed
Mesquita and Carvalho (2024)	International Journal of LCA	Common vegetarian and nonvegetarian dishes	1 kg/L of food and per 100 g of protein (for protein-rich foods)
Michalke et al. (2023)	Journal of Cleaner Production	22 food products (plant- and animal-based)	1 kg of product (for 22 different products)
Mitrovic et al. (2022)	Foods	Table eggs	1 kg of eggs
Mostert et al. (2022)	Poultry Science	Broiler chicken	1 kg broiler live weight
Nguyen et al. (2025)	Animals	Broiler chickens and poultry manure waste	1 kg of broiler chicken
Ogino et al. (2021)	Agriculture	Broiler chicken production systems	1 kg of live weight of broiler chicken
Payandeh et al. (2025)	Science of the Total Environment	Egg production	1 kg of eggs
Pelaracci et al. (2024)	Cleaner Environmental Systems	Egg production (organic and conventional systems)	1 kg of eggs

(Continued)

TABLE 15.2 (Continued)

Authors and year	Journal	Products analyzed	Functional/ declared unit
Portillo-Chávez & Vargas-Leitón (2025)	Agronomía Mesoamericana	Broiler chicken meat production in conventional systems	1 metric tonne (tM CP) of chicken meat at the slaughterhouse gate
Quevedo-Cascante et al. (2024)	Journal of Environmental Management	Eggs and apples	Per kg egg or kg apple; land occupation in m2a/kg class I product (apple or egg)
Sharifi et al. (2024)	Environment	Turkey meat production	1 tonne of live turkeys (farm stage); 1 tonne of processed turkey (slaughterhouse)
Sinke et al. (2023)	International Journal of LCA	Cultivated meat and conventional meats (beef, pork, chicken)	1 kg of meat (1.1 kg of final product for farmed meat)
Tukiainen et al. (2025)	International Journal of LCA	Meatballs (beef, chicken, pork, trout, perch, chickpeas, soy flour); full meals (meatballs with potatoes and salad)	Protein content (g); protein content corrected with DIAAS; NR-FIprot nutritional index (with and without DIAAS)
Turner et al. (2022a)	Journal of Cleaner Production	Egg production in various housing systems	1 tonne of eggs at the farm gate
Turner et al. (2022b)	Resources	Canadian egg production (conventional, enriched, cage-free, organic housing systems)	1 tonne of eggs
Turner et al. (2023)	Sustainable Production and Consumption	Laying hen systems (eggs)	1 tonne of eggs produced at farm gate
Usva et al. (2023)	Journal of Cleaner Production	Finnish broiler chicken meat	1 kg of CW

(Continued)

TABLE 15.2 (Continued)

Authors and year	Journal	Products analyzed	Functional/ declared unit
Vanbaelinghem et al. (2025)	Journal of Environmental Management	Eggs (and HVAC systems for poultry houses)	1 tonne of eggs; 1 kWh of heating and cooling energy, and electrical energy for ventilation
Vauterin et al. (2021)	Journal of Cleaner Production	Chicken meat; Insect proteins (as feed or direct food)	1 kg of protein
Wicaksono and Kaswara (2021)	Proceedings of the Int. Conf. on Industrial Engineering and Operations Management	Broiler chickens and poultry manure waste	1 kg of live chicken's weight
Yalçuk et al. (2025)	Journal of Food Process Engineering	Chicken meat production	1 kg of packaged chicken meat
Yayli and Kilic (2023)	International Journal of Environmental Science and Technology	Broiler chicken meat	1 kg of chicken meat
Yu et al. (2025)	Journal of Environmental Management	Beef, mutton, chicken, milk, pork, egg	1 tonne of animal products

of additives (citrus extracts) (Bui et al., 2023), or the adoption of circular economy approaches (Ghinea et al., 2024; Wicaksono & Kaswara, 2021).

A smaller number of publications (15 studies) focused on egg production, with particularly relevant contributions to the comparative assessment of different farming systems—conventional cages, enriched cages, barn systems, free-range, and organic systems—as well as the optimization of environmental performance through innovative technologies such as net-zero-energy barns (Li et al., 2021) or methodologies based on frontier efficiency analysis (Payandeh et al., 2025). Notable examples include Li et al. (2021), Pelaracci et al. (2024), and Payandeh et al. (2025). Finally, a group of studies simultaneously analyzed meat and egg production, offering integrated perspectives on the overall environmental impact of supply chains and protein

TABLE 15.3 Main methodological aspects: system boundaries and allocation.

Authors and year	System boundaries	Allocation
Alves et al. (2023)	From production to marketing, including all stages.	Substitution approach for waste valorization.
Araújo dos Santos et al. (2023)	Feed production, rearing, slaughter, and processing, up to packaging	The study evaluated and applied both economic allocation and mass allocation to distribute the environmental impacts due to the multifunctionality of the product system. It is noted that ISO 14040 recommends expanding the system.
Arrigoni et al. (2023)	From the production of materials and energy used in farming, through the final distribution of the packaged product, to final consumption.	Economic allocation, with an exception of dairy farming, where biophysical allocation is used for milk and meat.
Arulnathan et al. (2022)	Cradle-to-farm gate for Canadian conventional egg production. Excludes machinery, buildings, and infrastructure.	Various allocation methods are proposed and discussed, mainly allocation based on feed energy utilization (metabolizable energy). The subdivision and expansion of the system are also discussed as approaches to address multifunctionality or avoid allocation.
Arulnathan et al. (2025)	Cradle-to-farm gate, including pullets, feed, energy inputs, water, manure management, and transportation stages.	Allocation based on feed energy use in hens.
Azmi et al. (2023)	From cradle-to-grave, including feed production, rearing, slaughter, distribution, and consumer use.	Allocation approach to the main product. Allocation details for coproducts are not specified.
Beal et al. (2023)	Control volume, including upstream emissions, and on-site diesel and propane combustion	Economic allocation for all indicators.
Behboodi et al. (2025)	Life cycle assessment from raw material extraction to final disposal.	Allocation approach to the main product. Allocation details for coproducts are not specified.

(Continued)

TABLE 15.3 (Continued)

Authors and year	System boundaries	Allocation
Behboodi et al. (2023)	Environmental effects of the system under study.	Allocation approach to the main product. Allocation details for coproducts are not specified.
Berton et al. (2025)	From "cradle-to-farm-gate." Includes animal and manure management, feed production, and use of agricultural materials. Excludes land-use change.	Allocation approach to the main product. Allocation details for coproducts are not specified.
Bui et al. (2023)	Production of CEFA and its distribution to Thailand (main customer). For animal production, focus on the effects of the additive.	Allocation approaches are used for specific ingredients (e.g., a 40% juice/peel ratio for orange peel).
Chen and Qi (2025)	From "cradle-to-grave," divided into four stages: feed production, animal emissions, energy consumption in livestock farming, and processing of livestock products.	Allocation approach to the main product. Allocation details for coproducts are not specified.
Cheng et al. (2023)	Cradle-to-gate (maize feed production, broiler chicken farming, broiler chicken slaughter, and cutting). Biogenic CO_2 omitted.	Allocation approach to the main product. Allocation details for coproducts are not specified.
Ghinea et al. (2024)	"Gate-to-gate" approach. Slaughter and preparation of chicken meat products for human consumption.	Allocation approach to the main product. Allocation details for coproducts are not specified.
Giusti et al. (2022)	Cradle-to-retail, including fattening poultry stages and industrial slaughter phase. Wastewater treatment at slaughterhouse.	Allocation approach to the main product. Allocation details for coproducts are not specified.
Guðjónsdóttir et al. (2025)	From cradle-to-slaughterhouse gate for broiler chickens; from cradle-to-farm gate for egg production.	Allocation approach (based on protein content for sensitivity analysis).

(Continued)

TABLE 15.3 (Continued)

Authors and year	System boundaries	Allocation
Guillaume et al. (2022)	From cradle-to-gate (from hatchery to egg laying). The breeding phase, packaging, distribution, consumption, and maintenance of buildings/machinery are excluded.	Allocations were made based on mass and economics, considering eggs as the main product and the impact of end-of-life hens as insignificant compared to that of eggs.
Harasztiné Hargitai et al. (2024)	From water extraction to treatment and recycling, including product lines and water lines.	Allocation approach to the main product. Allocation details for coproducts are not specified.
Heines et al. (2022)	From the farm gate to the slaughterhouse gate (or egg production), with extensions for waste treatment.	Allocation was made based on prices for small quantities of goods.
Kang et al. (2025)	From feed production to chicken production, slaughter/meat processing, and transport for distribution (cradle-to-grave concept).	Not applicable.
Kheiralipour (2021)	Cradle-to-grave.	Not applicable.
Kolawole et al. (2025)	Typical conventional broiler farm in the Netherlands.	Allocation approach to the main product. Allocation details for coproducts are not specified.
Kyttä et al. (2023)	From cradle-to-plate, including primary production, processing, packaging, transport, and home cooking.	Allocation approach to the main product. Allocation details for coproducts are not specified.
Leinonen (2025)	"Gate-to-gate" approach. Excludes breeding generations.	Allocation approach to the main product. Allocation details for coproducts are not specified.
Li et al. (2025)	The footprint of the egg production chain, primarily feed production and manure management. LCC system boundary of layer production.	Allocation approach to the main product. Allocation details for coproducts are not specified.
Li et al. (2024)	From "cradle-to-grave", including cooking and food waste disposal.	Allocation approach to the main product. Allocation details for coproducts are not specified.

(Continued)

TABLE 15.3 (Continued)

Authors and year	System boundaries	Allocation
Li et al. (2021)	Cradle-to-farm gate production of one tonne of eggs. Barn models include material production, construction, use, and end-of-life stages.	Allocation based on gross energy was used in the baseline model of the Canadian egg supply chain (Pelletier, 2017).
Lu et al. (2024)	From cradle-to-farm gate.	A mass allocation approach was chosen to address the relative impact of the feed and animal product footprints.
Maciel et al. (2024)	From cradle-to-farm gate.	Economic allocation, considering eggs as the main product (94% allocation factor), organic compost (5%), and end-of-life hens (1%).
Maciel et al. (2023)	Cradle-to-gate. Includes feed production and manure management. Transports considered. Excludes infrastructure production/maintenance.	The study discusses allocation to address multifunctionality issues, referring to ISO 14040 and 14044 standards, but does not apply a specific method in this project context.
Mata et al. (2023)	"Gate-to-gate" approach. Focused on the slaughter and preparation of chicken meat.	Allocation approach to the main product. Allocation details for coproducts are not specified.
Matton et al. (2025)	Feed production.	It is based on Product Environmental Footprint (PEF) models.
Mesquita and Carvalho (2024)	From preproduction (input) to retail transport, excluding the postretail/consumer use stage.	Allocation approach to the main product. Allocation details for coproducts are not specified.
Michalke et al. (2023)	From cradle-to-farm gate in Germany.	For multiple output systems (e.g., cereal production generating straw), environmental impacts are allocated based on the economic value of the coproducts.

(Continued)

TABLE 15.3 (Continued)

Authors and year	System boundaries	Allocation
Mitrovic et al. (2022)	Entire egg supply chain, including farms, retail outlets, and households (cradle-to-consumer).	Allocation approach to the main product. Allocation details for coproducts are not specified.
Mostert et al. (2022)	Broiler chicken production chain, including breeding, hatchery, and chicken rearing stages.	Allocation approach to the main product. Allocation details for coproducts are not specified.
Nguyen et al. (2025)	"Entire study" implying cradle-to-farm gate for broiler production, focused on dietary treatments.	Not applicable.
Ogino et al. (2021)	Includes feed production, feed transport, animal housing, and manure management. Feed production includes CAA manufacturing.	Allocation approach to the main product. Allocation details for coproducts are not specified.
Payandeh et al. (2025)	Cradle-to-gate covers "production farms' inputs and outputs."	Allocation approach to the main product. Allocation details for coproducts are not specified.
Pelaracci et al. (2024)	Entire life cycle, including crop cultivation for feed, feed production, rearing stage, and disposal/recycling.	For agricultural feed coproducts (e.g., wheat/straw, soybean meal/oil, sunflower meal/oil), economic allocation was applied.
Portillo-Chávez & Vargas-Leitón (2025)	Cradle-to-gate (from raw material extraction to the slaughterhouse gate).	Allocation based on physical criteria (mass) for processes with multiple products (e.g., maize, soya, wheat). Chicken manure is allocated entirely to the broiler chicken production cycle, as it is considered a residue with low economic value.
Quevedo-Cascante et al. (2024)	Cradle-to-retail (food production [cradle-to-farm gate] and postharvest [cradle-to-retail for Class I products]).	Two modeling approaches were used: Model 1 (M1) employed economic allocation, while Model 2 (M2) subdivided the system into two subsystems. Biophysical allocation and system expansion for manure are also discussed.

(Continued)

TABLE 15.3 (Continued)

Authors and year	System boundaries	Allocation
Sharifi et al. (2024)	From farm to packaged meat (excludes sending to buyer/seller). Includes farm sector (turkey breeding), slaughter, and packaging.	Allocation approach to the main product. Allocation details for coproducts are not specified.
Sinke et al. (2023)	Cradle-to-gate, excluding packaging. Includes upstream processes and transport.	Allocation based on economic principles.
Tukiainen et al. (2025)	Based on Kyttä et al. (2023a) ; (2023b) ("cradle-to-plate"), at product and meal level.	Allocation approach to the main product. Allocation details for coproducts are not specified.
Turner et al. (2022a)	Cradle-to-farm gate.	Allocation approach to the main product. Allocation details for coproducts are not specified.
Turner et al. (2022b)	Cradle-to-farm gate egg supply chain.	The baseline model of the Canadian egg supply chain (Pelletier, 2017), to which this study refers as an update, used gross energy-based allocation.
Turner et al. (2023)	Laying phase of the hen's life cycle. Cradle-to-farm gate (excludes pullet, hatching, and breeder facilities).	Allocation approach to the main product. Allocation details for coproducts are not specified.
Usva et al. (2023)	Includes feed production, broiler breeding, egg laying, incubation, fattening, manure storage, and slaughter process. Excludes preparation, retail, and consumption of meat products.	Allocation according to PEF guidelines.
Vanbaelinghem et al. (2025)	Cradle-to-farm gate for eggs. Life cycle stages of NGB, GSHP, and NGB-EAHE HVAC scenarios' components.	Allocation approach at the point of substitution for modeled background systems (from Ecoinvent v3.9.1).
Vauterin et al. (2021)	Mainly from cradle-to-slaughterhouse/processing gate (or farm gate for insects).	Different allocation methods (economic, mass, energy) depending on the system.
Wicaksono and Kaswara (2021)	Gate-to-gate (limited to the closest processing activities, in particular the internal activities	Allocation approach to the main product. Allocation

(Continued)

TABLE 15.3 (Continued)

Authors and year	System boundaries	Allocation
	of chicken farming, from DOC farms to chickens ready for slaughter).	details for coproducts are not specified.
Yalçuk et al. (2025)	Cradle-to-grave, including slaughter, cutting, packaging, storage, distribution, and consumption.	Not applicable.
Yayli and Kilic (2023)	Within farm boundaries (from "gate-to-gate" within the shed).	Allocation approach to the main product. Allocation details for coproducts are not specified.
Yu et al. (2025)	Life cycle analysis framework, including feed production, feed processing, livestock and poultry management, manure management, and transportation stages.	Allocation approach to the main product. Allocation details for coproducts are not specified.

TABLE 15.4 Main methodological aspects: impact assessment method and data sources.

Data sources	Authors and year	Impact assessment method
Ecoinvent, agri-footprint, life cycle assessment (LCA) food DK, industry data, ULSCI and ELCD	Alves et al. (2023)	ReCiPe 2016 (H)
Ecoinvent, agri-footprint	Araújo dos Santos et al. (2023)	ReCiPe 2008, midpoint (H) v. 1.12
Ecoinvent, agri-footprint, ELCD	Arrigoni et al. (2023)	ReCiPe
Ecoinvent	Arulnathan et al. (2022)	CML-IA baseline
Ecoinvent, agri-footprint Global Feed LCA Institute (GFLI) database	Arulnathan et al. (2025)	Impact World+

(Continued)

TABLE 15.4 (Continued)

Data sources	Authors and year	Impact assessment method
Ecoinvent, agri-footprint	Azmi et al. (2023)	CML-IA baseline V3.06
Ecoinvent	Beal et al. (2023)	Intergovernmental Panel on Climate Change (IPCC) Tier 2 guidelines, ReCiPe midpoint (H), TRACI.
Ecoinvent	Behboodi et al. (2025)	ReCiPe 2016.
No life cycle inventory (LCI) database explicitly mentioned	Behboodi et al. (2023)	ReCiPe 2016.
Ecoinvent, agri-footprint	Berton et al. (2025)	Global warming potential (GWP100).
Agribalyse, Ecoinvent	Bui et al. (2023)	Environmental footprint (version 2.0)
China products carbon footprint factors database (CPCD, 2024)	Chen and Qi (2025)	Carbon footprint.
Ecoinvent	Cheng et al. (2023)	SDU (Shandong University) model for LCIA analysis, which includes 16 categories.
Ecoinvent e GaBi	Ghinea et al. (2024)	CML 2001
Ecoinvent	Giusti et al. (2022)	Available WAter REmaining (AWARE); IMPACT World+
Ecoinvent	Guðjónsdóttir et al. (2025)	CML-IA baseline (3.08-EU25).
Agri-footprint, GaBi, ecoinvent database	Guillaume et al. (2022)	Environmental footprint 3.0.
Hungarian energy mix, literature data, database (Sphera), ecoinvent database	Harasztiné Hargitai et al. (2024)	ReCiPe 2016.
Ecoinvent, agri-footprint	Heines et al. (2022)	IMPACT 2002+
No LCI database explicitly mentioned	Kang et al. (2025)	GWP IPCC (1996 e 2013).
GFLI	Kheiralipour (2021)	Not specified.

(Continued)

TABLE 15.4 (Continued)

Data sources	Authors and year	Impact assessment method
LCA agri-footprint e	Kolawole et al. (2025)	Sustell™
National food composition database in Finland, agri-footprint	Kyttä et al. (2023a)	IPCC 2013
Ecoinvent	Leinonen (2025)	GWP100
No LCI database explicitly mentioned.	Li et al. (2025)	N-footprint
National statistics database e "China Carbon Emissions Trading Network"	Li et al. (2024)	PAS 2050
(Pelletier, 2017) (modello LCI baseline), openLCA 1.7 software platform (per LCIA), pedigree matrix di OpenLCA 1.7 (per qualità dati)	Li et al. (2021)	Guinee (2002) for abiotic depletion and land use; IPCC (2006) 100-year baseline model for climate change
Livestock Environmental Assessment and Performance Partnership guidelines (per bovini, suini e pollame)	Lu et al. (2024)	Assessment of water footprint (WF) and nitrogen footprint; CRITIC method for weighting in the overall assessment
Ecoinvent	Maciel et al. (2024)	TRACI 2.1, IPCC 2021, CML 2016
Ecoinvent	Maciel et al. (2023)	Not specified
Literature	Mata et al. (2023)	ReCiPe 2016
No LCI database explicitly mentioned.	Matton et al. (2025)	Opteinics™
Metaanalysis	Mesquita and Carvalho (2024)	Carbon footprint
Agri-footprint	Michalke et al. (2023)	ReCiPe 2016 (H) v1.1
Ecoinvent	Mitrovic et al. (2022)	CML
No LCI database explicitly mentioned.	Mostert et al. (2022)	IPCC (2006)
No LCI database explicitly mentioned	Nguyen et al. (2025)	Opteinics™

(Continued)

TABLE 15.4 (Continued)

Data sources	Authors and year	Impact assessment method
Ecoinvent	Ogino et al. (2021)	CML
Ecoinvent, agri-Footprint, US LCI	Payandeh et al. (2025)	ReCiPe midpoint (H) v1.13
Ecoinvent	Pelaracci et al. (2024)	ReCiPe 2016, IPCC, 2006
Ecoinvent, agri-footprint	Portillo-Chávez & Vargas-Leitón (2025)	ReCiPe 2016 v.1.1
IPCC, DEFRA, Ecoinvent	Quevedo-Cascante et al. (2024)	CML-baseline
Ecoinvent	Sharifi et al. (2024)	CML 2 baseline 2000 v2.03
Ecoinvent, agri-footprint	Sinke et al. (2023)	ReCiPe 2016 midpoint v1.1 e cumulative energy demand
US Department of Agriculture FoodData Central (USDA 2023; Kyttä et al., 2023a, b)	Tukiainen et al. (2025)	IPCC 2013
No LCI database explicitly mentioned	Turner et al. (2022a,b)	CML 2 baseline 2000
Agribalyse	Turner et al. (2022a,b)	CML 2 baseline 2000
No LCI database explicitly mentioned	Turner et al. (2023)	Type 1 life cycle impact assessment method for animal welfare
Agribalyse, agri-footprint, Ecoinvent	Usva et al. (2023)	AWARE
Ecoinvent	Vanbaelinghem et al. (2025)	ReCiPe 2016 midpoint (H), cumulative energy demand
Various DB	Vauterin et al. (2021)	GWP
IPCC	Wicaksono and Kaswara (2021)	IMPACT 2002+
MiLCA	Yalçuk et al. (2025)	CML

(Continued)

TABLE 15.4 (Continued)

Data sources	Authors and year	Impact assessment method
Ecoinvent	Yayli and Kilic (2023)	Carbon footprint, WF, acidification, eutrofication, ozone depletion, photochemical oxidation potential, human toxicity potential
FAOSTAT e National Bureau of Statistics.	Yu et al. (2025)	Not specified

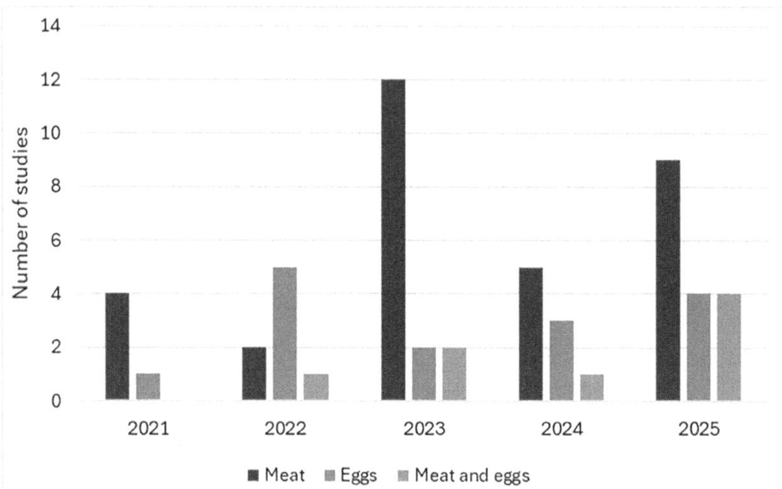

FIGURE 15.1 **Publication timeline of the reviewed life cycle assessment (LCA) studies.** Timeline of the reviewed LCA studies.

efficiency (Guðjónsdóttir et al., 2025) (see Fig. 15.1, including innovative approaches such as the use of insects in the diet (Heines et al., 2022) or the evaluation of integrated agroforestry systems (Quevedo-Cascante et al., 2024).

An analysis of the geographical distribution of studies in the literature showed a strong concentration in Europe, which accounted for 47.06% of the total. These data suggest significant scientific output in the poultry sector in Europe, with a variety of countries involved, notably Italy and Portugal (Heines et al., 2022; Mesquita & Carvalho, 2024; Pelaracci et al.,

2024). Asia followed with 22.35%, driven mainly by China and Iran, indicating growing interest and contributions from the Asian region to research (Behboodi et al., 2023; Cheng et al., 2023; Sharifi et al., 2024). North America contributed 14.12%, with Canada dominating over the United States of America and Mexico (Arulnathan et al., 2025; Beal et al., 2023; Portillo-Chávez & Vargas-Leitón, 2025; Turner et al., 2022b). South America was represented exclusively by Brazil (Araújo dos Santos et al., 2023; Maciel et al., 2023), while Africa (Barends-Jones & Pienaar, 2020) and Oceania showed a more marginal presence.

Standardization practices were commonly discussed across the reviewed literature. The ISO standards series (14040, 14044, 14045, and 14046) was explicitly cited in 43 studies, indicating widespread recognition of these frameworks (Maciel et al., 2023; Turner et al., 2023). In contrast, the LEAP guidelines (FAO, 2016) were referenced in only eight studies (approximately 14% of the total), reflecting a more limited adoption despite their relevance for livestock systems (Arulnathan et al., 2025; Kolawole et al., 2025).

LCA has emerged as the most widely adopted methodological framework for quantifying the environmental impacts associated with products and processes throughout their entire life cycle, from raw material extraction to end-of-life disposal. Its application within the poultry sector is particularly significant, given the industry's high efficiency in nutrient conversion and its substantial contribution to environmental burdens such as climate change, acidification, and eutrophication (Guðjónsdóttir et al., 2025). Researchers conducting LCA studies in the poultry sector frequently publish their work in a diverse range of scientific journals that span environmental science, agricultural engineering, and food systems. Among the most prominent are Journal of Cleaner Production, Journal of Environmental Management, and Poultry Science, which consistently feature studies on sustainability, resource efficiency, and environmental impact mitigation in poultry production. Other widely used journals include Animals, Science of the Total Environment, and Sustainable Production and Consumption, reflecting the interdisciplinary nature of LCA research. Specialized outlets such as Agriculture, Environmental Technology and Innovation, and Journal of Agricultural Engineering also serve as key platforms for methodological advancements and case studies. Additionally, proceedings from conferences highlight the growing interest in integrating LCA with technological innovation and system optimization in poultry farming.

Functional unit

A pivotal element in LCA studies is the definition of the FU, which serves as the reference output to which all inputs and environmental impacts are attributed. The choice of FU directly influences the comparability of systems and the interpretation of results. In poultry-related assessments, mass-based FUs—such as 1 kg of chicken meat at the slaughterhouse gate or 1 kg of eggs at the farm gate—are the most commonly employed. However, this approach may lead to underestimation of environmental impacts when critical stages such as industrial slaughtering and processing are simplified or excluded. For instance, studies such as Mata et al. (2023) and Ghinea et al. (2024) highlight the significant energy and water consumption associated with slaughtering, which is often overlooked in mass-based assessments.

In contrast, nutritional FUs have gained increasing attention for their ability to incorporate qualitative aspects of food products. These units consider not only the quantity but also the nutritional value of the output, including protein content, essential micronutrients, and dietary relevance. Research such as Kyttä et al. (2023a) and Tukiainen et al. (2025) exemplifies this shift. These studies propose indices that integrate protein, minerals, and vitamins, calibrated to national dietary models, thereby enabling more nuanced comparisons between animal- and plant-based products. For example, 1 kg of chicken meat containing 0.22 kg of protein may be evaluated alongside 1 kg of packaged eggs not only in terms of protein content but also with respect to micronutrients such as iron and vitamin B12, which are essential for a balanced diet (Alves et al., 2023). The integration of nutritional parameters into FU definitions represents a promising direction for future research, enabling more comprehensive evaluations of food systems that align with both environmental sustainability and human health objectives.

System boundaries

In LCA studies within the agri-food sector, the definition of system boundaries plays a critical role in determining the scope and accuracy of environmental impact evaluations. The reviewed literature reveals a diversity of boundary settings, ranging from "cradle-to-gate" and "gate-to-gate" to more comprehensive "cradle-to-grave" approaches. For instance, Azmi et al. (2023) adopt a cradle-to-grave perspective in assessing fried chicken meat in Indonesia, encompassing feed production, rearing, slaughter, distribution, and consumer use. Conversely, studies such as those by Mata et al. (2023) and Ghinea et al. (2024) focus on gate-to-gate boundaries, limiting the analysis to slaughter and preparation stages. More expansive boundaries, such as those employed by Arrigoni et al. (2023), include the entire chain from material and energy inputs in farming to final consumption, thereby capturing downstream impacts often overlooked in narrower scopes. The choice of system boundaries is frequently influenced by the study's objectives, data availability, and methodological standards, such as ISO 14040 and 14044, which recommend system expansion to address multifunctionality (Araújo dos Santos et al., 2023). Moreover, the inclusion or exclusion of infrastructure, machinery, and end-of-life stages varies significantly across studies, affecting comparability and the interpretation of results (Arulnathan et al., 2022; Beal et al., 2023). Thus establishing clear and consistent system boundaries is essential for ensuring methodological transparency and enabling meaningful comparisons across LCA studies in animal-based food production.

Allocation and system expansion

Multifunctionality in LCA arises when a process or system provides more than one function (i.e., eggs and meat). This creates a problem because the environmental burdens of that process must be assigned among the different outputs. One solution is allocation, in which impacts are partitioned among coproducts according to a chosen rule. Allocation can be based on physical relationships (e.g., mass or energy content) or economic value. However, results can vary significantly depending on the allocation criterion used. Another solution, suggested by the ISO standards 14040 and 14044, is system expansion, which avoids allocation by expanding the system boundary. In detail, under system expansion, the additional functions of coproducts are accounted for by subtracting the impacts of alternative products that would otherwise be produced (substitution approach).

Allocation methods—whether economic, mass-based, or energy-based—play a critical role in distributing environmental burdens among coproducts. For instance, Arulnathan et al. (2022) advocate for allocation based on metabolizable energy (ME) utilization, while Arrigoni et al. (2023) employ economic allocation to reflect market values. These choices significantly affect the outcomes and comparability of LCA studies.

One of the most critical methodological aspects in LCA of poultry systems concerns the management of manure and fertilizer. Poultry manure is typically treated as a by-product with negligible economic value, yet its environmental relevance is substantial due to its potential to substitute for synthetic mineral fertilizers. The most commonly adopted approach is system expansion, which attributes an environmental credit proportional to the nutrient content of the manure that is effectively utilizable in agricultural soils. This principle is exemplified in Arulnathan et al. (2025), where manure is credited for its nitrogen, phosphorus, and potassium content, reflecting its capacity to replace equivalent quantities of synthetic fertilizers. In contrast, the study by Guðjónsdóttir et al. (2025) illustrates the consequences of omitting leaching estimates in manure management. The absence of such data led to notably low results for acidification and eutrophication, suggesting that simplified assumptions may result in significant underestimation of environmental impacts (Guðjónsdóttir et al., 2025). GHG emissions from manure, particularly methane (CH_4) and nitrous oxide (N_2O), are generally estimated using Tier 2 methodologies from the Intergovernmental Panel on Climate Change (IPCC), while ammonia (NH_3) emissions are calculated using specific emission factors. It has been consistently observed that marine and freshwater eutrophication, as well as terrestrial acidification, are highly sensitive to manure management practices, both during storage and field application. However, variability in estimates remains considerable across studies.

Innovative manure management strategies, such as the integration of anaerobic digesters for biogas and biofertilizer production, have demonstrated substantial reductions in GHG emissions, fossil fuel consumption, and particulate matter formation. Nevertheless, these benefits are often accompanied by increased water demand. Araújo dos Santos et al. (2023) provide empirical evidence of such trade-offs, highlighting the complexity of balancing environmental gains across multiple impact categories. In China, the implementation of a national policy aiming to recycle 75% of poultry manure by 2020 has catalyzed the adoption of such technologies, underscoring the role of regulatory frameworks in enhancing the environmental performance of the poultry sector (Lu et al., 2024).

Another methodological challenge arises in the treatment of coproducts within multifunctional systems. In broiler meat production, carcass weight (CW) typically accounts for the vast majority of outputs, ranging from 98.5% to 99.2%. However, in egg production systems, meat from spent hens must also be considered. The allocation of environmental burdens among these outputs can be performed using various criteria. Economic allocation is the most prevalent method due to its applicability across complex systems, although the contribution of spent hens is generally minimal, often around 1%. Maciel et al. (2024) apply this approach, assigning 94% of impacts to egg production, 5% to compost, and 1% to spent hens.

Despite its widespread use, economic allocation has been critiqued for its limited reflection of biological processes. The LEAP guidelines advocate biophysical methods based on energy requirements for growth and egg production, which better capture the underlying metabolic functions. Arulnathan et al. (2022) introduce an advanced allocation model based on ME, which more accurately represents the distribution of nutrients within the animal's physiology. This approach yielded significantly different results compared with conventional methods, reinforcing the importance of methodological choices in LCA (Arulnathan et al., 2022). System expansion has also been proposed as an alternative to allocation, particularly in scenarios where spent hen meat is considered a substitute for broiler meat. However, this assumption is contentious due to the substantial differences in physicochemical properties and nutritional value between the two products, which undermine their functional equivalence.

Allocation issues are equally relevant in feed formulation, as feed ingredients often originate from multifunctional processes such as soybean or wheat processing. For instance, physical allocation based on mass assigns approximately 80.5% of impacts to soybean meal and 19.5% to soybean oil, while wheat processing allocates 77% to flour, 14% to bran, and 9% to other derivatives. The variability among allocation criteria necessitates sensitivity analyzes, a practice that remains underutilized despite its methodological importance. Studies such as Arrigoni et al. (2023); Michalke et al. (2023), and Heines et al. (2022) emphasize the need for such analyses to ensure robustness and transparency in LCA outcomes. Equally important is the treatment of manure as an organic fertilizer. Composting and anaerobic digestion can substantially reduce emissions and resource consumption, but the substitution model for synthetic fertilizers must be explicitly defined to avoid over or underestimation. In reducing the environmental impacts of Brazilian chicken meat production using different waste recovery strategies, the use of biodigesters was shown to reduce nitrogen emissions by up to 60.6%, although the modeling of such processes remains complex (Araújo dos Santos et al., 2023). This complexity is exacerbated in organic farming systems, where manure application rates often exceed those of conventional systems, challenging the validity of standard modeling assumptions.

In sum, the methodological choices surrounding manure management, coproduct allocation, and feed formulation significantly influence the results of LCA studies in poultry systems. The integration of biologically grounded allocation models and the adoption of system expansion, where appropriate, can enhance the accuracy and relevance of environmental assessments, particularly when supported by comprehensive sensitivity analyses and policy-driven implementation strategies.

Life cycle inventory and impact assessment method

A fourth critical dimension in the methodological framework of LCA concerns the quality and structure of inventory data (life cycle inventory [LCI]). The robustness of an LCA model is intrinsically linked to the reliability of both foreground and background data. Foreground data are typically obtained through direct surveys, interviews, and on-site measurements. For instance, in the study by Behboodi et al. (2023), data were collected through direct interviews with ostrich farmers, providing context-specific insights into production practices. Similarly, Beal et al. (2023) relied on farm-level data from broiler fattening cycles, ensuring that the environmental impacts reflected actual operational conditions. In Vanbaelinghem et al. (2025), the performance of HVAC systems was assessed through direct measurements, enhancing the precision of energy consumption estimates. Arulnathan et al. (2025) exemplify a structured approach to foreground data collection, integrating farmer-reported metrics on productivity, feed use, and manure management while supplementing gaps with national averages as proxies.

Background data, on the other hand, are sourced from bibliographic references and standardized databases. Among these, Ecoinvent remains the most widely used globally, with versions ranging from 3.4 to 3.9.1. Agri-footprint, particularly versions 4.0 and 5.0, and the Global Feed LCA Institute database are also frequently employed. These databases provide comprehensive datasets on crop production, fuel transformation, and energy supply chains. In several studies, such as Arrigoni et al. (2023), background datasets were adapted to reflect local agricultural conditions, with agri-footprint data calibrated to the specificities of Italian poultry farms. The distinction between foreground and background data, along with a clear articulation of assumptions and system boundaries, is essential to ensure the transparency and reproducibility of LCA results.

Impact assessment methods in LCA serve as essential tools for quantifying the environmental consequences of product systems, and their selection significantly influences the interpretation of results. Among the reviewed studies, the ReCiPe method—particularly the 2016 version with a hierarchical (H) perspective—is the most frequently employed, reflecting its comprehensive coverage of midpoint and endpoint categories and its alignment with ISO standards (Alves et al., 2023; Giusti et al., 2022; Portillo-Chávez & Vargas-Leitón, 2025). Other studies utilize the CML-IA baseline, especially in contexts where acidification, eutrophication, and global warming potential (GWP) are primary concerns (Arulnathan et al., 2022; Azmi et al., 2023). The environmental footprint method, including versions 2.0 and 3.0, is also applied in studies focusing on European regulatory frameworks and detailed toxicity assessments (Bui et al., 2023; Guillaume et al., 2022). Additionally, specialized models such as Available WAter REmaining for water scarcity and Sustell™ for integrated agricultural emissions are used to address specific impact categories (Harasztiné Hargitai et al., 2024; Kolawole et al., 2025). The diversity in methodological choices underscores the importance of context-specific selection, as different methods vary in their characterization factors, regionalization capabilities, and compatibility with background databases. Consequently, the comparability of LCA results across studies is contingent upon the harmonization of impact assessment frameworks and the transparency of methodological reporting.

The presentation of results constitutes another methodological concern, particularly in relation to land-use change (LUC). Guidelines such as those from LEAP and the PEF initiative recommend that impacts associated with LUC be reported separately due to their inherent uncertainty. This is especially relevant in the context of soybean production in South America, where deforestation linked to agricultural expansion can significantly alter GHG emission profiles. Studies such as Mostert et al. (2022) underscore the importance of distinguishing emissions from LUC, particularly when feed ingredients such as soy are sourced from regions with high deforestation risk.

Uncertainty analysis

Uncertainty analysis is a vital aspect of LCA, particularly in agri-food systems, where methodological choices and data variability can significantly influence results. Quantitative techniques such as Monte Carlo simulations are widely used to propagate uncertainty, often with 1000 or more iterations to generate probabilistic distributions and confidence intervals for impact categories such as GHG emissions (Turner et al., 2022b). Sensitivity analysis is frequently applied to test the robustness of results against variations in input parameters, allocation methods, and background datasets. For example, studies have examined the effects of changing feed sources, energy demand, and material inputs, as well as the influence of allocation strategies in multifunctional systems (Guðjónsdóttir et al., 2025; Vanbaelinghem et al., 2025). Data quality is commonly assessed using the ecoinvent pedigree matrix, which evaluates reliability, completeness, and contextual relevance, though reliance on generic or national average data can introduce additional uncertainty (Turner et al., 2022b). Statistical methods such as ANOVA, t-tests, and metaanalytic models are employed to determine the significance and variability of results (Vanbaelinghem et al., 2025). Scenario analysis is particularly relevant in exante studies of emerging technologies, where data scarcity is addressed through stakeholder-informed best and worst-case scenarios (Sinke et al., 2023). Expert elicitation methods, including Delphi processes, are used to manage scenario uncertainty and indicator weighting (Mitrovic et al., 2022). To address data gaps, some studies adopt conservative estimates or exclude highly uncertain categories such as animal welfare or occupational health. Advanced modeling techniques, including artificial neural networks and neuro-fuzzy systems, are also employed to simulate complex systems and identify influential variables (Behboodi et al., 2025). Despite these efforts, uncertainty remains underreported in many studies, underscoring the need for more rigorous and standardized approaches in future LCA research.

Discussion

Goal and scope

The breadth of research utilizing LCA in the poultry sector has expanded rapidly in recent years. The majority of studies focus on broiler chickens (meat production) and laying hens (egg production), with some exploring other species/products or integrated systems. Common focal points include the agricultural phase—particularly feed production and crop management—and on-farm operations, alongside the evaluation of postfarm processes such as processing and transportation.

Most reviewed LCA studies conducted on poultry systems primarily aim to describe environmental performance at different levels of the production chain or to compare various systems (e.g., in terms of housing or feeding strategies or manure management). These studies are predominantly aLCA, focusing on the impacts related to the physical input and output flows of the system under investigation. In contrast, cLCA studies are notably scarce. This scarcity is attributed to several factors, including the additional inventory data required for system expansion in aLCA, and the inherent greater uncertainty in cLCA studies, as consequential system changes in response to input and output flow alterations often need to be largely assumed. The reviewed literature also shows that LCA has been utilized as a tool for retrospective analysis. For example, Pelletier et al. (2014), Putman et al. (2017), and Pelletier (2018), employed LCA to analyze the evolution of environmental efficiency in the United States and Canadian poultry supply chains over the past 50 years.

Beyond environmental assessment, the integration of other considerations in poultry LCA studies is limited. For example, less than 20% of the reviewed studies combined economic assessment and LCA. The assessment of social performance was included in approximately 9% of LCA studies. When considered, this social pillar of sustainability was evaluated using multicriteria decision analysis and/or included specific indicators related to animal welfare. The lower inclusion of welfare indicators in LCA studies is an avenue to be addressed moving forward. Inclusion of complementary frameworks such as the welfare footprint framework (Schuck-Paim et al., 2025) could facilitate a more holistic sustainability assessment.

Functional unit

According to ISO 14044, the FU should be measurable, should quantify the function provided by the systems, and must be consistent with the goal and scope of the study and aligned with the considered system boundaries. Different FUs were considered when LCA was applied to agro-food products: area-based FUs, mass-based or volume-based FUs, energy-based FUs, and nutrition-based FUs (Kyttä et al., 2023a). Despite the heterogeneity of FUs adopted across the revised studies, there is a predominant use of mass-based FUs, both for production at the farm gate and after subsequent processing at different stages of the supply chain. This finding agrees with the previous reviews about LCA (Costantini et al., 2021; Ilemobayo et al., 2024). Alternative approaches for the selection of the FU, such as the use of the economic value or the protein content of the products (e.g., Tukiainen et al. (2025)), even if previously adopted (e.g.,Prudêncio da Silva et al. (2014)), who used €1000 of product as FU for comparing broiler meat production in Brazil and in France, were not considered in the LCA studies of the last 5 years.

With regard to mass-based FU, one identified inconsistency is reporting results for poultry meat in terms of expected CW without including impacts from the slaughter phase, instead assuming fixed yield values for slaughter and relating them to impacts up to the farm gate. This approach may lead to an underestimation of the impacts linked to carcass production, particularly for impact categories affected by energy consumption and water use.

System boundaries

Most LCA studies in the poultry sector have focused attention on the production phase, consequently limiting the system boundaries and often excluding retail, household, and end-of-life phases. This focus is justified by the fact that the production phase, and particularly the agricultural phase, has been identified as the most impactful for animal products (Notarnicola et al., 2017).

Three main types of system boundaries were identified in the reviewed studies:

- Cradle-to-farm gate: adopted by 34% of the total studies.
- Cradle-to-slaughterhouse gate: adopted by 13% of studies in the meat sector, although the slaughter phase was never considered for spent hens.
- Cradle-to-other systems: encompassing various downstream processes such as further processing, packaging, distribution to retailers, and household consumption.

Unfortunately, even studies adopting the same system boundaries may not consider the same types of input and output flows. LUC inclusion is a critical aspect that can strongly influence results, especially for GWP. The lack of a shared consensus on how to consider and calculate LUC can create distortions in interpretations and comparisons. Ideally, an LCA study should clearly report whether LUC is included and specify its computation method. Macleod et al. (2013) suggest LUC is responsible for 18% of global GHG emissions from chicken supply chains, underscoring its environmental relevance. However, most of the reviewed LCA studies explicitly exclude it due to great uncertainty, while others do not mention it.

Regarding the common use of manure as an organic fertilizer, system expansion has been frequently practiced. This involves accounting for avoided mineral fertilizer production as an environmental credit for the poultry production system, a common practice in LCA studies that incorporate residual handling leading to environmental benefits. However, there is a need for more clarity and transparency in the literature regarding how such substitution is assumed to occur. Key parameters such as the nutrient value of poultry manure and the estimation of emissions during storage and spreading are often not reported, making the LCA study not repeatable. Besides this, the system boundaries are not always clearly described when poultry manure is used for the fertilization of the crop used to produce feedstock for poultry feed. This can result in double accounting of the benefits related to poultry manure valorization: one for reducing the fertilization of the crops and one for avoiding mineral fertilizer production and the generation of environmental benefits.

A wide range of approaches is also observed in the consideration of capital goods and infrastructure. Their inclusion (though not always explicit) in upstream processes, starting from raw material extraction (e.g., transport, fuels, electricity consumption, and agricultural machinery for feed production), is common across most studies, utilizing background data from LCI databases. Conversely, the virtual consumption and maintenance of capital goods related to the growing phase (e.g., animal housing infrastructure, warehouses, silos, and tanks) and processing phases (e.g., slaughterhouse infrastructure) have almost always been excluded. Cleaning materials were only sporadically included within system boundaries, while veterinary products were consistently excluded. Regarding veterinary products, when clearly indicated within the system boundary, the authors justify this exclusion based on the lack of inventory data, as well as the low amount consumed.

Solving multifunctionality

For poultry meat production, the primary issue related to multifunctionality involves the carcass and secondary slaughter products (such as inedible organs, head, feet, blood, and feathers). Studies addressing the slaughter phase are generally divided into two categories: (1) studies considering secondary slaughter products as a residual products (i.e., having a possible subsequent use but zero economic value, per (FAO, 2016), thus assigning no environmental impact to them; (2) studies where slaughter products are considered as coproducts, applying allocation based on economic value. However, the economic value of these secondary products is generally low; consequently, more than 90% of the environmental impact is attributed to the produced meat.

In the egg production system, at least two coproducts are always produced: eggs and spent hens' meat. The LEAP guidelines recommend handling this coproduction using the "biophysical" method, which is based on the energy requirements for growth and egg production (FAO, 2016). Economic allocation was the most frequently used method, confirming that despite its limitations, it currently appears to be the most consistently applicable method for quantifying coproduct relationships in complex agricultural systems. Spent hen meat represents a minimal part of the system's coproduction, both in terms of quantity and value. Dekker et al. (2011), Leinonen et al. (2012), and Costantini et al. (2020) demonstrated that economic allocation for spent hens leads to limited changes (around 1%) in the environmental results of egg production.

A system expansion approach, treating spent hens' meat as an environmental credit for replacing an equivalent amount of specifically produced broiler meat, was not applied in the LCA studies of eggs published in the last 5 years. While avoiding allocation through system expansion is theoretically a priority practice according to ISO 14044, this solution is not applied frequently. Abín et al. (2018) applied this approach to solve multifunctionality, assuming that spent hens' meat is an equal replacement for broiler meat, which is not accurate due to significant differences in physicochemical characteristics and nutritional properties between the two products. Finally, the influence of different allocation choices on environmental results should be verified through sensitivity approaches within individual studies. Despite this, only a few studies have tested more than one type of allocation and/or compared allocation with system expansion.

Inventory data

Regarding the data sources, two different types of data can be used to build the inventory. Primary data were typically gathered from personal communications or by means of interviews with stakeholders across various stages of the productive chain, surveys, trials and experimental tests, as well as from automatic measurements (e.g., energy consumption reported in the electricity bills). Secondary data, which substitute for primary data, come from different literature sources such as a previously carried out study, databases representative of specific territories (e.g., regional, national, or international inventories), or are estimated using specific models.

The literature review in this chapter highlights that, in all the considered LCA studies, there is an integration of both primary and secondary data for their inventories. This agrees with the outcomes of previous reviews focused on livestock systems (Costantini et al., 2021; Ilemobayo et al., 2024). Most primary data refer to the amounts of the production factors consumed (e.g., feed, energy, fuels, water, etc.), while the secondary data are usually about the different emissions (e.g., from poultry manure management and application in the field, from animal rearing, etc.). A good practice involves reporting detailed and transparent information when presenting the inventory analysis. Data collection for LCA studies in the poultry sector should be as accurate as possible, especially for all factors concerning feed (its origin, composition, consumption, and possibly digestibility), as these factors, in turn, influence the characteristics and quantity of manure produced.

Regarding animal-related emissions, most studies adopt the emission factors proposed by the IPCC. However, these factors only cover manure-related emissions and do not account for enteric fermentation due to insufficient data for calculation. Consequently, the vast majority of reviewed studies consider methane emissions from enteric fermentation to be negligible. Nevertheless, the LEAP guidelines suggest including the enteric methane emission factors.

Key trends and future research perspectives

Key trends

This review analyzed the state of LCA applied to the poultry sector over the last 5 years, highlighting key environmental impacts, mitigation strategies, and future research needs.

The main findings are related to two different aspects. The first one refers to the LCA study results achieved in the reviewed studies (e.g., the identification of the main environmental hotspots, the quantification of the absolute environmental impacts, and the identification of potential mitigation strategies), while the second one focuses on how the LCA was applied to the poultry sector.

The results indicate that only a few conclusions can be drawn regarding the absolute impact due to different FUs and system boundaries, as well as the use of different impact assessment methods and/or the assumptions and modeling choices adopted by the authors. Based on the results, conclusions can be drawn about these different factors:

- Feed production and supply: Across all the reviewed studies, feed production and supply were found to be major contributors to the environmental impact of poultry production, including GWP, acidification potential (AP), and eutrophication potential (EP). Animal feeding is consistently identified as the main driver of environmental impact, regardless of the production system (cage, noncage, conventional, or organic). The agricultural phase, specifically the "cradle-to-farm gate" stage, is the primary environmental hotspot.
- Protein feeds: Among feed components, protein feeds are the most impactful, contributing significantly to GWP (mainly due to LUC) and other environmental categories, such as AP, EP, and particulate matter (due to nitrogen and phosphorus compound emissions during cultivation).
- Soybean: Soybean and its derived products are frequently identified as environmental hotspots because they are the main protein source. Concerns regarding soybean cultivation are primarily linked to LUC and the extensive use of pesticides, particularly when imported from South America.
- Animal performance and farm characteristics: Animal performance parameters related to feed, such as feed intake, daily weight gain, and hen-day egg production (and consequently, the feed conversion ratio [FCR]), significantly influence the environmental impact of final poultry products. Therefore continuous on-farm improvement of these parameters is crucial for more sustainable production. However, a low correlation was found between GWP and FCR across various studies. For broiler meat, the age at slaughter deeply affects environmental results. Other factors, such as building quality, equipment, and farm characteristics, also partially influence environmental performance variability.
- Slaughtering, processing, and packaging: These stages generally have a limited impact on most environmental categories, typically contributing only about 10%–15% to cumulative energy demand. While seeking continuous improvement here (e.g., energy efficiency, waste management, and recyclable packaging) is beneficial, the priority for mitigation must be sought in the agricultural phase due to its greater impact.
- Poultry manure management: Manure management is another important environmental hotspot, significantly influencing AP and EP. This is mainly due to ammonia volatilization across all phases (housing, storage, handling, and field spreading) and the loss of nutrients (nitrates and phosphates) into surface and groundwater after field application.

- Mitigation strategies: As expected, considering the results of the contribution analysis, most mitigation scenarios in the literature focus on the feed sector, encompassing both feed production and diet formulation.
- Diet formulation and protein sources: Given that selective breeding has not shown evidence of changed protein utilization efficiency in broiler chickens, mitigation interventions related to protein requirements must focus on diet formulation and/or the protein source itself for maximum influence.
- Partial soybean replacement: Replacing soybean, partially or entirely, with other protein crops, such as beans, peas, and sunflower, in broiler and laying hen diets can lead to GWP reduction trends. However, uncertainty exists due to LUC accounting methods and diet changes, as maintaining nutrient balance may require more synthetic amino acids or vegetable oils, which involve high unitary GHG emissions.
- Alternative protein sources: The poultry sector in Europe consumes more soybeans than other livestock, exposing the continent to food security risks due to low self-sufficiency. Novel ingredients are being explored to contain GWP and agricultural land occupation, including microalgae, yeast protein concentrate, bacterial protein meal, leaf protein concentrate, and insect meal.
- Manure energy valorization: This is explored as a mitigation option. Specifically, poultry manure can be valorized via anaerobic digestion and/or combustion. Poultry manure has high biogas production potential as a feedstock for anaerobic digestion, though it is currently underexploited. Regarding the valorization as biomass fuel, it should be considered that using poultry litter as fuel for biomass-burning power stations can bring environmental benefits, particularly for AP, EP, and energy demand. However, its GWP mitigation is slight because the loss of soil organic carbon (from not spreading litter on land) counterbalances the renewable energy benefit.
- Efficient management: Overall mitigation strategies also include efficient management of the supply chain, optimizing poultry housing conditions and rearing parameters, and reducing age at slaughter for poultry meat.

Concerning the comparison of different farming systems, drawing unambiguous conclusions when comparing different farming systems is challenging, and the results vary when different poultry products are considered:

- Eggs:
  - Conventional systems often show lower environmental burdens than organic systems.
  - For GWP, cage farming systems appeared less impactful, with smaller, often statistically insignificant, differences between organic, barn, and free-range systems.
  - Organic production typically shows higher land use per kg of eggs compared to conventional systems.

- However, some studies, such as Pelletier (2017) in Canada, found opposite results, potentially due to the use of animal-derived feed components in conventional systems.
- Chicken meat:
  - Conventional rearing systems were found to have a slightly lower impact on GWP than organic systems.
  - The difference is more marked with standard indoor systems, while free-range systems fall in between the latter and organic systems.
  - Organic systems generally show significantly higher AP, EP, and land use impacts, largely due to greater feed consumption and manure production linked to longer cycle durations.
  - Stocking density has a minor role in environmental performance, despite the industry's shift towards lower densities to meet animal welfare expectations, which is a potential indicator for social LCA.

Regarding the environmental impact per unit of product, and, in particular, the results for the climate change impact categories, a general trend can be identified. Considering the findings of Costantini et al. (2021) and the LCA studies reviewed in this chapter, over time, a significant decrease in the environmental impact per unit of product was observed across all evaluated impact categories. This improvement in environmental performance is related to different factors, with the most relevant being:

(i) Improvements in animal performance: driven by advancements in selective breeding and genetics, as well as enhanced feeding, housing techniques, and overall management. In general, the poultry supply chain is becoming more and more efficient year after year. Consequently, the recent LCA studies have reported lower environmental impact because they analyzed poultry production processes characterized by lower feed conversion rate, higher slaughter yield, and less energy-demanding processes.

(ii) Improvements in background processes: including increased yields in feed production, greater energy efficiency, changes in electricity mixes, optimization of the supply chain for the different inputs, and reduced emissions related to transport.

Future research perspectives

The review identified several critical areas for future research:

- Methodological unification: A unified application of the LCA method is needed, especially concerning system boundaries (e.g., manure management, LUC inclusion, and accounting) and the allocation of environmental burdens. System boundaries must be clearly defined and consistent with the adopted FU.
- Food-oriented FUs: Future studies should adopt more food-oriented FUs that consider the nutritional value and quality aspects of products, moving

beyond simple mass or volume. This is crucial for avoiding misinterpretation and enabling consistent comparisons, as mass-based FUs do not reflect the true function of food.

- cLCA: Given the dynamic and multifunctional nature of the poultry sector, cLCA should be explored to understand environmental impacts from a broader cause-and-effect chain perspective, especially as production and associated by-products/waste streams increase.
- Integration of sustainability aspects: Current LCA studies have primarily focused on environmental aspects. Future research must integrate economic considerations to enable feasible mitigation strategies and address social aspects of sustainability, including animal welfare, which are largely neglected. Sustainability is a multilayered concept requiring greater integration of these different aspects.
- Geographical coverage: There are significant shortcomings in geographical coverage. Asia, the largest producer of eggs and poultry meat, has relatively few representative studies, and continental Africa is largely unaddressed, despite the sector's development there.
- Alternative protein sources: The potential of alternative protein sources for environmental mitigation in poultry production is still poorly documented despite their expected benefits.
- Feed–food competition: The competition between feed and food ingredients in poultry diets should be increasingly considered and evaluated from an environmental perspective.
- Unexplored environmental effects: Several environmental effects remain largely unexplored by current LCA studies, including odor impact, consequences on biodiversity, and the release of antimicrobials, estrogens, and heavy metals through manure. These represent serious environmental concerns for livestock production.

In conclusion, while significant attention has been given to the environmental impact of the poultry sector using LCA, many gaps remain. The agricultural phase, particularly feed consumption, is the main environmental hotspot, and diverse mitigation strategies, including efficient management, improved diet formulation, and enhanced manure management, offer potential for impact reduction. Future research must address methodological unification, adopt more comprehensive FUs, explore cLCA, and integrate economic and social aspects to gain a more holistic understanding of sustainability in the poultry sector.

Conclusions

A comprehensive review of the existing literature was conducted with the primary objective of evaluating the current state of LCA applications within the poultry production sector. Over the past decade, numerous studies have been published, reflecting ongoing interest and advances in this research area.

Most of these investigations adopt a cradle-to-farm gate perspective, focusing predominantly on the agricultural phase of poultry production. Within this phase, feed consumption consistently emerges as a critical environmental hotspot, significantly contributing to the overall environmental impacts associated with poultry farming.

Studies that extend their analysis beyond the farm gate to include processing stages—such as egg or meat processing—indicate that these downstream activities generally exert limited influence on the total environmental burden across most impact categories. Notably, the literature reveals considerable variability in the reported results, underscoring a need for greater standardization and harmonization of methodological approaches in future research. This variability is partly attributable to differences in system boundaries, FUs, data quality, and assessment methods, which hinder direct comparisons and consensus building within the field.

Despite the variability, the potential for reducing the environmental footprint of the poultry sector remains substantial. Numerous mitigation strategies have been explored and demonstrate promise for environmental improvements. These strategies encompass optimizing supply chain operations, enhancing poultry housing conditions, and refining management/husbandry parameters. Specific attention has been given to diet formulation and feeding strategies, particularly the adoption of alternative protein sources and feeds that do not compete with human food resources—such as insect meal, algae, or by-products (for more discussion on alternative feeds, see Chapter 14). Additionally, reducing the age at slaughter to improve resource-use efficiency and implementing best practices for manure management—potentially augmented by energy recovery technologies—are seen as viable avenues for impact mitigation.

Nevertheless, further research is essential to deepen the understanding of the sector's environmental performance, particularly in areas where current knowledge remains limited. Beyond environmental considerations, integrating economic and social dimensions into the life cycle perspective is increasingly recognized as vital for comprehensive sustainability assessment. Future studies should therefore aim to incorporate these aspects to provide a more holistic evaluation of the poultry sector's sustainability profile. Continued investigation and methodological refinement will be critical for identifying effective solutions to mitigate environmental impacts while ensuring social acceptability and economic viability.

References

Abín, R., Laca, A., Laca, A., & Díaz, M. (2018). Environmental assesment of intensive egg production: A spanish case study. *Journal of Cleaner Production, 179*, 160–168. https://doi.org/10.1016/j.jclepro.2018.01.067.

Alves, E. C., dos Santos Alves, I. H., Soares, B. B., Borges, A. F., Jalal, A., Jani, A. D., Abreu-Junior, C. H., Capra, G. F., & Rodrigues Nogueira, T. A. (2023). Resource recovery of biological residues from the Brazilian poultry industry in mitigating environmental impacts: A life cycle assessment (LCA) approach. *Journal of Cleaner Production, 416*, 137895. https://doi.org/10.1016/j.jclepro.2023.137895.

Araújo dos Santos, R., Silva da Costa, J., Maranduba, H. L., Almeida Neto, J. Ad, & Rodrigues, L. B. (2023). Reducing the environmental impacts of Brazilian chicken meat production using different waste recovery strategies. *Journal of Environmental Management, 341*, 118021. https://doi.org/10.1016/j.jenvman.2023.118021.

Arrigoni, A., Marveggio, D., Allievi, F., Dotelli, G., & Scaccabarozzi, G. (2023). Environmental and health-related external costs of meat consumption in Italy: Estimations and recommendations through life cycle assessment. *Science of the Total Environment, 869*, 161773. https://doi.org/10.1016/j.scitotenv.2023.161773.

Arulnathan, V., Heidari, M. D., & Pelletier, N. (2022). Internal causality in agri-food Life Cycle Assessments: Solving allocation problems based on feed energy utilization in egg production. *Journal of Environmental Management, 309*, 114673. https://doi.org/10.1016/j.jenvman.2022.114673.

Arulnathan, V., Turner, I., Doyon, M., Li, E., & Pelletier, N. (2025). NESTT – Development of an online, life cycle-based sustainability assessment and management platform for Canadian egg farmers. *Journal of Cleaner Production, 493*, 144954. https://doi.org/10.1016/j.jclepro.2025.144954.

Azmi, S., Suprihatin, S., Indrasti, N. S., & Romli, M. (2023). The assessment of environmental impact of the chicken meat agroindustry in Indonesia: Life cycle assessment (LCA) perspective. *Tropical Animal Science Journal, 46*(2), 249–260. https://doi.org/10.5398/tasj.2023.46.2.249.

Baldini, C., Gardoni, D., & Guarino, M. (2017). A critical review of the recent evolution of life cycle assessment applied to milk production. *Journal of Cleaner Production, 140*, 421–435. https://doi.org/10.1016/j.jclepro.2016.06.078.

Barends-Jones, V., & Pienaar, L. (2020). Western Cape Department of Agriculture Division for Macro & Resource Economics. The South African Ostrich Industry Footprint.

Beal, C. M., Robinson, D. M., Smith, J., Gerber Van Doren, L., Tabler, G. T., Rochell, S. J., Kidd, M. T., Bottje, W. G., & Lei, X. (2023). Economic and environmental assessment of U.S. broiler production: Opportunities to improve sustainability. *Poultry Science, 102*(10), 102887. https://doi.org/10.1016/j.psj.2023.102887.

Behboodi, B., Parashkoohi, M. G., Zamani, D. M., & Firouzi, S. (2023). A comparison of the energy use and environmental impacts of raising meat and egg production of ostrich. *Environmental Resources Research, 11*(2), 305–316. https://doi.org/10.22069/ijerr.2024.21958.1418.

Behboodi, B., Parashkoohi, M. G., Zamani, D. M., & Firouz, S. (2025). A study comparing energy consumption and environmental emissions in ostrich meat and egg production. *Journal of Agricultural Engineering, 56*(1). https://doi.org/10.4081/jae.2025.1718.

Berton, M., Sturaro, E., Schiavon, S., Bittante, G., Cecchinato, A., Xiccato, G., & Gallo, L. (2025). Biogenic and fossil main greenhouse gas emissions of dairy, beef, pig and poultry systems. *Animal, 19*(7), 101562. https://doi.org/10.1016/j.animal.2025.101562.

Bui, H., Cisse, S. H., Ceccaldi, M., Perrin, A., Benarbia, M. E. A., & Chicoteau, P. (2023). Mitigating the environmental impacts from pig and broiler chicken productions: Case study on a citrus extract feed additive. *Animals, 13*(23), 3702. https://doi.org/10.3390/ani13233702.

Chen, Y., & Qi, L. (2025). Carbon emissions of animal-based food can be reduced by adjusting production and consumption of residents in China. *Environmental Technology & Innovation, 37*, 103966. https://doi.org/10.1016/j.eti.2024.103966.

Cheng, Z., Jia, Y., Bai, Y., Zhang, T., Ren, K., Zhou, X., Zhai, Y., Shen, X., & Hong, J. (2023). Intensifying the environmental performance of chicken meat production in China: From perspective of life cycle assessment. *Journal of Cleaner Production, 384*, 135603. https://doi.org/10.1016/j.jclepro.2022.135603.

Costantini, M., Lovarelli, D., Orsi, L., Ganzaroli, A., Ferrante, V., Febo, P., Guarino, M., & Bacenetti, J. (2020). Investigating on the environmental sustainability of animal products: The case of organic eggs. *Journal of Cleaner Production, 274*, 123046. https://doi.org/10.1016/j.jclepro.2020.123046.

Costantini, M., Ferrante, V., Guarino, M., & Bacenetti, J. (2021). Environmental sustainability assessment of poultry productions through life cycle approaches: A critical review. *Trends in Food Science & Technology, 110*, 201–212. https://doi.org/10.1016/j.tifs.2021.01.086.

de Vries, van Middelaar, & de Boer (2015). Comparing environmental impacts of beef production systems: A review of life cycle assessments. *Livestock Science, 178*, 279–288. https://doi.org/10.1016/j.livsci.2015.06.020.

Dekker, S. E. M., de Boer, I. J. M., Vermeij, I., Aarnink, A. J. A., & Groot Koerkamp, P. W. G. (2011). Ecological and economic evaluation of Dutch egg production systems. *Livestock Science, 139*(1-2), 109–121. https://doi.org/10.1016/j.livsci.2011.03.011.

FAO, (2016). Greenhouse gas emissions and fossil energy demand from poultry supply chains: Guidelines for assessment. Livestock Environmental Assessment and Performance Partnership. FAO, Rome, Italy. https://openknowledge.fao.org/handle/20.500.14283/i6421en.

FAO Food Outlook: Global output of key food commodity crops on course for new records. (2025). https://www.fao.org/newsroom/detail/fao-food-outlook–global-output-of-key-food-commodity-crops-on-course-for-new-records/en#:~:text=Avian%20influenza,-The%20FAO%20Food&text=Global%20hen%20egg%20production%20reached,raised%20in%20closed%20housing%20systems.

Ghinea, C., Prisacaru, A.E., & Leahu A. (2024). Romania. Life cycle assessment of chicken meat production: A Romanian case study. 12th E-health and bioengineering conference, EHB 2024, pp. 1-4. https://doi.org/10.1109/EHB64556.2024.10805692.

Giusti, G., Marques, T. L., Figueirêdo, M. C. Bd, & Silva, D. A. L. (2022). Integrating water footprint in the eco-efficiency assessment of Brazilian chilled chicken. *Sustainable Production and Consumption, 33*, 331–342. https://doi.org/10.1016/j.spc.2022.07.009.

Goldáraz-Salamero, N., Blanc, S., Sierra-Perez, J., & Brun, F. (2025). From food loss and waste to feed: A systematic review of life cycle perspectives in livestock systems. *International Journal of Life Cycle Assessment, 30*(7), 1586–1606. https://doi.org/10.1007/s11367-025-02464-5.

Guillaume, A., Hubatová-Vacková, A., & Kočí, V. (2022). Environmental impacts of egg production from a life cycle perspective. *Agriculture, 12*(3), 355. https://doi.org/10.3390/agriculture12030355.

Guðjónsdóttir, S. B., Vásquez-Mejía, C. M., Shrivastava, S., & Ögmundarson, Ó. (2025). A life cycle assessment of broiler chicken meat and egg production in Iceland. *Poultry Science, 104*(6). https://doi.org/10.1016/j.psj.2025.105072.

Harasztiné Hargitai, R., Sebestyén, V., & Somogyi, V. (2024). Potential water reuse pathways from a life cycle analysis perspective in the poultry industry. *Journal of Water Process Engineering, 64*, 105577. https://doi.org/10.1016/j.jwpe.2024.105577.

Heines, W., Ristic, D., Rosenberger, S., Coudron, C., Gai, F., Schiavone, A., & Smetana, S. (2022). Eggs or meat? Environmental impact and efficiency assessment of chicken protein production with potential of *Hermetia illucens* use in feed. *Resources, Conservation & Recycling Advances, 16*, 200121. https://doi.org/10.1016/j.rcradv.2022.200121.

Ilemobayo, J., Durodola, O., Akindipe, S., Hayford, D., Hamid, T. M., Anyibama, B. J., Habeeb, S., Falana, O., Iledare, A., Akinrinde, E., & Edu, O. (2024). A review on life cycle assessment of broiler production systems. *International Journal of Multidisciplinary Research and Growth Evaluation, 5*(3), 535–541.

Kang, H. Y., Hwang, Y. W., Lee, J. H., Cho, S. J., Jeon, Y. S., Kim, N. S., & Kim, J. (2025). Evaluating the greenhouse gas emissions footprint of chicken meat production in South Korea: A life cycle perspective. *Food and Bioproducts Processing, 150*, 230–239. https://doi.org/10.1016/j.fbp.2025.01.017.

Kheiralipour, K. (2021). Environmental life cycle assessment of poultry production systems. In *Interdisciplinary applications of the life cycle assessment tool*. Iran: Nova Science Publishers, Inc, 45–65. https://novapublishers.com/shop/interdisciplinary-applications-of-the-life-cycle-assessment-tool.

Kolawole, O., Valliere, C., Krska, R., David, M., Lorran, G., Schatzmayr, D., Schatzmayr, G., Eskola, M., Petchkongkaew, A., & Elliott, C. (2025). Potential economic and environmental impacts of mycotoxins in poultry: A meta-analysis and life cycle assessment approach. *Environmental Impact Assessment Review, 115*, 107989. https://doi.org/10.1016/j.eiar.2025.107989.

Kyttä, V., Kårlund, A., Pellinen, T., Pietiläinen, O., Tuomisto, H. L., Kolehmainen, M., Pajari, A. M., & Saarinen, M. (2023a). Product-group-specific nutrient index as a nutritional functional unit for the Life Cycle Assessment of protein-rich foods. *International Journal of Life Cycle Assessment, 28*(12), 1672–1688. https://doi.org/10.1007/s11367-023-02217-2.

Kyttä, V., Kårlund, A., Pellinen, T., Tuomisto, H. L., Kolehmainen, M., Pajari, A.-M., & Saarinen, M. (2023b). Extending the product-group-specific approach in nutritional life cycle assessment. *The International Journal of Life Cycle Assessment, 30*, 93–109. https://doi.org/10.1007/s11367-023-02235-0.

Leinonen, I. (2025). Potential climate performance of modern fast- and slow-growing broiler genotypes. *Animal, 19*(5), 101499. https://doi.org/10.1016/j.animal.2025.101499.

Leinonen, I., Williams, A. G., Wiseman, J., Guy, J., & Kyriazakis, I. (2012). Predicting the environmental impacts of chicken systems in the United Kingdom through a life cycle assessment: Egg production systems. *Poultry Science, 91*(1), 26–40. https://doi.org/10.3382/ps.2011-01635.

Li, D., Zhang, X., Zhao, Z., Wang, S., Wang, J., & Wang, H. (2025). Integrated assessment of productive, environmental, and social performances of adopting low-protein diets technology for laying hens. *Animals, 15*(2), 146. https://doi.org/10.3390/ani15020146.

Li, H., Hu, S., & Tong, H. (2024). Carbon footprint of household meat consumption in China: A life-cycle-based perspective. *Applied Geography, 169*, 103325. https://doi.org/10.1016/j.apgeog.2024.103325.

Li, Y., Allacker, K., Feng, H., Heidari, M. D., & Pelletier, N. (2021). Net zero energy barns for industrial egg production: An effective sustainable intensification strategy? *Journal of Cleaner Production, 316*, 128014. https://doi.org/10.1016/j.jclepro.2021.128014.

Lu, Y., Ma, W., & Shao, L. (2024). Strategies to mitigate the environmental footprints of meat, egg and milk production in northern China. *Journal of Cleaner Production, 443*, 141027. https://doi.org/10.1016/j.jclepro.2024.141027.

Maciel, Fd. F., Gates, R. S., Tinôco, Id. F. F., Sousa, F. Cd, Pelletier, N., Ibarburu-Blanc, M. A., & Oliveira, C. E. A. (2023). Life cycle assessment project for the Brazilian egg industry. *Animals, 13*(9), 1479. https://doi.org/10.3390/ani13091479.

Maciel, Fd. F., Gates, R. S., Tinôco, Id. F. F., Pelletier, N., Ibarburu-Blanc, M. A., Renato, Nd. S., Sousa, F. Cd, Andrade, R. R., Silva, G. Md. M., & Becciolini, V. (2024). Environmental impacts of the brazilian egg industry: Life cycle assessment of the battery cage production system. *Animals, 14*(6), 861. https://doi.org/10.3390/ani14060861.

Macleod, M., Gerber, P. J., Mottet, A., Tempio, G., Falcucci, A., Opio, C. I., Vellinga, T. V., Henderson, B., & Steinfeld, H. (2013). *Greenhouse gas emissions from pig and chicken supply chains – A global life cycle assessment.* Food and Agriculture Organization of the United Nations (FAO), Rome.

Mata, T. M., Rodrigues, J. N. F. G., da Silva, J. C. G. E., & Martins, A. A. (2023). Life cycle energy and climate change impacts of a chicken slaughtering process. *Environmental Science and Engineering,* 215–222. https://doi.org/10.1007/978-3-031-43559-1_21.

Matton, B., Verhelle, A., Vlerick, L., Keirsschieter, K., & Saremi, B. (2025). Feeding broilers with reduced dietary crude protein or reduced soybean meal diets has no negative impact on the performance and carcass traits while reducing the feed global warming potential. *Animals, 15*(12), 1753. https://doi.org/10.3390/ani15121753.

Mesquita, C., & Carvalho, M. (2024). The carbon footprint of common vegetarian and non-vegetarian meals in Portugal: An estimate, comparison, and analysis. *International Journal of Life Cycle Assessment, 29*(12), 2169–2183. https://doi.org/10.1007/s11367-023-02161-1.

Michalke, A., Köhler, S., Messmann, L., Thorenz, A., Tuma, A., & Gaugler, T. (2023). True cost accounting of organic and conventional food production. *Journal of Cleaner Production, 408,* 137134. https://doi.org/10.1016/j.jclepro.2023.137134.

Mitrovic, M., Tomasevic, I., & Djekic, I. (2022). Assessment of environmental impacts from different perspectives—Case study of egg value chain system in serbia. *Foods, 11*(12), 1697. https://doi.org/10.3390/foods11121697.

Mostert, P. F., Bos, A. P., van Harn, J., & de Jong, I. C. (2022). The impact of changing toward higher welfare broiler production systems on greenhouse gas emissions: A Dutch case study using life cycle assessment. *Poultry Science, 101*(12), 102151. https://doi.org/10.1016/j.psj.2022.102151.

Nguyen, H. D., Moss, A. F., Yan, F., Romero-Sanchez, H., & Dao, T. H. (2025). Effects of feeding methionine hydroxyl analogue chelated zinc, copper, and manganese on growth performance, nutrient digestibility, mineral excretion, and welfare conditions of broiler chickens: Part 2: Sustainability and welfare aspects. *Animals, 15*(3), 419. https://doi.org/10.3390/ani15030419.

Notarnicola, B., Tassielli, G., Renzulli, P. A., Castellani, V., & Sala, S. (2017). Environmental impacts of food consumption in Europe. *Journal of Cleaner Production, 140,* 753–765. https://doi.org/10.1016/j.jclepro.2016.06.080.

OECD/FAO. (2025). OECD-FAO Agricultural Outlook 2025-2034. OECD Publishing. https://doi.org/10.1787/601276cd-en.

Ogino, A., Oishi, K., Setoguchi, A., & Osada, T. (2021). Life cycle assessment of sustainable broiler production systems: Effects of low-protein diet and litter incineration. *Agriculture, 11*(10), 921. https://doi.org/10.3390/agriculture11100921.

Payandeh, Z., Mesri-Gundoshmian, T., Jahanbakhshi, A., Shahgholi, G. H., Rossi, L., & Bacenetti, J. (2025). Optimization of environmental and energy performance of egg production using data envelopment analysis (DEA) and life cycle assessment (LCA). *Science of the Total Environment, 963,* 178493. https://doi.org/10.1016/j.scitotenv.2025.178493.

Pelaracci, S., Paolotti, L., Rocchi, L., Boggia, A., & Castellini, C. (2024). Life cycle assessment of organic and conventional egg production: A case study in northern Italy. *Cleaner Environmental Systems, 15,* 100226. https://doi.org/10.1016/j.cesys.2024.100226.

Pelletier, N. (2017). Life cycle assessment of Canadian egg products, with differentiation by hen housing system type. *Journal of Cleaner Production, 152,* 167–180. https://doi.org/10.1016/j.jclepro.2017.03.050.

Pelletier, N. (2018). Changes in the life cycle environmental footprint of egg production in Canada from 1962 to 2012. *Journal of Cleaner Production, 176*, 1144–1153. https://doi.org/10.1016/j.jclepro.2017.11.212.

Pelletier, N., Ibarburu, M., & Xin, H. (2014). Comparison of the environmental footprint of the egg industry in the United States in 1960 and 2010. *Poultry Science, 93*(2), 241–255. https://doi.org/10.3382/ps.2013-03390.

Portillo-Chávez, F., & Vargas-Leitón, B. (2025). Environmental impact of the broiler production chain under conventional systems. *Agronomia Mesoamericana, 36*, 61097. https://doi.org/10.15517/am.2025.61097.

Prudêncio da Silva, V., van der Werf, H. M. G., Soares, S. R., & Corson, M. S. (2014). Environmental impacts of French and Brazilian broiler chicken production scenarios: An LCA approach. *Journal of Environmental Management, 133*, 222–231. https://doi.org/10.1016/j.jenvman.2013.12.011.

Putman, B., Thoma, G., Burek, J., & Matlock, M. (2017). A retrospective analysis of the United States poultry industry: 1965 compared with 2010. *Agricultural Systems, 157*, 107–117. https://doi.org/10.1016/j.agsy.2017.07.008.

Quevedo-Cascante, M., Dorca-Preda, T., Mogensen, L., Zollitsch, W., Waqas, M. A., Hörtenhuber, S., Geßl, R., Kongsted, A. G., & Knudsen, M. T. (2024). Life cycle assessment and modeling approaches in silvopastoral systems: A case study of egg production integrated in an organic apple orchard. *Journal of Environmental Management, 372*, 123377. https://doi.org/10.1016/j.jenvman.2024.123377.

Schuck-Paim, C., Alonso, W. J., Verkuijl, C., Hegwood, M., & Hartcher, K. (2025). The welfare footprint framework can help balance animal welfare with other food system priorities. *Nature Food, 6*(8), 737–739. https://doi.org/10.1038/s43016-025-01213-z.

Sharifi, M., Soodmand-Moghaddam, S., & Moloudi, H. (2024). Investigation of environmental, energy and economic indicators of the turkey breeding farms: A case study in West Azarbaijan and Zanjan, Iran. *Environment, Development and Sustainability, 26*(9), 24221–24245. https://doi.org/10.1007/s10668-023-03642-x.

Sinke, P., Swartz, E., Sanctorum, H., van der Giesen, C., & Odegard, I. (2023). Ex-ante life cycle assessment of commercial-scale cultivated meat production in 2030. *International Journal of Life Cycle Assessment, 28*(3), 234–254. https://doi.org/10.1007/s11367-022-02128-8.

Tukiainen, K., Kyttä, V., Gómez-Gallego, C., Kolehmainen, M., Pajari, A. M., Tuomisto, H. L., Saarinen, M., & Kårlund, A. (2025). Re-evaluating the importance of protein quality: Insights on its limited role in multi-nutrient functional units. *International Journal of Life Cycle Assessment, 30*(5), 928–938. https://doi.org/10.1007/s11367-025-02451-w.

Turner, I., Heidari, D., & Pelletier, N. (2022a). Environmental impact mitigation potential of increased resource use efficiency in industrial egg production systems. *Journal of Cleaner Production, 354*, 131743. https://doi.org/10.1016/j.jclepro.2022.131743.

Turner, I., Heidari, D., & Pelletier, N. (2022b). Life cycle assessment of contemporary Canadian egg production systems during the transition from conventional cage to alternative housing systems: Update and analysis of trends and conditions. *Resources, Conservation and Recycling, 176*, 105907. https://doi.org/10.1016/j.resconrec.2021.105907.

Turner, I., Heidari, D., Widowski, T., & Pelletier, N. (2023). Development of a life cycle impact assessment methodology for animal welfare with an application in the poultry industry. *Sustainable Production and Consumption, 40*, 30–47. https://doi.org/10.1016/j.spc.2023.06.010.

UNI. (2010). Ente Nazionale Italiano di Unificazione (UNI) Italy UNI EN ISO 14025:2010 – Environmental labels and declarations – Type III environmental declarations – Principles and procedures.

Usva, K., Hietala, S., Nousiainen, J., Vorne, V., Vieraankivi, M.-L., Jallinoja, M., & Leinonen, I. (2023). Environmental life cycle assessment of finnish broiler chicken production – Focus on climate change and water scarcity impacts. *Journal of Cleaner Production, 410*, 137097. https://doi.org/10.1016/j.jclepro.2023.137097.

Vanbaelinghem, L., Arulnathan, V., Costantino, A., & Pelletier, N. (2025). Comparative life cycle assessment of alternative heating, ventilation and air-conditioning (HVAC) systems for poultry houses. *Journal of Environmental Management, 383*, 125541. https://doi.org/10.1016/j.jenvman.2025.125541.

Vauterin, A., Steiner, B., Sillman, J., & Kahiluoto, H. (2021). The potential of insect protein to reduce food-based carbon footprints in Europe: The case of broiler meat production. *Journal of Cleaner Production, 320*, 128799. https://doi.org/10.1016/j.jclepro.2021.128799.

Wicaksono, P. A. & Kaswara, D. (2021). Environmental impact reduction in broiler chicken farm using life cycle assessment in the implementation of circular economy. Proceedings of the International Conference on Industrial Engineering and Operations Management. IEOM Society International. https://doi.org/10.46254/SA02.20210643.

Yalçuk, A., Ongun, G. K., & Üçtuğ, F. G. (2025). Life cycle assessment of white meat supply to final consumer: A case study from türkiye. *Journal of Food Process Engineering, 48*(7). https://doi.org/10.1111/jfpe.70170.

Yayli, B., & Kilic, I. (2023). Assessment of environmental impacts of broiler farms using different indicators. *International Journal of Environmental Science and Technology, 20*(1), 125–134. https://doi.org/10.1007/s13762-022-03918-1.

Yu, S., Yue, W., Li, S., Chen, S., & Wu, P. (2025). Characteristics identification and mitigating potentials of provincial gaseous reactive nitrogen emissions from livestock and poultry breeding systems in China. *Journal of Environmental Management, 374*, 124126. https://doi.org/10.1016/j.jenvman.2025.124126.

Zumsteg, J. M., Cooper, J. S., & Noon, M. S. (2012). Systematic review checklist: A standardized technique for assessing and reporting reviews of life cycle assessment data. *Journal of Industrial Ecology, 16*(1), S12–S21. https://doi.org/10.1111/j.1530-9290.2012.00476.x.

Chapter 16

Consumer perceptions

Michelle Sinclair
A World of Good Initiative Inc., Santa Fe, NM, United States

Consumers and why their perceptions matter

Much can, and has been, said about consumers—their knowledge (or lack of knowledge), their expectations, and their willingness (or lack of willingness) to pay for higher welfare animal products. Known as the "father of advertising," and the reported inspiration behind the popular series "Mad Men," David Ogilvy once famously stated that "Consumers don't think how they feel, they don't say what they think, and they don't do what they say" (Ogilvy, 1969). Considering the poultry industry specifically and to further confound the situation, researchers recently conducting focus groups with poultry consumers in Minnesota, United States of America (USA) stated "Consumers are critical stakeholders, yet their growing disconnection from agricultural production practices complicates efforts to achieve more sustainable systems" (Dewi et al., 2024). All aside, one thing can be said with certainty: consumers drive the market. When they choose what to consume and therefore where to send their hard-earned dollars, they make markets. They also break markets, the very same markets they made with their earlier choices. Consumers can be demanding, unrealistic, and at times, a bit confusing—but ultimately, they're incredibly powerful. Understanding the way consumers think, or more importantly, will think in the future, is the key to the economic sustainability of any industry.

Thankfully, we have a significant tool at our disposal—a vintage crystal ball just sitting on its too-tiny wooden stand waiting to be prompted by data: the age-old art of market research. Contrary to popular belief, market research is not just about anticipating and planning for the survivability and success of commercial enterprises. Market research is a largely useful activity in policy creation for government bodies, and a greatly underutilized activity for building sustainable strategies in nongovernment sectors. This includes those organizations working to increase welfare provisions and protections for poultry. From whichever position one sits, consumer buy-in is integral for economic sustainability, and for an industry or practice to remain economically sustainable and commercially viable in the long term, we need to consider

Integrating Welfare and Sustainability in Poultry Systems. DOI: https://doi.org/10.1016/B978-0-443-40594-5.00012-2

social landscapes and trends, including the extent to which animal welfare and sustainability are important to consumers of the future.

In this chapter, we will explore what we know about consumers' perceptions of poultry and the systems in which poultry are farmed around the world. We will dip a toe in the complicated realm of consumer psychology and take a look at what the consumer of the future is prioritizing in different geopolitical regions. With that, we will discuss perceptions around environmental sustainability and how it lives in the minds of young people across cultures—in tandem with animal welfare and protection, because ultimately, "Consumers don't buy products, they buy better versions of themselves". This chapter will mostly center on chickens, due to the increased focus on this species in the research literature and the number of chickens in farming systems as compared to other species. Where possible, we will include some research on consumer perceptions around turkeys and ducks.

Poultry around the world: culture and perceptions

As humans, we have a general tendency to try to understand the world and our interactions with others according to how we personally experience the world. We often have a tendency to assume others think the way we think. That is, unless they don't look like us or act like us in our estimation. In fact, the more "human-like" an animal is deemed, or how "cute" we think animals are, the more charismatic or "beautiful" and the more intelligence we tend to associate with them, the more concern we have for their suffering and habitat conservation (Batt, 2009; Woods, 2000). One of the major measures of "cuteness," as far as our minds are concerned, is the largeness of neonate-looking eyes positioned at the front of the head. Where does that leave poultry?

Inserting the ubiquitous chicken as the ambassador species for poultry more generally, we set about an international study to assess just that—how important do consumers consider the welfare of species with vastly different phylogenetic traits, found in different societal contexts and across cultures? Although exact results varied a little by culture, the general aggregate ranking remained relatively consistent; participants proceeded to rank the species in accordance with the phenomena described above (Sinclair et al., 2022). The welfare of humans was deemed most important, followed by other mammalian species (e.g., dogs and cattle), then followed by birds (chickens), fish, and then lastly, reptiles (see Fig. 16.1 for full findings by country). Still, the associated average importance of chicken welfare does not fall below 5.5 out of 7 for all countries. Summarizing simply, although deemed less important than the welfare of humans and dogs, consumers around the world express that the welfare of chickens is important to them (Sinclair et al., 2022).

This sentiment is continued when looking specifically at broiler chickens. A European study with consumers in Denmark, France, the Netherlands and the United Kingdom found that in all of these countries, "animal welfare" was

	Human animals	Companion animals	Frequently farmed animals					Mostly wild animals	
		Dogs	*Cattle*	*Pigs*	*Chickens*	*Fish*	*Soft-shelled turtle*	*Kangaroo*	*Koala*
Australia	6.62	6.70	6.49	6.52	6.46	6.19	6.48	6.53	6.63
Bangladesh	6.70	5.72	6.09	5.05	5.88	5.51	4.83	4.94	4.65
Brazil	6.64	6.72	6.59	6.56	6.49	6.31	6.53	6.55	6.56
Chile	6.60	6.89	6.62	6.46	6.51	6.35	6.57	6.57	6.65
China	6.74	6.13	5.82	5.67	5.49	5.24	5.24	5.81	5.81
India	6.32	6.17	6.14	5.70	5.74	5.75	5.74	5.94	5.83
Malaysia	6.58	6.19	5.89	5.66	5.76	5.62	5.79	5.99	6.04
Nigeria	6.62	5.93	5.97	5.65	6.03	5.80	4.87	4.97	4.85
Pakistan	6.54	4.15	6.27	2.33	6.28	6.12	3.72	3.90	3.63
Philippines	6.61	6.47	6.14	6.15	6.05	5.95	6.14	6.20	6.26
Sudan	6.19	5.11	5.50	4.19	5.62	5.47	4.63	4.83	4.62
Thailand	6.72	6.41	6.19	6.11	6.00	5.88	5.86	6.20	6.20
United Kingdom	6.64	6.39	6.05	6.01	5.93	5.74	5.98	6.07	6.05
United States	6.08	5.98	5.53	5.54	5.51	5.39	5.46	5.64	5.54

FIGURE 16.1 Perceived importance of welfare by species and country, as presented by means of a scale from 1 (very unimportant) to 7 (extremely important).
From Sinclair, M., Lee, N.Y.P., Hötzel, M.J., de Luna, M.C.T., Sharma, A., Idris, M., Derkley, T., Li, C., Islam, M.A., Iyasere, O.S., Navarro, G., Ahmed, A.A., Khruapradab, C., Curry, M., Burns, G.L. & Marchant, J.N. (2022). International perceptions of animals and the importance of their welfare. Frontiers in Animal Science, 3, 960379. https://doi.org/10.3389/fanim.2022.960379.

amongst the most frequently mentioned purchase criteria, with one respondent specifically articulating that "If there is no chicken in the fridge that had a good life, then I am not having chicken that day" (Lehmann & Christoph-Schulz, 2022).

Shifting to ducks, a Taiwanese study found that consumers care deeply about the welfare of ducks when they consider buying duck eggs (Yang & Wang, 2025). Likewise, another study with consumers in Uganda found that the quality attribute of "welfare" of ducks was among the most important considerations for consumers and restaurant users (Higenyi, 2014).

The next logical question for the consumer becomes, what constitutes animal welfare infringements for poultry? If we choose to follow the scientific logic of sentience, that would depend on which species has the capacity to experience/suffer (or not experience/suffer). We know, through scientific research, that chickens have the capacity to suffer considerably—both physiologically and psychologically, but what do consumers think?

In the same study described above, we also asked participants the level to which they agree chickens feel pain. In aggregate across countries, 79.3% of participants agreed to some level that chickens experience pain. Likewise, 71.4% of participants believed chickens have emotions, and on a practical level, 73.5% believed that chickens need space to explore (Sinclair et al., 2022). Likewise, another study in the USA found that 89% of the general public believes that turkeys feel pain (Bir et al., 2019). This study also found that for turkey welfare attributes, consumers were the most concerned that the birds had "room to move around" (Bir et al., 2019). Interestingly, this study also identified that "pet owners" were far more likely to care about the welfare of the birds to the point of seeking animal welfare-friendly labeling on their turkey product purchases (Bir et al., 2019). Worded in the reverse, similar results were also seen considering broiler chickens, where 70% of participants in Australia disagreed that "chickens don't feel pain like humans" (Coleman, 2007). Another study in Taiwan found that most consumers, especially younger consumers, felt the welfare of the ducks was important enough to pay a premium price for eggs from ducks with higher welfare, should the opportunity be presented (Yang & Wang, 2025).

By way of simple summary, consumers across cultural contexts tend to believe that the welfare of poultry species matters; that poultry have the capacity for pain and emotion, and specifically that they need space to explore and resources to express normal behaviors.

Farming system diversity: perceptions, preferences and (lack of) knowledge

The farming systems in which poultry are reared and kept vary all over the world. When considering chickens for egg production, however, we can very broadly draw a line between systems that are cage-based and systems that are not cage-based. Cage-based systems incorporate "battery" or conventional cages which are frequently criticized for causing levels of confinement that cause the birds severe restriction in expressing any normal behaviors, for example, sitting on eggs, stretching their wings, perching, forage-pecking and dust bathing (Appleby, 2003; EFSA Panel on Animal Health and Welfare, 2005; Lay et al., 2011; Appleby et al., 2004; Webster, 2004; Fraser et al., 2001; Duncan, 2004; European Food Safety Authority, 2015). Due to the welfare concerns prevalent with these systems, the demand of consumers in some countries has led to a ban on these egg systems, including in Australia, Canada, New Zealand, and some states in the USA and across the Eurozone. In some areas of the world, as a step better than conventional cages, eggs are also produced in "enriched" cages. These cages are slightly bigger in size and usually allow the birds enrichment opportunities, such as to perch or to scratch. Although readily accepted as an improvement, these systems receive some of the similar criticisms as their rudimentary cousin, the battery cage—still not enough space to express normal behaviors (Nicol, 2015; Tauson, 2005; Lay et al., 2011). For this reason, driven by consumer demand for hens to be out of cages entirely, countries such as Austria, Luxembourg, and Switzerland have also banned "enriched" cages, soon to be followed by Germany and Sweden.

Looking at non-cage-based systems, they can broadly be described as "barn" —out of cages inside barns—and "free-range" (either entirely or partially), outside on pasture for some of their day. Although the scientific literature vastly favors the non-cage-based systems for reasons of improved hen welfare, neither system is entirely perfect. Barn systems are criticized for the inability of the hens to go outside, or that barns are overstocked and therefore cramped, and free-range systems for the possibility that the birds could be predated upon. The ill-favor that both noncage systems receive in relation to the welfare of the birds, primarily from industry rather than consumers, is the reduced ability to identify and treat diseases in individual birds as they occur.

In our 2022 study with consumers around the world, a vast majority of our 4292 participants (72.3%) stated that they would prefer to buy eggs from chickens that have not been kept in cages at all. This percentage varied across countries, with 31.6% of the participants in Nigeria (a divergent outlier) at the lowest, to 93.6% of the participants in Australia at the highest (see Table 16.1). When we sought to ascertain if consumers knew what production system was prevalent in their country, however, the results became even more interesting. We asked participants to rank their agreement with the statement "most eggs in my country are produced in cages," then analyzed their responses against the realities of current egg production in their country. Participants in Australia and the UK were mostly wrong (63% and 54% respectively) in believing most eggs in their country were from cage-based systems, when these two countries are

TABLE 16.1 **Agreement with the statement "I would prefer to buy eggs from chickens that have not been kept in cages" by country.**

	%
Australia	93.6
Bangladesh	46.2
Brazil	92.1
Chile	94.1
China	65.5
India	64.0
Malaysia	69.31
Nigeria	31.6
Pakistan	89.8
Philippines	71.9
Sudan	58.4
Thailand	68.6
United Kingdom	90.2
United States	77.7
International aggregate	72.3

Source: From Sinclair, M., Lee, N.Y.P., Hötzel, M.J., de Luna, M.C.T., Sharma, A., Idris, M., Derkley, T., Li, C., Islam, M.A., Iyasere, O.S., Navarro, G., Ahmed, A.A., Khruapradab, C., Curry, M., Burns, G.L. & Marchant, J.N. (2022). International perceptions of animals and the importance of their welfare. Frontiers in Animal Science, 3, 960379. https://doi.org/10.3389/fanim.2022.960379.

both now predominantly cage-free. Participants from the rest of the countries were mostly correct in identifying that most eggs in their country are produced in cages (Bangladesh 63%, Brazil 73%, Chile 78%, China 77%, India 65%, Malaysia 70%, Pakistan 87%, Philippines 79%, Sudan 55%, Thailand 75%, and the United States of America 82%). Participants in the USA were the most likely to confidently (and accurately) state that most eggs in their country were produced in cages, with 71.5% of eggs produced in cages in 2020. Interestingly, participants in the USA also reported lower levels of concern for chicken welfare, and less preference for cage-free eggs as compared to the other high-income countries in the study. 74% of US-American participants stated that they cared that chickens producing eggs should not suffer, with 78% stating a preference for cage-free eggs; however, that is significantly less compared to Australian participants (93%/94%) and British participants (88%/90%). Still, these findings demonstrate a gap between consumer perception/stated preference, and the current realities of the egg industry in the USA. Not only are most eggs still produced in cages in the USA, with the exception of some state-based legislation—such as that found in Californian Proposition 12—the USA is mostly without legal restriction in terms of hen welfare requirements in egg production. Put simply, 78% of consumers in the USA state that they would prefer to buy eggs from chickens not kept in cages, yet over 71% of US-American eggs are produced in cages, and minimal laws are currently in place (Sinclair et al., 2022).

Public knowledge of turkey production was also deemed to be low in the USA, despite a similarly large percentage of the population purchasing turkey products (Bir et al., 2019). Turning to broiler production, at the onset of public workshops conducted by a team in York, United Kingdom, a majority of participants admitted how little they knew about the process of broiler production and expressed shock at some of the facts presented to them (Hall & Sandilands, 2007). Looking at waterfowl, an unpublished survey conducted by the Royal Society for the Prevention of Cruelty to Animals in the United Kingdom found that most consumers believed that ducks raised for meat have access to water, when in reality, most ducks only have access to water for drinking through nipple drinkers, with no ability to swim (Babington & Campbell, 2022).

In general, it is entirely reasonable to summarize that while consumers do not have overflowing knowledge about poultry farming systems, they would far prefer for hens to be in production systems outside of cages, for chickens and turkeys to be raised humanely for meat, and for ducks to be raised with access to water.

What do egg producers think about this? In 2022 (de Luna et al., 2022), we made contact with 224 egg producers in Asia (China, Indonesia, Malaysia, the Philippines, and Thailand). Asia is responsible for ~60% of global egg production. As in most of the world, nearly all the egg-producing hens are housed in cages. To investigate closing the gap between consumer perception and preference and current practices, we asked producers using cages to share their perceptions with us. Why do they favor cages? Their unsurprising answer is: for their efficiency and ease of management. The primary reason they would consider switching to non-cage-based systems is their perception that they consider them better for the welfare of the birds. They reported that the challenges in making the switch to cage-free systems rest mostly with perceived reduced profitability, higher running costs, and an increased risk of biosecurity and disease events. Most of the producers (72%) said that more support would be needed for them to feel comfortable transitioning, and the support they envisaged needing took the form of market development, technical expertise, training, and provisions (see Fig. 16.2). Ultimately, 65% of the cage-based producers in this study said that they believed cage-free systems to be a viable option in their country (de Luna et al., 2022).

In an earlier study, we surveyed 146 meat chicken producers in China, and approximately 66% stated they were using barn systems, 31% of which provided no enrichment. More producers in this study agreed than disagreed that intensive indoor farming was better for food safety (47%). A total of 30%

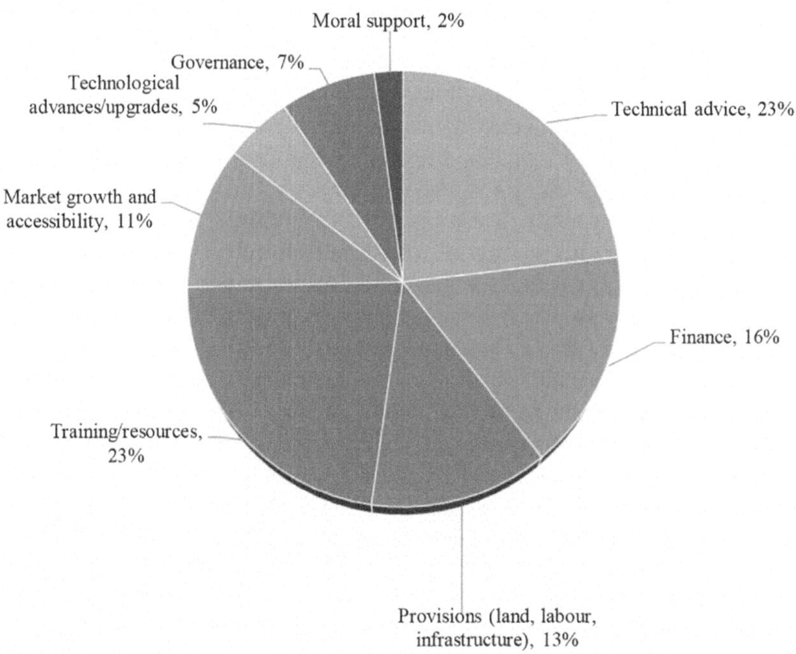

FIGURE 16.2 Egg producers' perception of the support needed to transition to cage-free systems, by emerging theme.

From de Luna, M.C.T., Yang, Q., Agus, A., Ito, S., Idrus, Z., Iman, R.H.S., Jattuchai, J., Lane, E., Nuggehalli, J., Hartcher, K. & Sinclair, M. (2022). Cage egg producers' perspectives on the adoption of cage-free systems in China, Japan, Indonesia, Malaysia, Philippines, and Thailand. Frontiers in Veterinary Science, 9, 1038362. https://doi.org/10.3389/fvets.2022.1038362.

of these producers were also of the opinion that intensive indoor systems were better for the welfare of the chickens, although only 4% disagreed that outdoor farming systems, where chickens are in groups and able to roam, are better for the welfare of the chickens. In all instances, however, a majority of respondents opted for neutrality (Sinclair et al., 2019). This may indicate a lack of knowledge or confidence regarding welfare in the different poultry farming systems on behalf of the producers in some areas of the world—not just of consumers.

Future consumers: the role of environment, sustainable agriculture and animal welfare

Real sustainability requires securing solutions that work for the future, and in the context of this chapter, that means understanding what the perceptions are expected to be of future leaders, producers, retailers, and, of course, consumers.

In many societies, the future consumer, the current youth, is frequently cited as the reason to be concerned about the environment. "We do not inherit the earth from our ancestors, we borrow it from our children"—Native American Proverb. With Greta Thunberg front and center, young people are also frequently associated with increased ethical concern for the future of the environment, so much so that many individuals and institutions openly attribute their hopes for rectifying long-term environmental concern firmly into their hands. "What gives me hope are young people: They know it and they call it."—Inger Andersen, under-secretary-general of the United Nations and executive director of the United Nations Environment Program. However unfairly deferred any action on environmental protection may be, are these common perceptions of younger generations accurate and consistent around the world? Do younger people who care about the environment also care about animal welfare? How about sustainable development?

In a bid to understand which social issues mattered to young people in different cultures of the world, we published survey data from universities situated across a broad spectrum of global cultures, including China, the Czech Republic, Estonia, Ireland, Iran, Portugal, Korea, the Republic of Macedonia, Norway, Serbia, Spain, and Sweden (Sinclair & Phillips, 2017). Within the survey, they were asked to rank the importance of 13 major world social issues according to their own priorities. In alphabetical order, these social issues included animal protection, capital punishment, environmental protection, equality for the LGBT community, genetic engineering, human cloning, human euthanasia, peace and security, professional ethics, racial equality, reduction of poverty, sustainable development, and women's rights. What we found was a complete surprise at the time. Environmental protection, sustainable development, and animal protection were rated with the greatest importance across all surveyed nations collectively. With few exceptions, these social issues also dominated the rankings by country.

Chinese youth, for example, despite living in one of the two nations that contribute the most annual greenhouse gases, gave these three social issues an equal perfect score in terms of how important they are to them. In recent years, the notion of "ecological civilization," that is, promoting harmonious coexistence between humans and nature, has been increasingly integrated into national policy in China, in recognition of public demand for improved environmental quality and the importance of sustainable development (Dong et al., 2025). This national policy agenda is planned to be continued into the future (The State Council Information Office of the People's Republic of China, 2023).

Looking across all the countries, some deviation was present, including that for youth in the Czech Republic, Estonia, Serbia and Sweden "women's rights" were rated just above "animal protection"; in Norway, Sweden and Serbia "peace and security" was rated just above "animal protection," in the Republic of Macedonia "animal protection" was rated significantly lowly (as was "racial equality") as compared to other countries, and for youth in Iran "peace and security" was rated above all else. In general, however, it remains very clear that the "nature trifecta" of social issues—"environmental protection," "sustainable development" and "animal protection"—remain of paramount importance to young human beings, despite culture, and for the most part, despite geopolitical challenges of the time (see Table 16.2).

More recently, we embarked on a study for Faunalytics to investigate in more detail the perceptions and behaviors of "Generation Z" (young adults born between 1999 and 2007) towards the environment and animal protection and welfare (Sinclair et al., 2025). Selected for the scale of their agricultural trade, regional importance and diversity of political and cultural contexts, in this study we focussed on youth in the USA, China, Thailand, and Indonesia. The major findings? Most of our participants across all countries (93%) expressed positive attitudes towards environmental protection and animal protection. A total of 86% of them indicated that they had preferences for purchasing environment and animal conscious products, and 84% reported that their behaviors reflected these preferences—that they had made alterations to their life in support of environmental and animal protection efforts and were able to tell us about those actions (Sinclair et al., 2025). Unlike studies with earlier generations that demonstrated buyer behavior does not always align with reported buyer values or intention (Yang, 2023; Napolitano et al., 2008; White et al., 2012), this suggests that up-and-coming generations may be more inclined to put their "money" and "efforts" where their mouth is.

Closing gaps through consumer and stakeholder engagement

In any commercial industry, the consumer, or rather, consumers as a collective group, are the big-ticket stakeholder that have a powerful say over the thrive or demise of any business or production system within it. This is often termed the "social license to operate." Being aware of consumer perceptions and keeping

TABLE 16.2 Least squares means for the effects of country on importance value for 13 world social issues, weighted according to relative ratings within country, together with coefficients of variation between countries for each issue.

	CN	CZ	ES	IE	IR	KR	MK	NO	PT	SE	SR	UK	CV
Environmental protection	100	100	100	100	99.0	100	100	89.78	100	91.2	100	99.97	3.8
Sustainable development	100	77.73	100	91.69	24.03	94.67	87.98	100	89.77	100	81.38	100	24.5
Animal protection	100	82.99	83.33	81.88	99.03	95.81	35.33	81.70	85.79	76.72	75.91	75.36	20.7
Peace and security	85.18	80.97	84.21	81.88	100	25.85	80.91	85.53	74.43	80.99	88.56	75.77	22.7
Women's rights	77.77	86.23	95.51	72.45	91.82	56.27	87.27	71.06	70.45	82.42	77	51.98	17.3
Reducing poverty	81.14	51.82	82.45	81.50	76.92	53.23	90.10	63.82	78.97	73.97	70.43	71.8	16.0
Racial equality	78.11	51.41	77.63	84.15	85.57	72.62	35.33	77.02	76.70	73.87	63.50	62.99	20.8
Professional ethics	85.18	42.91	56.14	60.75	15.86	54.75	59.36	49.78	44.31	46.55	72.26	65.19	31.6
Genetic engineering	53.53	61.53	28.50	15.09	14.42	56.27	52.29	5.95	0	23.75	45.25	4.94	74.8
Human euthanasia	1.68	28.74	14.47	13.96	46.63	20.53	9.54	0	9.09	27.55	23.72	39.20	73.1
LGBT equality	0.67	0	28.50	56.22	22.59	0.38	0	20.42	10.79	54.86	0	22.02	113.5
Capital punishment	26.26	30.76	31.14	27.92	43.75	0	22.26	5.95	23.29	0	25.91	0	74.0
Human cloning	11.11	72.46	0	0	0	42.58	36.04	7.65	15.90	19.95	30.65	14.09	103.3

Source: From Sinclair, M. & Phillips, C.J.C. (2017). The Cross-Cultural Importance of Animal Protection and Other World Social Issues. Journal of Agricultural and Environmental Ethics, 30(3), 439–455. https://doi.org/10.1007/s10806-017-9676-5.

a finger on the pulse of consumer sentiment is, however, only really the first stage of consumer engagement. One way to ensure continued consumer engagement (or engagement with any key stakeholder, for that matter) is to co-create future products alongside them. Not just to keep an ear open, but to actively ask, to listen, and to respond. In 2004, the Netherlands, threatened with the first round of bird flu, encouraged the development of options that delivered more sustainable futures for Dutch poultry farmers. With the help of Wageningen University, egg producers Rondeel collaborated with consumers to co-develop a round barn (esthetically preferred by consumers), that prioritizes bird welfare and environmental sustainability, hosts an integrated landscape that is transparent (live cameras are available on their website for over two decades later), and importantly, is economically viable (Rondeel Eieren, 2025). Still going strong, this producer is unlikely to be concerned about upcoming consumer-driven legislative changes in Europe that are likely to see enriched cages become a thing of the past, with its future-proof system. Likewise, responding to early consumer voices who were becoming aware of the growth rates of broiler hens and the associated hen welfare concerns, a coalition of farmers and retail partners in the Netherlands evolved. In 2007, they launched "Volwaard," a slower growing breed of chicken that was housed in lower stocking densities and higher welfare systems. By 2012, Plukon Food Group estimated "Volwaard" captured 90% of the market share of animal-friendly chicken meat in Dutch supermarkets (Plukon Free-range Chicken chain to experience 20% growth, 2025). Having made the transition and necessary adaptations long ago, it is again unlikely these producers and processors are concerned with the growing number of commitments that consumer-mindful industry big players (such as KFC, Nestlé, Unilever, and Nando's) are making to international conventions such as the "Better Chicken Commitment" (The Better Chicken Commitment [US], 2025)—who are pledging to only source their poultry meat from slower growing breeds in the near future.

If waiting for legislative change was ever a good business practice in animal agriculture, it is no longer. Understanding the forces that drive legislative change (i.e., the citizen consumer, en masse) before it gets to that point is a far smarter, more sustainable, and economically efficient approach to business in an evolving industry. Good business practice is entrepreneurial. It is identifying existing societal tensions, considering them the first indicators of a coming trend or market spaces due for significant attention. Good business practice dons a collaborative hat; it meets stakeholders, including citizen consumers, with respect and genuine interest. Consumers always tend to want more for less, so good business practice is also about working out how to maximize their perceptions in sustainable ways. Identifying existing societal tensions includes reading studies, paying some note to polls and the activities of NGOs and politicians, watching market changes carefully, and seeking case studies from other areas of the world.

At the time of writing, in a time of increased "misinformation," one such tension in multiple societies around the world is trust; truth, authenticity, naturalness, and a priority on personal and professional ethics. In investigating consumer behavior and purchase intention, a team of researchers in Colombia recently assessed a variety of known factors of influence and their strength in relation to the purchase intention of poultry products. "Trust," partially presented in the study as "I view the company as a brand responsive to the needs and interests of the community," was affirmed to be one of only a couple of highly significant factors influencing poultry product consumer behavior (Hernandez-Rey et al., 2024). Kipster, "the farm that gives a cluck," a high-welfare egg-producing brand originating in Europe, now also expanded to the United States of America, prides itself on complete transparency. Kipster farms are open to the public, and for those who cannot make the journey to one of their farms, they livestream their hen-houses 24/7 (Kipster, 2025). Hailed for producing the world's first carbon-neutral egg by the Washington Post, labeled the most sustainable egg in the world by Innovation Origins, and presented as the "egg farm of the future" by the Poultry Site (Here's what the egg farm of the future looks like, 2018), the accolades for Kipster do not come in short supply. In addition to their commitment to trust through transparency, Kipster also clearly attributes their award-winning innovative practices to their "revolutionary collaborative approach to animal welfare and sustainability," cocreating future-proofed systems with thought-leaders and consumer advocate groups.

The summary message here goes beyond "don't anger the masses by hiding poor practices and getting caught out misleading consumers."

The key takeaways that bridge consumer perceptions and sustainability in the poultry sector include;

1. Understand consumer perceptions. Critically, consumers across geopolitically diverse cultures care about chicken welfare, would prefer to buy higher welfare products where they are available, and consumers of the future value animal protection, environmental protection, and sustainable development (a "nature trifector") as critically important.
2. Engage with consumers. Take it one step further and engage with consumers by asking, listening, and cocreating future products and systems with their perceptions in mind.
3. Look for societal tensions and trends. The pinnacle of entrepreneurship is to identify societal tensions and trends and innovate. At a minimum, respond to ethical challenges and be an early adopter.
4. Be transparent with consumers. The societal tension around the values of truth and authenticity currently seen in many societies around the world can only lead to good things on "profit and loss" lines for producers who are working towards improvement and keeping it real.

References

Appleby, M. C., Mench, J. A., & Hughes, B. O. (2004). *Poultry behaviour and welfare*. CABI Publishing. http://doi.org/10.1079/9780851996677.0000.

Appleby, M. C. (2003). The European Union ban on conventional cages for laying hens: History and prospects. *Journal of Applied Animal Welfare Science, 6*(2), 103–121. https://doi.org/10.1207/S15327604JAWS0602_03.

Babington, S., & Campbell, D. L. M. (2022). Water for domestic ducks: the benefits and challenges in commercial production. Frontiers in Animal Science, 3, 782507. https://doi.org/10.3389/fanim.2022.782507.

Batt, S. (2009). Human attitudes towards animals in relation to species similarity to humans: A multivariate approach. *Bioscience Horizons, 2*(2), 180–190. https://doi.org/10.1093/biohorizons/hzp021.

Bir, C., Davis, M., Widmar, N., Zuelly, S., & Erasmus, M. (2019). Perceptions of animal welfare with a special focus on turkeys. *Frontiers in Veterinary Science, 6*, 413. https://doi.org/10.3389/fvets.2019.00413.

Coleman, G. (2007). Public perceptions of animal pain and animal welfare. *Proceedings of the Australian Animal Welfare Strategy Science Summit on Pain and Pain Management, 27*(1), 8.

de Luna, M. C. T., Yang, Q., Agus, A., Ito, S., Idrus, Z., Iman, R. H. S., Jattuchai, J., Lane, E., Nuggehalli, J., Hartcher, K., & Sinclair, M. (2022). Cage egg producers' perspectives on the adoption of cage-free systems in China, Japan, Indonesia, Malaysia, Philippines, and Thailand. *Frontiers in Veterinary Science, 9*, 1038362. https://doi.org/10.3389/fvets.2022.1038362.

Dewi, G., Smith, C., Martin, W., Venkitanarayanan, K., & Kollanoor Johny, A. (2024). Focus groups exploring American consumer perspectives on contemporary poultry production reveal critical insights to educate sustainable practices for producers. *Frontiers in Sustainable Food Systems, 8*, 1416167. https://doi.org/10.3389/fsufs.2024.1416167.

Dong, Z., Wang, B., & Shao, C. (2025). The historical evolution and modernization path of China's ecological and environmental governance. *Energy & Environmental Sustainability, 1*(2), 100014. https://doi.org/10.1016/j.eesus.2025.100014.

Duncan, I. J. H. (2004). Welfare problems of poultry. In G. J. Benson, & B. E. Rollin (Eds.). *The well-being of farm animals: Challenges and solutions* (pp. 307–323). Ames: Blackwell, Ames Blackwell. https://doi.org/10.1002/9780470344859.ch14.

EFSA Panel on Animal Health and Welfare (2005). Opinion of the Scientific Panel on Animal Health and Welfare (AHAW) on the impact of the current housing and husbandry systems on the health and welfare of domestic fowl (*Gallus gallus*). *EFSA Journal, 197*, 1–23. https://doi.org/10.2903/j.efsa.2005.197.

European Food Safety Authority. (2015). Scientific opinion on welfare aspects of the use of perches for laying hens. EFSA Journal, 13, 1–70, 4131. https://doi.org/10.2903/j.efsa.2015.4131.

Fraser, D., Mench, J. A., & Millman, S. T. (2001). Farm animals and their welfare in 2000. In D.J. Salem & A.N. Rowan (Eds.), The state of the animals 2001 (pp. 87-99). Washington, DC: Humane Society Press. https://www.wellbeingintlstudiesrepository.org/sota_2001/9/.

Hall, C, & Sandilands, V (2007). Public attitudes to the welfare of broiler chickens. *Animal Welfare, 16*(4), 499–512. https://doi.org/10.1017/s0962728600027433.

Here's what the egg farm of the future looks like. (2018). https://www.thepoultrysite.com/articles/heres-what-the-egg-farm-of-the-future-looks-like.

Hernandez-Rey, J. W., Casadiego-Alzate, R., Silva Cortés, A., Villa Enciso, E. M., & Marín Correa, J. F. (2024). Influences on consumer behaviour regarding the purchase intention of poultry products. *Revista de Gestão Social e Ambiental, 18*(11), e09247. https://doi.org/10.24857/rgsa.v18n11-030.

Higenyi, J. (2014). Social and quality attributes influencing consumption of native poultry in Eastern Uganda. *Animal and Veterinary Sciences, 2*(2), 42. https://doi.org/10.11648/j.avs.20140202.15.

Kipster (2025). Available at: https://kipster.farm/. Accessed June 6, 2026.

Lay, D. C., Fulton, R. M., Hester, P. Y., Karcher, D. M., Kjaer, J. B., Mench, J. A., Mullens, B. A., Newberry, R. C., Nicol, C. J., O'Sullivan, N. P., & Porter, R. E. (2011). Hen welfare in different housing systems. *Poultry Science, 90*(1), 278–294. https://doi.org/10.3382/ps.2010-00962.

Lehmann, C., & Christoph-Schulz, I. (2022). Relevance of animal welfare for poultry meat consumers – results of multinational focus groups. System Dynamics and Innovation in Food Networks, 2022, 2–30. https://doi.org/10.18461/pfsd.2022.2202.

Napolitano, Pacelli, Girolami, & Braghieri (2008). Effect of information about animal welfare on consumer willingness to pay for yogurt. *Journal of Dairy Science, 91*(3), 910–917. https://doi.org/10.3168/jds.2007-0709.

Nicol, C. J. (2015). *The behavioural biology of chickens.* CABI. http://doi.org/10.1079/9781780642499.0000.

Ogilvy, D. (1969). *Confessions of an advertising man.* London: Southbank Publishing.

Plukon Free-range Chicken chain to experience 20% growth. (2025). https://www.plukon.com/plukon-free-range-chicken-chain-to-experience-20-growth/.

Rondeel Eieren. (2025). https://www.rondeeleieren.nl/en/home/.

Sinclair, M., Lee, N. Y. P., Hötzel, M. J., de Luna, M. C. T., Sharma, A., Idris, M., Derkley, T., Li, C., Islam, M. A., Iyasere, O. S., Navarro, G., Ahmed, A. A., Khruapradab, C., Curry, M., Burns, G. L., & Marchant, J. N. (2022). International perceptions of animals and the importance of their welfare. *Frontiers in Animal Science, 3*, 960379. https://doi.org/10.3389/fanim.2022.960379.

Sinclair, M., & Phillips, C. J. C. (2017). The cross-cultural importance of animal protection and other world social issues. *Journal of Agricultural and Environmental Ethics, 30*(3), 439–455. https://doi.org/10.1007/s10806-017-9676-5.

Sinclair, M., Stennett, J., & Chung, J. Y. (2025). Exploring Gen Z's attitudes towards animals and the environment. *Faunalytics.* https://faunalytics.org/exploring-gen-zs-attitudes-towards-animals-and-the-environment/.

Sinclair, M., Yan, W., & Phillips, C. J. C. (2019). Attitudes of pig and poultry industry stakeholders in Guangdong Province, China, to animal welfare and farming systems. *Animals, 9*(11), 860. https://doi.org/10.3390/ani9110860.

State Council Information Office of the People's Republic of China, 2023. China's green development in the new era. [online] Available at: http://www.scio.gov.cn/zfbps/zfbps_2279/202303/t20230320_707666.html [Accessed 18 August 2025].

Tauson, R. (2005). Management and housing systems for layers – Effects on welfare and production. *World's Poultry Science Journal, 61*(3), 477–490. https://doi.org/10.1079/wps200569.

The Better Chicken Commitment [US]. (2025). https://betterchickencommitment.com/us/.

Webster, A. B. (2004). Welfare implications of avian osteoporosis. *Poultry Science, 83*(2), 184–192. https://doi.org/10.1093/ps/83.2.184.

Woods, B. (2000). Beauty and the beast: Preferences for animals in Australia. *Journal of Tourism Studies, 11*(2), 25–35.

White, K., MacDonnell, R., & Ellard, J. H. (2012). Belief in a just world: Consumer intentions and behaviors toward ethical products. *Journal of Marketing, 76*(1), 103–118. https://doi.org/10.1509/jm.09.0581.

Yang, J. M. (2023). Willingness to buy animal welfare products: A study on the conflict between moral attitudes and perceived higher prices. *Journal of Applied Animal Research, 51*(1), 447–455. https://doi.org/10.1080/09712119.2023.2225586.

Yang, Y.-C., & Wang, C.-C. (2025). The determinants of willingness to pay for improving the animal welfare of laying ducks: The case of Taiwan. *International Food and Agribusiness Management Review, 28*(1), 61–73. https://doi.org/10.22434/ifamr1018.

Chapter 17

Conclusion: welfare and sustainability for poultry's future

Marisa A. Erasmus[1] and Dana L.M. Campbell[2]
[1]*Department of Animal Sciences, Purdue University, West Lafayette, IN, United States,*
[2]*Agriculture and Food, CSIRO (Commonwealth Scientific and Industrial Research Organisation), Armidale, NSW, Australia*

The question of poultry welfare and sustainability

This book began by highlighting the global significance of poultry production and the growing demand for poultry products. The main question posed in this book is how poultry can be produced in ways that achieve optimal animal welfare, minimize environmental impacts, and remain economically viable. The chapters in this book explored this question through an examination of different poultry species, production systems, and frameworks, and there is no simple answer. The relationship between animal welfare and sustainability is characterized by both synergies and trade-offs that are discussed within the chapters.

Chapter 2 discussed laying hens housed in indoor systems with a focus on cage-free aviaries. These systems are increasing in many global regions as a result of consumer-driven pressures, as they provide hens with more behavioral freedom. Hens can engage in foraging and dustbathing in a littered area, and a greater degree of movement reduces the osteoporosis risks that are seen in caged housing systems. However, keel bone damage remains a prevalent welfare issue along with damaging behaviors that are seen in large groups, such as feather pecking, toe pecking, and smothering. Aviary systems can be more expensive to operate as they require more space and increased labor to manage. Higher bird mortality can reduce the overall system production output as well as increase greenhouse gas (GHG) emissions. Higher ammonia emissions relative to caged systems can put both birds and workers at health risk. However, breeding and management strategies could contribute toward optimizing bird welfare in aviary systems while simultaneously improving the system's efficiency. For example, genetic and nutritional interventions may contribute to reducing hens' susceptibility to keel bone damage, and emission reduction technologies

Integrating Welfare and Sustainability in Poultry Systems. DOI: https://doi.org/10.1016/B978-0-443-40594-5.00010-9

could reduce the GHG footprint of aviary housing. Aviary systems bring welfare benefits to hens, but the birds themselves and management practices need to be further optimized for these systems to be a sustainable housing solution.

Chapter 3 focused on grower broiler chickens, discussing the differences and trade-offs between fast-growing and slower-growing strains. While fast-growing birds have been selected over generations to be one of the most efficient protein sources available, with both high growth rate and feed conversion efficiency, they experience a multitude of welfare issues as a consequence. Research shows that raising slower-growing strains alleviates some of the welfare compromise through reduced growth-related physical and physiological demand on the bird, alongside an improved ability to exhibit natural behaviors. However, slower growing means reduced feed efficiency. Producing more birds to provide the equivalent protein equates to more land, more GHG emissions, more manure, and an increase in the associated environmental impacts of growing chicken feed crops and transporting them around the world. These increased environmental impacts associated with slower-growing birds can partially be offset through lower protein requirements of the birds and reductions in mortality and carcass condemnations through improved welfare status. Fewer breeder birds may also be needed as a result of greater fertility. Consumer demand for higher welfare meat chickens is likely to necessitate an industry change in the future. Whether this is in the form of increased use of slower-growing strains or reduced densities in fast-growing birds remains to be seen. Integrating dwindling planetary resources and higher bird welfare in broiler production may need to challenge human protein consumption habits for the industry to be sustainable into the future.

Chapter 4 continued the discussion on meat chickens but focused on the breeder rather than the grower birds. Breeder birds that supply the fast-growing broilers must be feed-restricted for most of their life to curb their growth and ensure they remain fertile. This restriction, and its associated behavioral issues, is the primary welfare concern for these birds, but remains necessary to avoid obesity-related welfare issues that would result from ad libitum feeding. This paradox is a major challenge for the chicken meat industry. Another challenge is the genetic selection for leaner growth, contributing to decreased breeder fertility, with a 60% fertility decline predicted by 2050. A significant increase in the number of breeders required, or a decrease in the amount of chicken meat produced, is predicted, bringing into question whether sustainability is possible along the current trajectory. While different types of feed restriction and different management modifications, such as provision of environmental enrichment, are applied in practice, there is currently no solution to the broiler breeder paradox. The introduction of slower-growing strains brings similar environmental impacts at the breeder and grower level, as outlined in Chapter 3. Remaining with the status quo puts longer-term sustainability at risk and will require significant advancements in poultry science to find a balance between bird welfare and all three pillars of sustainability.

Chapter 5 dove into a global overview of duck production, which is growing in some countries. Similar to geese production (Chapter 6), research into duck welfare and sustainability is limited, and several key research gaps include the influence of outdoor access on duck welfare and productivity, breeder duck welfare, optimal lighting programs, and waste management, to name a few. Ducks have some of the same ethical and welfare issues as other meat poultry; for example, foot and leg health issues and handling and transportation challenges; however, issues that are unique to duck (and goose) production is that of foie gras (fatty liver) and a need to consider the strong motivation that waterfowl have for performing water-related behavior. A similar theme emerged to that of Chapter 15, which is that duck welfare needs to be incorporated into life cycle assessment to evaluate sustainability aspects of duck production. Other emerging issues that need to be considered include alternative protein sources and how sustainability aspects related to duck housing systems are impacted by the scale of production.

Chapter 6 characterized the major goose breeds and production systems across the globe, with China being the global leader in goose production. In addition to foie gras production methods, another unique ethical consideration for goose production compared to other poultry includes how feathers are harvested for their use in clothing, bedding, and other products. From a sustainability standpoint, geese have the potential to reduce the carbon footprint when housed in well-managed pasture-based systems due to their ability to graze, but the impacts of goose manure on water and air quality need to be considered and managed. A corresponding trade-off is therefore between the potential welfare benefits and environmental impacts of housing geese in outdoor systems. Another trade-off that was identified is between the economic costs and welfare benefits associated with housing systems that provide resources such as water access for swimming and environmental enrichment. Similar to ducks, there is a critical need for additional research into the effects of housing and management on goose welfare, as well as developing policies that can support ethical and sustainable goose production.

Chapter 7 discussed the challenges that are experienced in another poultry meat industry—turkeys. This industry is smaller than the chicken meat industry, and consequently, the available body of research on bird welfare is also smaller. Many recommendations for turkey practices that might optimize bird welfare are adapted from broiler research, which does not always translate, as turkeys have their own species-specific requirements. Across all stages of turkey production from hatchery and grower bird management through to transportation of market-age birds, this chapter considered the social, economic, and environmental sustainability challenges where applicable. Much more research on turkeys is needed, particularly with current genetic lines, focusing on multiple areas, including lighting, enrichment, shed environment, stocking density, transport, and almost all factors in breeder turkey management. An increase in science in this understudied species will inform how each management area can be optimized for both bird welfare and industry sustainability.

Chapter 8 stepped away from commercial high-production chicken strains to focus on dual-purpose and specialty, or local breeds. Dual-purpose birds are a step toward more ethical production through growing out male chicks that would otherwise be killed at hatch. Local strains have high genetic value for the diversity they bring to the chicken industry for both eggs and meat. These birds are often adapted to local regions, which may make them more resilient and, thus, they may experience improved welfare. However, welfare research on the numerous local strains worldwide is fragmented and limited, restricting conclusions that can be drawn for individual breeds. These local strains typically have lower production output compared with the intensely selected genetic lines, reducing their production efficiency. However, their diversity, local adaptation, and cultural significance highlight their unique value. Local strains hold importance in niche markets and smaller-scale production systems. Through crossbreeding, they may also bring the genetic innovation that commercial hybrids need to improve bird welfare, alongside strengthening industry sustainability. A greater research investment and marketing focus on local strains could drive locally-adapted and welfare-friendly poultry production on a larger scale.

Chapter 9 focused on laying hens again, but delving into pasture-based systems, also termed free-range, where hens have access to an outdoor area. This includes small flocks at a backyard scale through to intensive free-range farms that house thousands of hens. Pasture-based systems are popular with consumers as the outdoor space is viewed as providing a more natural living environment for the birds, allowing greater behavioral freedom than exclusively indoor systems. From a welfare perspective, there are documented benefits of outdoor access, including increased exhibition of motivated behaviors such as foraging and improved plumage condition. However, the outdoor resource is not always utilized to its full potential, with ranging influenced by factors such as the quality of the range area, flock size, and individual bird traits. Outdoor access can also expose birds to diseases from wild bird populations. Manure on the range can increase nutrient loading in the soil beyond acceptable limits, and ventilation systems are not always designed to cope with the shed opening created by popholes. Pasture-based systems require more resources to produce the same number of eggs, but costs can often be passed along to the consumer. Ultimately, the welfare benefits of pasture-based systems will depend on individual flock management. A One Health approach views healthy food as coming from healthy animals. The variation and complexities of pasture-based systems require a careful approach to sustainably manage the land and the animals, balancing potential trade-offs in these systems.

Chapter 10 provided a discussion of animal welfare and sustainability within organic production systems for broiler chickens and laying hens, and noted similarities and differences among organic standards in the European Union, United States and Australia. Similar to pasture-based systems, outdoor access is a major aspect of organic production, and in this regard, organic systems have similar welfare implications as pasture-based systems. However, organic systems have additional requirements and standards. Most notably, organic systems restrict the use of antibiotics and coccidiostats for preventive use, which may have implications for bird health and the specific breeds or genetic strains most suited to organic production. Compared to conventional production systems, organic production is generally associated with lower stocking densities, which leads to a trade-off between the welfare benefits from lower stocking densities and the potential for higher economic costs. These higher costs could, in part, be recouped when the products are sold at a higher price. While organic production is viewed as providing poultry with welfare benefits, the expansion of organic production in the future to meet demand raises questions about organic production being able to stay true to its original definition.

Chapter 11 examined silvopastoral systems, where animal production is intentionally integrated with trees and forage on the same land. One of the noted challenges with assessing silvopastures in terms of poultry welfare, productivity, and sustainability is that these systems are not very common in the United States and Europe, and the types of trees, forages, and even poultry that are used vary from operation to operation. Nonetheless, compared with other outdoor systems, the trees and vegetation present on silvopasture provide behavioral enrichment and may provide poultry with some protection (from certain diseases and aerial predators) and opportunities for thermal comfort or choice. From a sustainability standpoint, trees are beneficial for soil health, and poultry excreta may have both positive and negative effects on soil health, depending on the number of poultry and the specific vegetation on the land. Silvopasture may offer economic resilience by allowing producers to diversify where and how income is generated, but there is as yet little information available to inform questions about the economic costs and benefits of these systems.

Chapter 12 detailed the growth and intensification of the poultry sector in India, one of the world's largest producers of eggs and one of the fastest-growing poultry sectors. In India, growing public awareness and evolving regulations about animal welfare are prompting a shift from caged to cage-free systems to meet rising ethical standards. Key welfare focuses discussed in this chapter included transportation stress in hot climates and optimizing lighting to reduce mortality and provide poultry with behavioral opportunities. Alongside the animal welfare challenges and opportunities, some sustainability challenges include feed security, particularly diverting maize (corn) for ethanol production, and the need to adopt water-efficient agricultural practices. Considering technical capacity alongside regulatory frameworks and consumer awareness of animal welfare will continue to be important as poultry production continues to grow in India and globally.

Chapter 13 discussed moving beyond efficiency toward robustness and replenishing natural resources. The chapter argued that focusing on robustness can have trade-offs in terms of short-term productivity, but can lead to greater health and resiliency of the system in the long term. Some of the environmental impacts of intensive poultry production were discussed, including impacts on soil and water, biodiversity, and waste management, among others. The chapter further proposed using agroecological processes, such as nutrient cycling, renewable energy, and mutualistic relationships as part of ecological practices. Another key theme was that animal welfare is not a separate concern from sustainability, because animal welfare considerations often align with ecological health and social equity. In this context, building a food system that incorporates respect for both the animals and the people is essential for maintaining a social license to operate.

Chapter 14 addressed the environmental and economic concerns with traditional corn, soy, and wheat as the main ingredients of poultry diets and discussed the factors affecting the long-term viability of feeding poultry using these main ingredients. Some concerns with continuing to rely on corn, wheat, and soy-based diets include competition between human food production and feed crops for resources such as water and land, the environmental footprint, and loss of biodiversity. Additional concerns include increasing costs of transportation and the associated greater carbon emissions, because most countries need to import feed ingredients from elsewhere. The chapter explored various potential alternative feed ingredients that are locally sourced; for example, using insects and algae as protein, using drought-resilient crops like amaranth and buckwheat, and combining pasture-based systems with trees and other crops (silvopasture), among others. There are challenges with implementing these alternatives, including the nutritional quality of the plant-based alternatives, which often have higher fiber; the investment in infrastructure and labor that is needed to implement these alternatives; and scalability—producing enough of these alternatives to feed the growing poultry population.

Chapter 15 provided a review of 55 studies that evaluated the poultry supply chain using life cycle assessment (LCA), which is a standardized methodology for evaluating environmental impacts of a product across the entirety of the product's life cycle. The chapter emphasized that while poultry production is very efficient compared to other livestock production, the scale of poultry production requires continuous improvement for sustainability. The chapter discussed feed production and manure management as primary sources of environmental impact in the poultry industry. Possible strategies that can be used to reduce the poultry industry's footprint include using alternative protein sources in poultry diets, converting manure into energy, and using precision management to improve animal performance. The chapter noted that there is a general lack of consistency in how sustainability is measured, and calls for a unification of LCA methodologies, particularly as applied to system boundaries (which stages of production are included within the LCA). Further, future research on LCA should strive to incorporate animal welfare as a formal metric.

Chapter 16 explored the drivers of consumer perceptions and choice. Understanding consumer perceptions is critical for maintaining the poultry industry's social license to operate. Globally, concern for animal welfare is generally high, with consumers in various countries stating that they consider animal welfare when purchasing poultry products; however, the value that people place on animals depends on the species of animal being considered and other characteristics, such as the perceived intelligence of the animal. Demographic factors influence people's perceptions. Surveys conducted with youth—as the influential group of future consumers—identified the social issues of "environmental protection, sustainable development, and animal protection" as being important across several countries. However, consumer knowledge of production practices remains limited, with transparency and the perceived "naturalness" of systems being key factors influencing consumer trust. The chapter concluded that engaging with consumers, understanding their perceptions, identifying trends, and being transparent are important for connecting consumer perceptions with the sustainability of housing system design and management practices.

Key themes that emerged from these chapters include

Trade-offs. There is a complex trade-off between animal welfare and sustainability, with both animal welfare and sustainability being influenced by a multitude of factors. Generally, production systems that provide more space and opportunities for poultry to express their natural behavior are associated with positive animal welfare outcomes and improve the birds' quality of life; however, these systems require more land, feed, and labor inputs, which can increase their overall carbon and resource footprints. The crucial impact of management on both poultry welfare and sustainability factors cannot be negated—any given system can have great poultry welfare and sustainability potential, and both welfare and sustainability are highly dependent on the management at the farm.

Transitions and transformations. Cage-free and pasture-based systems are gaining increasing attention across the globe, and consumer perceptions continue to be important drivers of transitions within the poultry sector. The transition from caged to cage-free housing is a global phenomenon, not limited to Europe and North America, but becoming a focus in rapidly developing poultry sectors like that of India. However, context is important because optimal approaches vary with climate, market conditions, cultural norms, and regulatory environments. In addition to transitions of housing systems, the poultry sector is experiencing a transformation in the form of growing demand for poultry, partly driven by consumer preference changing from red meat to poultry (OECD/FAO, 2025).

Circularity and systems thinking. The focus on circularity and systems thinking in production is growing, with increasing emphasis on alternative feeds and integrated systems that aim to maximize efficient use of resources while minimizing waste. Changes in one aspect of the system can have One Health and One Welfare implications for the entire system, affecting animal welfare, environmental impacts, human health, economics, and labor requirements. Broom (2019) argues that the evaluation of a system necessitates the precise measurement of all of the components of sustainability, which includes the welfare of both humans and animals within the system. In addition, innovation is critical for optimizing poultry production, and new technologies, housing designs, and management approaches need to be evaluated holistically before being widely adopted.

Technology and innovations for addressing animal welfare and sustainability

Technology has been changing rapidly, leading to improvements, and sometimes unintended consequences, for both poultry welfare and sustainability. There are a host of factors, technologies, techniques, and tools, such as the gut microbiome, proteomics, and epigenomics, to name a few, which have not been discussed in this book, that may provide insights and advances into poultry welfare and sustainability that will be important for the future of poultry. Below, we discuss a few technologies and innovations that have already impacted, or have the potential to greatly impact, poultry welfare and sustainability.

Breeding and genomics. Several chapters mentioned bird genetics or breeding, highlighting the concept of finding the bird best suited to the particular production system and environment—this is not a new concept; in 2008, Jensen et al. (2008) stated that a future challenge would be to find "farming practices where the genotype and environment harmonize," and this remains a challenge. Advances in genetic technologies offer opportunities to address challenges that cannot be addressed using management approaches alone. For example, selectively breeding against traits such as high fearfulness has yielded positive results (Jensen et al., 2008), the sequencing of the chicken genome (International Chicken Genome Sequencing Consortium, 2004) was postulated to offer new opportunities for understanding the genetic bases of breed differences and chicken biology (Burt, 2004) and whole genome resequencing has provided opportunities for studying economically important traits (Xiong et al., 2023). Advances in gene and genome editing (see (Khwatenge & Nahashon, 2021) and (Wang et al., 2024) for reviews) also offer opportunities for addressing poultry health and welfare challenges, such as disease resistance (e.g., poultry that are resistant to avian influenza; Idoko-Akoh et al., 2023), reducing the need for culling male day-old chicks (Tizard et al., 2019), and potentially reducing skeletal abnormalities or improving walking ability (for example, Mulim et al., 2024 reported that some aspects of walking ability are moderately heritable in turkeys)—all of which have the potential to alleviate suffering and reduce mortality, and contribute to more efficient production and sustainability. In their overview of meat poultry breeding practices, Neeteson et al., 2023 state that holistic breeding can "contribute to sustainability" because the gains made through breeding for particular traits are permanent and accumulate across generations. However, considering the ethical, social, and regulatory implications of breeding practices and technology broadly is extremely important. Genetic selection has had unintended negative consequences for poultry welfare, such as higher incidences of musculoskeletal disorders and cardiovascular and metabolic diseases, among others (reviewed in Hocking, 2014). In addition, public acceptance of gene editing technologies, such as CRISPR-Cas9 for food-production animals varies among countries (e.g., Borgdorf et al., 2026); nonetheless, people generally had positive views about using gene editing for improving animal welfare (Borgdorf et al., 2026; Naab et al., 2021). Debates around the acceptability and effects of technologies and breeding practices that alter animals will continue to impact the development and application of technology in animal agriculture, especially considering the rapid pace at which disruptive technologies such as artificial intelligence have been changing the world around us.

Precision livestock farming and artificial intelligence. Similar to most other animal production industries, precision livestock farming and the application of artificial intelligence technologies are steadily increasing, and with this comes the potential for research and management innovation (Neethirajan, 2025). Technologies that can monitor individual birds, such as on-animal radio-frequency identification (RFID) tags, or flock-level remote artificial intelligence using computer vision, have the ability to provide insights that were not previously possible. For example, RFID tracking of individual aviary-housed hens allowed researchers to match behavioral profiles with potential genes to better understand the underlying mechanisms for consistent variation among individual hens (Hoeksema et al., 2026). These technologies can also advance the way birds are managed, allowing more continuous monitoring to facilitate timely interventions (Engelbrecht et al., 2026). For example, a computer-vision model was trained to detect feather pecking behavior in cage-free hens in an experimental setting (Subedi et al., 2023), which could eventually be applied in a commercial context to alert producers as this behavior increases in frequency.

Challenges do exist when it comes to poultry due to the sheer number of animals housed at any given time and the value of the individual bird compared with larger livestock species. As a result, on-animal devices become less viable in a commercial setting, but this does not detract from the value that can still be gained from precision monitoring in a research context that advances our understanding and guides commercial management. Continual technological advances in monitoring devices could one day make tagging or micro-chipping of flocks a viable option (Kazanskiy et al., 2025). As computer vision algorithms for detecting individual bird behaviors within a flock become more reliable and widely available, this could open up the potential for sophisticated bird management and early detection of behavioral issues. However, the interpretation of any monitoring output, and the setting of relevant thresholds for intervention, is a new challenge to be addressed to ensure the data produced by artificial intelligence systems is meaningful for welfare assessment and improvement (Fuentes et al., 2026). Rollout to commercial farms, while no easy feat (Bist et al., 2026), will ensure the potential of these systems is realized for actual bird-welfare optimization (Rowe et al., 2019).

Housing system innovations. The chapters have collectively discussed the challenges associated with designing and managing housing systems for poultry that are sustainable while also providing resources that aim to optimize bird welfare, including opportunities for positive experiences. To be able to meet these criteria requires innovation, and there are some examples of this already occurring. The Rondeel™ system (Rondeel, n.d.) for laying hens was developed in the Netherlands as part of a government-funded initiative to develop a welfare-friendly and sustainable system for egg production (Klerkx et al., 2012; Spoelstra et al., 2013). This circular-shaped system combined resource-heavy areas for hens to behaviorally engage with, covered outdoor areas, natural light, no beak trimming, and areas for public viewing. While this system can be viewed as successful for meeting animal welfare needs and facilitates public engagement, the eggs incur a premium and thus do not yet represent a majority of the egg production systems available in the Netherlands. Another example is the Kipster system (Kipster, n.d.), also based in the Netherlands. This system houses laying hens and also rears male chicks for meat that would otherwise be culled. The system, while aiming for higher bird welfare, also applies multiple practices to reduce environmental footprints, such as feeding hens human feed waste (e.g., bakery/candy waste), and the use of solar panels (van Hal et al., 2019). While the system is more efficient in some aspects, there are trade-offs in others. For example, a life cycle assessment showed that while raising male chicks was beneficial from a social-sustainability perspective, there was an environmental cost to the practice (van Hal et al., 2019). These are only two examples of potential innovation in housing-system design where collaboration among scientists of different disciplines, as well as other stakeholders, can drive change and provide solutions for animals, the environment, and people (de Olde et al., 2020).

Collaboration for poultry welfare and sustainability

Optimizing poultry welfare and sustainability requires coordination and collaboration among all stakeholder groups within the poultry sector. The poultry value chain is characterized by interconnectedness: breeding company decisions influence producers; producers' practices impact the products available to consumers; consumers' preferences and willingness to pay influence companies' social license to operate, what producers can economically implement, and, in some cases, regulatory frameworks; and research helps inform production practices, poultry-welfare standards, and policy decisions.

Future poultry production requires the sustained development and validation of tools that can integrate animal welfare measures, life cycle assessment metrics, and economic indicators. Including measures associated with worker health and safety, animal disease resilience, community impacts, and food security in the development of these tools will be valuable. Ultimately, technical analyses will still need to be integrated with value judgements to determine acceptable trade-offs in different contexts, with input from diverse stakeholder groups.

Science is going to play a key role in system innovation as well as continuing to tackle persistent challenges across different industry sectors, but real change will require buy-in from all stakeholders. Animals are an integral part of people's way of life, yet many indicators suggest that the current trajectories of increased animal production may not be infinite. The respective roles of societal influence and political drivers in a changing global landscape are likely to be key determinants of what shape poultry industries of the future take—how we care for the birds while feeding the people and considering the associated environmental impacts.

References

Bist, R. B., Wang, D., Chai, L., & Xiong, Y. (2026). Precision farming technologies for monitoring livestock and poultry. *AgriEngineering, 8*(2), 64. https://doi.org/10.3390/agriengineering8020064.

Borgdorf, L., Delanoue, E., Candek-Potokar, M., Haas, V. P., Kramer, K., & Meijboom, F. L. B. (2026). Attitudes towards genome editing in farmed animals – A cross-cultural study. *Food Ethics, 11*(1), 23. https://doi.org/10.1007/s41055-026-00205-4.

Broom, D. M. (2019). Animal welfare complementing or conflicting with other sustainability issues. *Applied Animal Behaviour Science, 219*, 104829. https://doi.org/10.1016/j.applanim.2019.06.010.

Burt, D. W. (2004). The chicken genome and the developmental biologist. *Mechanisms of Development, 121*(9), 1129–1135. https://doi.org/10.1016/j.mod.2004.04.020.

Engelbrecht, D., Steyn, N., Djouani, K., & Bosman, H. (2026). Precision farming: A review of artificial intelligence applications in broiler poultry farming. *Emirates Journal of Food and Agriculture, 38*, 1–13. https://doi.org/10.3897/ejfa.2026.172240.

Fuentes, A., Han, S., Liu, J., Park, J., Yoon, S., & Park, D. S. (2026). Review: Understanding cattle social behaviour in modern penned production systems with AI technology: Are we tracking welfare indicators? *Animal, 20*(4), 101788. https://doi.org/10.1016/j.animal.2026.101788.

van Hal, O., Weijenberg, A. A. A., de Boer, I. J. M., & van Zanten, H. H. E. (2019). Accounting for feed-food competition in environmental impact assessment: Towards a resource efficient food-system. *Journal of Cleaner Production, 240*, 118241. https://doi.org/10.1016/j.jclepro.2019.118241.

International Chicken Genome Sequencing Consortium (2004). Sequence and comparative analysis of the chicken genome provide unique perspectives on vertebrate evolution. *Nature, 432*, 695–716. https://doi.org/10.1038/nature03154.

Hocking, P. M. (2014). Unexpected consequences of genetic selection in broilers and turkeys: Problems and solutions. *British Poultry Science, 55*(1), 1–12. https://doi.org/10.1080/00071668.2014.877692.

Hoeksema, K., Baes, C. F., Gebhardt-Henrich, S. G., Petelle, M. B., Toscano, M. J., & Makanjuola, B. O. (2026). Genome-wide association study of durations in zones of an aviary for laying hens. *Poultry Science, 105*(6), 106719. https://doi.org/10.1016/j.psj.2026.106719.

Idoko-Akoh, A., Goldhill, D. H., Sheppard, C. M., Bialy, D., Quantrill, J. L., Sukhova, K., Brown, J. C., Richardson, S., Campbell, C., Taylor, L., Sherman, A., Nazki, S., Long, J. S., Skinner, M. A., Shelton, H., Sang, H. M., Barclay, W. S., & McGrew, M. J. (2023). Creating resistance to avian influenza infection through genome editing of the ANP32 gene family. *Nature Communications, 14*(1), 6136. https://doi.org/10.1038/s41467-023-41476-3.

Jensen, P., Buitenhuis, B., Kjaer, J., Zanella, A., Mormède, P., & Pizzari, T. (2008). Genetics and genomics of animal behaviour and welfare—Challenges and possibilities. *Applied Animal Behaviour Science, 113*(4), 383–403. https://doi.org/10.1016/j.applanim.2008.01.012.

Kazanskiy, N. L., Khorin, P. A., & Khonina, S. N. (2025). Biochips on the move: Emerging trends in wearable and implantable lab-on-chip health monitors. *Electronics, 14*(16), 3224. https://doi.org/10.3390/electronics14163224.

Khwatenge, C. N., & Nahashon, S. N. (2021). Recent advances in the application of crispr/cas9 gene editing system in poultry species. *Frontiers in Genetics, 12*, 627714. https://doi.org/10.3389/fgene.2021.627714.

Kipster. (n.d.). https://kipster.nl/

Klerkx, L., van Bommel, S., Bos, B., Holster, H., Zwartkruis, J. V., & Aarts, N. (2012). Design process outputs as boundary objects in agricultural innovation projects: Functions and limitations. *Agricultural Systems, 113*, 39–49. https://doi.org/10.1016/j.agsy.2012.07.006.

Mulim, H. A., Hernandez, R. O., Vanderhout, R., Bai, X., Willems, O., Regmi, P., Erasmus, M. A., & Brito, L. F. (2024). Genetic background of walking ability and its relationship with leg defects, mortality, and performance traits in turkeys (*Meleagris gallopavo*). *Poultry Science, 103*(7), 103779. https://doi.org/10.1016/j.psj.2024.103779.

Naab, F. Z., Coles, D., Goddard, E., & Frewer, L. J. (2021). Public perceptions regarding genomic technologies applied to breeding farm animals: A qualitative study. *BioTech, 10*(4), 28. https://doi.org/10.3390/biotech10040028.

Neeteson, A.-M., Avendaño, S., Koerhuis, A., Duggan, B., Souza, E., Mason, J., Ralph, J., Rohlf, P., Burnside, T., Kranis, A., & Bailey, R. (2023). Evolutions in commercial meat poultry breeding. *Animals, 13*(19), 3150. https://doi.org/10.3390/ani13193150.

Neethirajan, S. (2025). Rethinking poultry welfare—integrating behavioral science and digital innovations for enhanced animal well-being. *Poultry, 4*(2), 20. https://doi.org/10.3390/poultry4020020.

OECD/FAO. (2025). *OECD-FAO agricultural outlook 2025–2034*. OECD Publishing. https://doi.org/10.1787/601276cd-en.

de Olde, E. M., van der Linden, A., Bolhaar, L. D.olde, & de Boer, I. J. M. (2020). Sustainability challenges and innovations in the Dutch egg sector. *Journal of Cleaner Production, 258*, 120974. https://doi.org/10.1016/j.jclepro.2020.120974.

Rondeel. (n.d.). https://www.rondeeleieren.nl/en/about-us/

Rowe, E., Dawkins, M. S., & Gebhardt-Henrich, S. G. (2019). A systematic review of precision livestock farming in the poultry sector: Is technology focussed on improving bird welfare? *Animals, 9*(9), 614. https://doi.org/10.3390/ani9090614.

Spoelstra, Koerkamp, G., Bos, Elzen, & Leenstra (2013). Innovation for sustainable egg production: Realigning production with societal demands in The Netherlands. *World's Poultry Science Journal, 69*(2), 279–298. https://doi.org/10.1017/S0043933913000305.

Subedi, S., Bist, R., Yang, X., & Chai, L. (2023). Tracking pecking behaviors and damages of cage-free laying hens with machine vision technologies. *Computers and Electronics in Agriculture, 204*, 107545. https://doi.org/10.1016/j.compag.2022.107545.

Tizard, M. L., Jenkins, K. A., Cooper, C. A., Woodcock, M. E., Challagulla, A., & Doran, T. J. (2019). Potential benefits of gene editing for the future of poultry farming. *Transgenic Research, 28*(S2), 87–92. https://doi.org/10.1007/s11248-019-00139-0.

Wang, Y., Zhou, Q., Chen, J., et al. (2024). A review of gene editing technologies in poultry breeding: Focus on CRISPR/Cas9. *Animal Molecular Breeding, 14*(6), 345–353. https://doi.org/10.5376/amb.2024.14.0036.

Xiong, X., Liu, J., & Rao, Y. (2023). Whole genome resequencing helps study important traits in chickens. *Genes, 14*(6), 1198. https://doi.org/10.3390/genes14061198.

Index

Note: Page numbers followed by "*f*," "*t*," and "*b*" refer to figures, tables, and boxes, respectively.

C

D

E

I

K

L

M

N

O

Printed and bound by CPI Group (UK) Ltd, Croydon, CR0 4YY
17/09/2026
02218538-0004